Principles and Problems in

Physical Chemistry for Biochemists

THIRD EDITION

Nicholas C. Price

Department of Biochemistry and Molecular Biology, University of Glasgow

Raymond A. Dwek

Oxford Glycobiology Institute, Department of Biochemistry, University of Oxford

R. George Ratcliffe

Department of Plant Sciences, University of Oxford

Mark R. Wormald

Oxford Glycobiology Institute, Department of Biochemistry, University of Oxford

Oxford University Press, Great Clarendon Street, Oxford OX2 6DP

Oxford New York
Athens Auckland Bangkok Bogotá Buenos Aires Calcutta
Cape Town Chennai Dar es Salaam Delhi Florence Hong Kong Istanbul
Karachi Kuala Lumpur Madrid Melbourne Mexico City Mumbai
Nairobi Paris São Paulo Singapore Taipei Tokyo Toronto Warsaw

and associated companies in
Berlin Ibadan

Oxford is a trade mark of Oxford University Press

Published in the United States
by Oxford University Press, Inc., New York

First published 2001

Library of Congress Cataloging in Publication Data
(Data applied for)
1 3 5 7 9 10 8 6 4 2

ISBN 0 19 879281 6 (Pbk)

Typeset by Newgen Imaging Systems (P) Ltd., Chennai, India

Printed in Great Britain on acid free paper by
Bath Press Ltd., Bath, Avon

"Biology is the search for the chemistry that works"

Prof. R.J.P. Williams

Preface to the third edition

It is now twenty years since this text was last updated and, whilst the basic principles of physical chemistry have not changed, the field of biochemistry has moved on very considerably since then. Advances in technology have resulted in a wider variety of physical techniques being applied to the study of biological systems. Such advances are making it possible to obtain more quantitative information about these systems. In addition, advances in biochemical knowledge and techniques are making it possible to study more and more complex systems. As a result, the detailed analysis and interpretation of quantitative data has become much more important. A good example of this is in the study of flux through metabolic pathways, where the ability to manipulate such pathways *in vivo* has provided evidence counter to the original "intuitive" approach to control in such systems of "flux generating steps" and has lead to the much more rigorous and mathematical approach of metabolic control analysis.

The underlying philosophy of the book remains just as valid today as it was twenty years ago. We feel that the use of biologically relevant examples and problems is the most effective way of conveying and reinforcing the principles of physical chemistry and their application to biological systems. While the principles are often perceived as "difficult topics" for students, there is increasing recognition that a sound appreciation of them will be necessary to take full advantage of the ever-increasing amount of structural information on biological systems being generated in large international projects.

The text of the 3rd edition has been completely rewritten and widened in scope in order to provide a more rounded coverage of the topics at the core of biochemistry and related areas of the molecular biosciences, namely molecular structure, thermodynamics and kinetics. In each of these topics new chapters and sections have been added to provide the connection between the basic principles and the biological systems. For example, the principles of atomic and molecular structure are linked to the structure of macromolecules and spectroscopy. The section on thermodynamics is linked via the concept of chemical potential to membrane bioenergetics. Chemical kinetics is linked to the kinetics of enzyme-catalysed reactions and to metabolic control analysis.

The expansion of these areas has necessitated the omission of some topics covered in the 2nd edition, including the sections on the ultracentrifuge and the use of radioactive isotopes. We believe that these topics are less relevant to the core of physical chemistry in the early years of an undergraduate programme, and already well covered in more advanced texts.

We hope that the 3rd edition of *Principles and Problems in Physical Chemistry for Biochemists* will build on the success enjoyed by the earlier editions of the book and will prove to be a valuable educational tool for students and teachers alike.

Glasgow and Oxford
June 2001

Nicholas C. Price
Raymond A. Dwek
R. George Ratcliffe
Mark R. Wormald

Preface to the second edition

In preparing the second edition of this book we have drawn on the experience of those using the first edition, and we are grateful for all the helpful suggestions and comments which have been made. A number of new problems have been included and we have made alterations to every chapter. The major changes have been to the chapter on binding studies (where an account of co-operative behaviour has been included), to the chapter on electrochemical cells (where a section on oxidative phosphorylation has been added), to the chapter o chemical kinetics (where greater emphasis is placed on the transition-state theory), and to the chapter on enzyme kinetics (which has been expanded to include a discussion of two-substrate reactions). A new chapter on macromolecules includes some of the previous material but a new section on the ultracentrifuge has been added.

In this edition we have changed to SI units as far as possible. While we realize that some of these units may not be familiar to some teachers, we feel that we should follow the advice of bodies such as the Biochemical Society who recommend that SI units should be used. A note on these units is included.

Stirling and Oxford
August 1978

Nicholas C. Price
Raymond A. Dwek

Preface to the first edition

There is a widespread belief, to which we subscribe, that the teaching of physical biochemistry is best accomplished by the students solving problems. However, all too frequently the examples chosen, while being ideal for chemistry students, have little relevance for the student of biochemistry who often has difficulty in seeing the application to his subject. We have tried to set problems which we think ought to be within the capabilities of a first or second year student, and which illustrate some of the more important present-day ideas and methods. The emphasis is on principles rather that any sophisticated mathematical manipulation (unfortunately all too common in many physical chemistry textbooks). In this light we have chosen our text specifically to provide the background for the problems at the end of each chapter, and the worked examples form an integral and important part of the text – serving to illustrate several new points. The emphasis in the treatment in the book is on equilibria and rates for we believe that it is an understanding of these phenomena that provides a secure basis for more advanced topics (such as the physical chemistry of macromolecules) to be developed.

In the first part of the book we have attempted to emphasize the universal applicability of thermodynamic equations to systems in equilibrium, not only in dealing with reactions but also with properties of solutions, acids and bases and oxidation-reduction processes. The second part of the book deals with rates of reactions, and here it is shown how many of the basic principles of chemical kinetics can be carried over into the kinetics of enzyme-catalysed reactions. For completeness we have included two short chapters of spectrophotometry and the uses of isotopes because these are of considerable importance in biochemistry. The section dealing with solution to problems contains not only the numerical answers but sufficient comment and working to enable individual students to see if, or where, they may have made errors or misunderstood certain principles. This should enable students to work through the problems, to a large extent, on their own. Where appropriate we have tried to suggest the importance of the results. There are several appendices containing material which can be omitted at a first reading without creating difficulties in understanding the text.

Finally, it should be noted that biological systems are in general much more complicated than those dealt with by chemists. It is therefore often necessary to make drastic simplifications to perform any calculations from fundamental principles, at least at the level of this book.

Oxford
November 1973

Nicholas C Price
Raymond A. Dwek

Acknowledgments

First edition

In writing this text we have drawn freely on the help and advice of our colleagues in the Department. In particular we should like to thank Drs Keith Dalziel, David Brooks, John Griffiths and Simon van Heyningen. Professor H. Gutfeund made some useful criticisms of the manuscript.

We are also indebted to Mrs Shirley Greenslade who carefully and patiently typed the original manuscript.

Second edition

We should like to thank those friends and colleagues who have given help and advice in the preparation of this edition and in particular Drs Simon Easterbrook-Smith, Lewis Stevens, Stuart Ferguson, and Peter Zavodszky.

Second edition – 1982 reprint

We have made a number of changes to the text, notably in Chapter 7 where we have attempted to make the treatment of acid/base chemistry thermodynamically more rigorous. Nearly all the changes in the text were inspired by Charles Eliot and we thank him for his enthusiasm.

Third edition

Many of our students and colleagues have continued to give us helpful comments on the previous editions of this text. In particular, MRW would like to thank Dr Peter Hore for many helpful discussions and much advice and Professor Iain Campbell who read many of the new chapters at an early stage. We also thank Susan Ratcliffe for proof-reading the entire text.

Contents

Notes to the reader

This book is written with the intention of providing a comprehensive coverage, sufficient for a biochemistry degree, of the basic principles of physical chemistry. We have not attempted to cover in any detail the wide range of biophysical techniques that are now used, although the principles covered should provide a solid grounding for understanding such techniques.

Some of the sections dealing with applications of these principles can also be treated as introductions to more advanced texts. These include the sections on the use of electrochemical gradients as biological energy stores (Chapter 7), metabolic control analysis (Chapter 13), spectroscopy (Chapter 17) and macromolecular structure (Chapter 18).

Some of the material is also rather more advanced and may be beyond the scope of some courses. This includes the treatments of multiple ligand binding (Chapter 4), strong ion difference (Chapter 5) and multiple substrate enzyme kinetics (Chapter 12).

Worked examples are an integral part of the book and are often used to clarify and explain further the principles discussed in the main text. The text can be read omitting the worked examples but we would strongly encourage the reader not to do this. The problems, the worked answers to which are given in appendix 3, are also an integral part of the book. Whilst these do not introduce any new concepts or principles, they do provide additional applications of such principles and give a wider range of examples of how quantitative analysis can give insight into the behaviour of complex systems.

There are two short appendices, 1 and 2, at the end of the text. These are intended to make the book easier to use. The first deals with the definitions of the units used throughout the text (SI units), the conversion factors between different units and the values of fundamental constants. It is very important during calculations that consistent units are used for all quantities. The second is a brief summary of mathematical techniques. The purpose of this appendix is not to teach the maths, but to act as a checklist of the tools that you need to be able to follow the text and tackle the problems. For those unfamiliar with SI units or who may be intimidated by the thought of the maths, it is probably a good idea to have a quick look through these appendices first.

Symbols

The following is a list of the symbols used in this text. For units and the values for the fundamental constants, see Appendix 1. Unfortunately, some standard symbols have two distinct meanings (such as *h* and S) but we have retained these to be consistent with other texts.

[x]	Concentration of a compound x
$[x]_{eq}$	Concentration of a compound x at equilibrium
a_x	Activity of a compound x
m_x	Stoichiometry coefficient of x
n_x	Number of moles of a compound x
N_x	Mole fraction of a compound x, equals n_x divided by total number of moles present
Δy	Large change in a quantity y
δy	Small change in a quantity y
Δy°	Change in a quantity y with all components in their standard states
$\Delta y^{\circ\prime}$	Change in a quantity y with all components in their biochemical standard states
$\Delta y^{\ddagger}$	Change in quantity y between reactants and the transition state
$\mathcal{A}$	Pre-exponential factor in the Arrhenius equation
A	Absorbance, or optical density
A, *B*	Constants in the Debye–Hückel equation
β	Buffering capacity
c	Speed of light in a vacuum
$C^{j}_{E_i}$	Flux control coefficient for enzyme E_i in a metabolic pathway
C_f	Cooperativity factor in folding of macromolecules
C_p	Heat capacity at constant pressure
Δ	Crystal field splitting
Δp	Proton motive force
ϵ	Dielectric constant, permittivity
$\epsilon^{V_i}_{X}$	Elasticity coefficient for an effector molecule X in a metabolic pathway
ϵ_o	Vacuum permittivity

ϵ_λ	Extinction coefficient at wavelength λ
e	Charge on a proton
E	Energy
E_a	Activation energy, used in reaction kinetics
Φ	Electric potential
F	Faraday constant
γ	Activity coefficient of an ion
$\gamma_\pm$	Mean activity coefficient of an ionic solution
Γ	Mass action ratio
g	Acceleration due to gravity
G	Gibbs free energy
h	Hill coefficient for cooperative binding (the symbol n_H is used in some texts)
h	Planck's constant, $\hbar = h/2\pi$
h	height
H	Enthalpy
I	Ionic strength
J	Flux through a metabolic pathway
κ	Transmission coefficient in transition state theory
k	Boltzmann constant
k	Rate constant
K	Equilibrium constant
K^{app}	Apparent equilibrium constant, determined using concentrations
$K^{\ddagger}$	Equilibrium constant between reactants and the transition state
K_a	Dissociation constant for an acid
K_b	Dissociation constant for a base
K_d	Dissociation constant for ligand/substrate binding to a macromolecule (macroscopic dissociation constant)
κ_d	Dissociation constant for a ligand binding to a specific site on a multisite macromolecule (microscopic dissociation constant)
K_m	Michaelis constant for an enzyme reaction
K_w	Dissociation constant for water
λ	Wavelength of a wave
l	Angular quantum number for an electron in an atomic orbital (see n and m)
L	Avogadro number
μ	Dipole moment
μ	Chemical potential
$\bar{\mu}$	Electrochemical potential (chemical potential in an electric field)
m	Angular quantum number for an electron in an atomic orbital (see n and l)
m	Mass
m_w	Molar mass of water
n	Radial quantum number for an electron in an atomic orbital (see l and m)

n	Number of ligand binding sites on a macromolecule
$\bar{n}$	Average number of ligands bound to a macromolecule
N	Mole fraction
Π	Osmotic pressure
p	Momentum
P	Pressure of a gas
P_X	Partial pressure of a component X of a gas
P^*	Vapour pressure of a gas in equilibrium with the liquid phase
P_i	Inorganic phosphate ion
θ	Fraction of binding sites occupied on a protein
q	Heat energy
ρ_w	Density of water
r	Internuclear distance
R	Gas constant
s	Spin quantum number for an electron in an atomic orbital
S	Net spin of all the electrons in an atom or a molecule
S	Entropy
t	Time
T	Temperature
U	Internal energy
ν	Frequency of a wave
v_R	Relative velocity of two molecules
v	Velocity or velocity of a reaction
V	Volume
$\bar{V}_w$	Partial molar volume of water
V_{max}	Limiting rate for an enzyme reaction
W	Work energy
$\{X\}_{effective}$	Effective concentration of two groups on a macromolecule relative to each other
Ψ	Water potential

INTRODUCTION

The consequences of physics and chemistry for life

1

All biological systems, varying from single cells to multicellular organisms, like the reader, are a complex collection of interlinked chemical reactions. These chemical reactions are responsible for growth, which involves making more chemical compounds and assembling them; reproduction, creating new independent biological systems just like the parent; movement, exerting a force on the outside world, like turning a page; and even awareness and intelligence, involved in reading this book. The chemical reactions involved in these processes are very carefully controlled, both temporally, that is when and how fast they are allowed to occur, and spatially, that is where in a cell or organism they are allowed to occur. However, despite the vast complexity of the multitude of chemical reactions that occur and the great variety of chemical compounds that can be made, living systems cannot just do anything that they want. They have to work within very stringent restrictions and limitations.

The constraints imposed on a biological system by its environment

Biological systems do not exist in isolation but interact in a complex way with their external environment. In general, biological systems can only change their environment in limited ways. Thus, biological systems have evolved in such a way as to survive and grow within the limits imposed by their environment. These limits include

- Natural abundance of the elements. Biological systems cannot synthesise elements, thus they must make do with what they can find. For instance, evolving extensive life based on boron chemistry would not be possible on this planet.
- Chemical availability of elements. Even if an element is abundant in the environment, it has to be extracted by the biological system. Elements that are only available in inert forms are very difficult to utilize.
- External physical conditions, including gravity, pressure, radiation levels, temperature. Biological systems have to withstand whatever the environment

throws at them. For instance, the difference in pressure between the inside of a cell and its environment must be such that it does not implode or explode.

- External chemical environment. As well as the physical impact of the environment, the biological system has to cope with the chemical reactivity of the species that make up the environment. For instance, there is no point in having a cell wall that is water soluble if you want to survive in the ocean.
- Available sources of energy. As well as building materials, biological systems need energy to survive and grow and this must also come from the environment, either directly from light and heat, or indirectly by eating other living, or once living, things.

Thus the environment places considerable limits on how life can evolve; change the environment and you will change the nature of the biological systems that will develop.

The constraints imposed on a biological system by physics and chemistry

As well as environmental constraints, there are further constraints imposed on biological systems that apply in *all* environments. These are very basic constraints that have their origin in the fundamental way that matter and energy behave.

- Chemical properties of the elements and their compounds. The chemical properties of the elements limit the kinds of compounds that can be made from them. The physical and chemical properties of these compounds, determined by the elements that make them, limit their uses and the reactions that they will undergo.
- Energetic requirements of chemical reactions. In order to make chemical reactions occur, it is necessary to supply energy. Thus a balance has to be maintained between the energy available from the environment and the reactions that need to be done.
- Rates of chemical reactions. Even if chemical reactions can be made to occur, they have to be done at the right speed to fit in with all the other reactions that are taking place. It is usually very difficult to slow down chemical reactions, except by physically separating the reactants, and speeding them up involves providing either additional energy or a catalyst, such as an enzyme.

Physical chemistry and 'Biochemists'

The above constraints limit both what a living system can do, that is what reactions it can carry out, and how it can do them. Unless we understand how these constraints operate, we cannot understand how biological systems carry

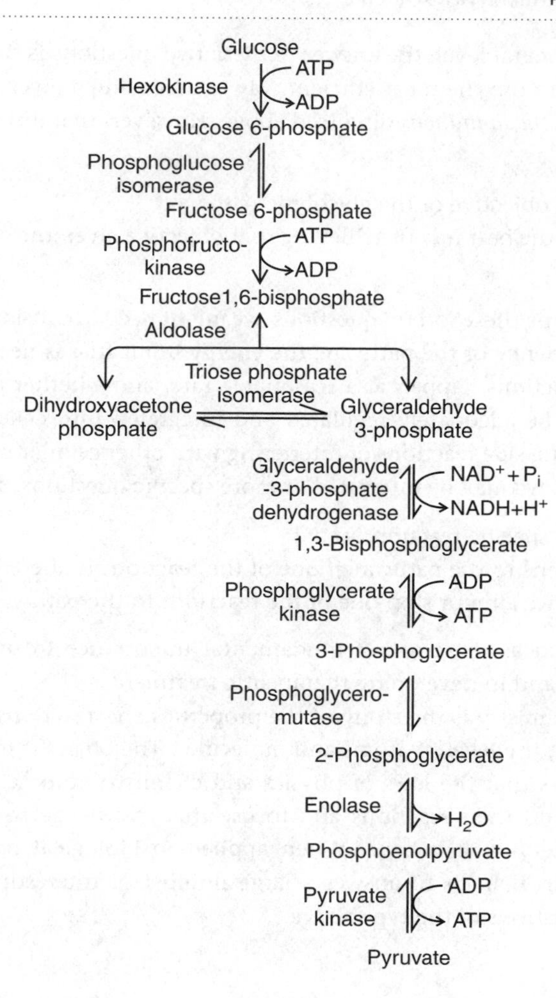

Fig. 1.1 The glycolytic pathway breaks down glucose obtained from the environment into pyruvate and the chemical energy released in this process is stored as ATP.

out and control reactions and, more generally, why they do things in the way that they do.

For example, the glycolytic pathway (figure 1.1) converts glucose to pyruvate, producing ATP. This is a complex series of coupled chemical reactions. The situation becomes more complex when you realize that many of the chemicals are also involved in other processes as well. Although the figure tells us *what happens* in the glycolytic pathway, it does not answer some of the more fundamental questions

- Why has evolution selected for this method of making ATP by converting glucose to pyruvate, rather than by using a different energy-producing reaction?
- Why is this precise route for the conversion of glucose to pyruvate used and not any other potential route, indeed why is it not just done in one single step?

At the most general level, the answer to these two questions is that biological systems tend to use the most efficient way of achieving a given target while *working within the limitations* discussed above. However, that just recasts these two questions as

- What is the objective of the glycolytic pathway?
- Why is this the best way of achieving that objective given the constraints, if indeed it is?

When answering these sorts of questions, we might need to consider the overall energetic efficiency of the pathway, the energy input that is needed to make individual reactions happen at a reasonable rate, and whether the reactions involved can be adequately regulated and integrated into cellular function without causing side reactions or interfering with other cellular processes.

In addition, we may be interested in more specific questions, such as

- How does a specific reaction occur?
- What happens to the pathway if one of the reactions is altered?
- How could we alter or stop one of the reactions in the pathway?

Questions such as these are of fundamental importance in understanding disease states and in developing therapeutic treatments.

Physical chemistry is the study of the properties and reactions of chemical compounds at the level of atoms and molecules. The object is to understand the constraints that the laws of physics and chemistry impose on chemical compounds and their reactions and to use this knowledge to predict how chemical systems will behave. When applied to biological processes, this knowledge can help us to answer a large number of interesting questions, including questions of the type above.

Scope of this book

This book is divided into three main sections. These deal with the major constraints that the laws of physics and chemistry impose on biological systems

• The energetics of chemical reactions (Thermodynamics)	Deals with the energy requirements of chemical processes.
• The rates of chemical reactions (Chemical kinetics)	Deals with the rates at which processes occur.
• Atomic and molecular structure (Quantum mechanics)	Deals with how the atomic structures of molecules determine their chemical properties.

The last section also provides an introduction to molecular spectroscopy, the way in which chemical compounds interact with electromagnetic radiation.

This is an important area of biochemistry, concerned with such systems as photosynthesis and vision, and also one of the main techniques for studying molecular and macromolecular structures and interactions.

This book is not intended to be a general physical chemistry textbook, and we shall limit ourselves to those areas that have a direct application to biology. The emphasis throughout will be on understanding the basic principles involved and applying them to biochemical systems.

FURTHER READING

P.R. Bergethon, 1998, 'The physical basis of biochemistry', Springer-Verlag—*a book that you should aspire to read after this one.*

R.J.P. Williams and J.J.R.F. da Silva, 1999, 'Bringing chemistry to life', Oxford University Press—*a wide ranging account of the doctrine that 'biology is the search for the chemistry that works'.*

THE ENERGETICS OF CHEMICAL REACTIONS

Basic thermodynamics

2

KEY POINTS:

- Thermodynamics is the study of energy and how it is distributed.
- The total energy in a system can be divided into *work energy* and *heat energy*.
- The first law of thermodynamics is the law of conservation of energy. This allows energy to be converted from one form to another but not to be created or destroyed.
- The second law of thermodynamics deals with the change in work energy during a process. This says that during a spontaneous process the amount of work energy must decrease.
- *Enthalpy* (H) is a measure of the total energy of a system, after allowing for expansion. *Entropy* (S) is a measure of the disorder in a system, and entropy multiplied by temperature (TS) is a measure of the heat energy. *Free energy* (G) is a measure of the work energy in a system. These are related by $\Delta G = \Delta H - T\Delta S$.
- *Standard states* of compounds are the reference points from which changes in H, S and G can be measured.
- A process will proceed spontaneously if G decreases, i.e. if $\Delta G < 0$. At equilibrium $\Delta G = 0$.
- The values of ΔH and ΔS for a reaction can be related to the molecular changes that occur during the reaction. However, it is important to remember that ΔH and ΔS are properties of the entire system and include contributions from all changes that occur.

What is thermodynamics?

In everyday life we are familiar with the idea that energy is important in making things happen, for example energy is clearly involved in lifting weights, moving about or generating heat, light and electricity. Moreover, it is also apparent that the right amount of energy needs to be used to achieve a particular objective, for example there is a minimum amount of energy that will be sufficient to knock a nail into a wall but there is no point in using so much energy that the wall collapses as the nail is driven home. The availability of energy is equally critical for the behaviour of molecular systems at the microscopic level. Thus the extent to which a collection of molecules, such as those comprising a living cell, can change in some way, for example as a result of a chemical reaction or as a result of a transport process that redistributes the molecules, depends on the energy requirement of the change and the availability of sufficient energy to enable it to happen.

Considerations of this kind lead to a series of fundamental questions that can be asked of any molecular system

- Is the system stable?
- If not, then which rearrangements of the system would lead to an increase in stability?
- How long will it take to reach the more stable state?

In order to answer these questions it is necessary to have a framework for analysing the availability of energy in the system and the way in which it can be utilised and redistributed. This is the province of thermodynamics—the study of energy, its distribution and the way in which it is converted from one form to another. Thermodynamics is central to any understanding of the way in which systems change and it can be used to answer a host of biochemically relevant questions, such as

- Are living systems stable?
- What are the energetic costs of cellular functions?
- How can energy be stored and made available when required?
- To what extent do cells equilibrate with their surroundings?

As we shall see in Chapter 9, the availability of energy is also critical for determining the rate at which changes occur and so thermodynamic concepts are also needed to answer such questions as

- On what time scales do different biochemical reactions occur?

Thermodynamics is the subject of this and the next six chapters, and before proceeding it is worth making a few general points.

- Thermodynamics is based on simple empirical laws, derived from observation of the real world. We do not need any knowledge about the nature of matter or the universe in order to use these laws, we do not even have to know that atoms or molecules exist. We use thermodynamics because it works, not because it is an elegant theory.
- As the laws of thermodynamics are not based on any specific physical model, they can be applied to an enormous range of problems, varying from mechanical engineering (performance of engines, refrigerators, etc.) and large-scale biological systems (efficiency of muscle action) through to chemical reactions and the stability of molecular systems.

- In order to relate thermodynamic quantities to the properties of a system, we do need a physical model. If we then predict the wrong behaviour for our system, it is because the model we have chosen is wrong.

Basic definitions

Thermodynamic ideas depend critically on the distinction between a system and its surroundings.

A *system* consists of matter. A complete description of a system would include defining the composition, as well as the pressure, volume and temperature. There are three different types of system (figure 2.1)

- Isolated system — a system that cannot gain or lose matter or energy
- Closed system — a system that can gain or lose energy but not matter
- Open system — a system that can gain or lose matter and energy

Chemical reactions occurring within cells, or performed in the laboratory, can often be treated as closed systems. Processes involving transfer across membranes usually have to be treated as open systems. We shall deal initially with closed and isolated systems, as these are the more straightforward to analyse.

The *surroundings* are anything in contact with the system that can influence its state. In a closed system energy can be exchanged with the surroundings, while in an open system energy *and* matter can be exchanged with the surroundings. An open or closed system plus its surroundings makes an isolated system.

Any property of a system that only depends on its current state, and not on how it got that way, is called a *state function*. There are two different kinds of state function

- Intensive state function — independent of the size of the system, e.g. temperature, pressure, density, etc.
- Extensive state function — depends on the size of the system, e.g. mass, volume, etc.

If the size of a system is doubled, the value of an intensive state function remains the same whereas the value of an extensive state function doubles. Most thermodynamic functions are extensive state functions.

When talking about changes to state functions, we shall use Δ to represent a large, measurable change and δ to represent a very small change (see Appendix 2). Thus, ΔT would be a large change in temperature and δT would be a very small change in temperature. As the temperature of the initial and final states of a system are independent of the path taken to get from one to the other, the change in temperature must be

$$\Delta T = T_{final} - T_{initial}$$

This type of equation (final – initial) applies to *all* other state functions.

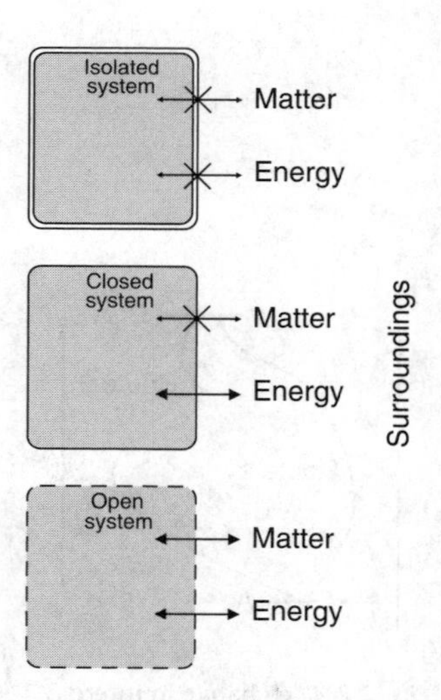

Fig. 2.1 A diagrammatic representation of the classification of systems used in thermodynamics.

Statement of the first law of thermodynamics

The first law of thermodynamics is the law of conservation of energy.

> **FIRST LAW**
> The algebraic sum of all the energy changes in an isolated system is zero.

Thus, energy can be moved from one place to another, as in heat flow, or it can change form, as in the conversion of chemical energy to electrical energy in a battery, but it cannot be created or destroyed.

It is useful to define a quantity called the *internal energy*, U. This is the total energy of a system, regardless of type. If energy is converted from one form to another, for instance chemical energy to electrical energy, the internal energy remains unchanged. In an isolated system, which cannot gain or lose any energy to the surroundings, the first law of thermodynamics states that the change in internal energy during any process must be zero, i.e. for an isolated system

$$\Delta U = 0$$

For a closed or open system, energy can be exchanged with the surroundings. ΔU for the system need no longer be zero, but any energy lost or gained by the system must be balanced *exactly* by a gain or loss of energy by the surroundings because the system and its surroundings together comprise an isolated system, i.e. for a closed system

$$\Delta U(\text{system}) = -\Delta U(\text{surroundings})$$

Thus ΔU can be used as an accounting tool, it lets us keep track of energy irrespective of the form that it takes. It also follows from the first law that U must be a state function.

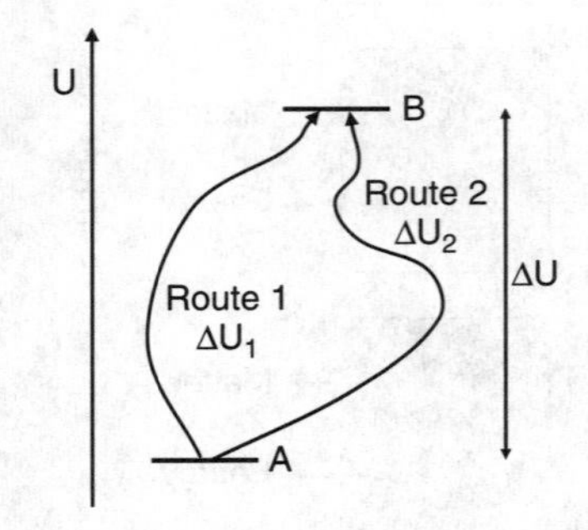

Fig. 2.2 The change in internal energy, ΔU, going from A to B must be the same regardless of the path taken. This is the definition of a state function.

> **WORKED EXAMPLE**
>
> **Prove that U is a state function.**
>
> **SOLUTION:** Consider a process going from state A to state B that can be done by two different routes, 1 and 2, with changes of internal energy of ΔU_1 and ΔU_2 respectively (figure 2.2).
>
> If we go from A to B by route 1 and back from B to A by route 2, the total change in U on going from A to A will be $\Delta U_1 - \Delta U_2$.
>
> If ΔU_1 did not equal ΔU_2, we would have either created or destroyed energy. Hence ΔU_1 must equal ΔU_2, which is the definition of a state function.

Work and heat and internal energy

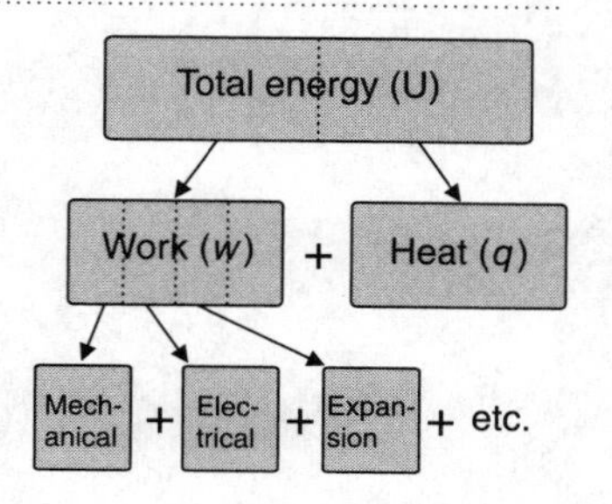

Fig. 2.3 The total internal energy in a system can be divided into work energy and heat energy. There are also many different forms of work energy. The total energy in an isolated system is constant but work energy can be converted into heat energy or into another form of work energy.

As we have seen, U is an accounting tool that we can use to track the movement of energy. However, it is also convenient to distinguish between two different forms of energy, *work*, w, and *heat*, q (figure 2.3). We are used to distinguishing between work and heat in everyday life and similar definitions can be used in molecular systems. Work is the energy associated with the orderly movement of bodies, for example pushing back a boundary (expanding a piston, or increasing volume). Heat is the energy associated with the disorderly movements of bodies, for example the molecular motion in liquids and gases (when we heat something up, the molecules move faster).

The work done on a system by the surroundings is defined as Δw and the heat absorbed by the system from the surroundings as Δq. The internal energy of the system must increase by

$$\Delta \text{U} = \Delta q + \Delta w \qquad \text{or} \qquad \delta \text{U} = \delta q + \delta w$$

Similarly, the internal energy of the surroundings must decrease by the same amount. The total amounts of work and heat energy need not be constant, unlike U, as energy can be converted from one form to the other.

Work energy in chemical systems

In some biochemical systems several forms of work energy can be carried out, such as the mechanical work performed by muscles or the electrical work required to move an ion through an electric field gradient. These work terms are often of great interest to biochemists. However, there is one form of work that almost all chemical reactions perform at constant pressure and that is of almost no interest, apart from removing energy from the system. This is the work of expansion.

If the volume of a system increases by ΔV during a chemical reaction then the system must be doing work on the surroundings, it is pushing back its own boundaries and making the surroundings smaller. Δw, which is defined as the work done on the system, is therefore negative. At constant pressure P, Δw is given by

$$\Delta w_{\text{P}}(\text{expansion}) = -\text{P}\Delta \text{V}$$

where the subscript $_{\text{P}}$ is used to denote a process occurring at constant pressure. If expansion is the only form of work energy exchanged between the system and the surroundings, as is the case for most chemical reactions in solution, then from the first law

$$\Delta \text{U}_{\text{P}} = \Delta q_{\text{P}} - \text{P}\Delta \text{V}$$

For reactions in solution, ΔV is very small and so the work associated with expansion is also very small.

The concept of enthalpy

For reactions at constant pressure, it is convenient to define an energy function *enthalpy*, H, such that

$$H = U + PV$$

At constant pressure,

$$\Delta H = \Delta U + P\Delta V$$

which can be rewritten as

$$\Delta H = \text{Change in total energy} - \text{Energy required for expansion}$$

Thus,

H can be viewed as a corrected form of U that allows for the work of expansion at constant pressure.

If expansion is the only form of work energy exchanged between the system and the surroundings, then

$$\Delta U_P = \Delta q_P + \Delta w_P = \Delta q_P - P\Delta V$$

and so

$$\Delta q_P = \Delta U_P + P\Delta V = \Delta H$$

Thus ΔH equals the heat absorbed at constant pressure, but only if the system is doing no work other than expansion.

Standard states

Every system has an absolute value of U and H that depends on the state of the system, i.e. composition, temperature, etc. However, these values can almost never be measured. The values that can be measured are the changes in U and H as the system changes from one state to another, ΔU and ΔH. It is useful to have a common reference point for such changes.

Standard states are the reference points from which energy changes are measured.

Standard states are defined as follows

- Solid pure solid
- Liquid pure liquid
- Solute concentration of 1 M (1 mole per litre of solution)
- Gas pure gas at a pressure of 1 atmosphere

Standard states can be defined at any temperature, although if a temperature is not specified it is usually assumed to be 25°C or 298 K.

If the solute is H^+, then an aqueous solution in the standard state would correspond to a proton concentration of 1 M, or pH = 0 (see Chapter 5). This standard state is of less interest to biochemists who are often concerned with solutions at pH ≈ 7. So a *biochemical standard state* is defined where all components are present in their standard states except for H^+ which is present at 10^{-7} M.

When 1 mole of reactants in their standard states is converted into 1 mole of products in their standard states, thermodynamic quantities relating to this process are denoted by a superscript ° (ΔU°, ΔH°). When the reactants and products are in the biochemical standard state, a superscript °′($\Delta H^{\circ\prime}$) is used.

Enthalpy as a state function

For many chemical reactions, the change in enthalpy can be measured directly because $\Delta H = \Delta q_P$ and Δq_P can be determined by measuring the change in temperature of the system caused by the reaction (a technique called calorimetry). We may also be interested in ΔH of a reaction that is difficult to do. However, because enthalpy is an extensive state function we can do a reaction in a series of steps and then add all the resultant enthalpy terms together. This is known as Hess's law of constant heat summation.

WORKED EXAMPLE

The enthalpy change when 1 mole of a substance is made from its elements is called the enthalpy of formation, ΔH_f°. How could ΔH_f° be determined for glucose?

SOLUTION: The formation of glucose from its elements is given by the following equation.

$$6C(graphite) + 6H_2(g) + 3O_2(g) \rightleftharpoons C_6H_{12}O_6(s)$$

We cannot do this reaction and make direct experimental observations. However, we can measure the enthalpy of combustion, the enthalpy change when 1 mole of a substance is fully oxidised by O_2, for each of the components involved. As enthalpy is an *extensive* state function, we can add and subtract enthalpy changes directly.

(1) $C(graphite) + O_2(g) \rightleftharpoons CO_2(g)$ $\Delta H^\circ = -393{\cdot}1\ kJ\ mol^{-1}$
(2) $H_2(g) + \frac{1}{2}O_2(g) \rightleftharpoons H_2O(l)$ $\Delta H^\circ = -285{\cdot}5\ kJ\ mol^{-1}$
(3) $C_6H_{12}O_6(s) + 6O_2(g) \rightleftharpoons 6CO_2(g) + 6H_2O(l)$ $\Delta H^\circ = -2821{\cdot}5\ kJ\ mol^{-1}$

The reaction for the formation of glucose is given by

$$6 \times (1) + 6 \times (2) - (3)$$

Thus

$$\Delta H_f^\circ = 6(-393{\cdot}1) + 6(-285{\cdot}5) - (-2821{\cdot}5)$$
$$\Delta H_f^\circ = -1250{\cdot}1\ kJ\ mol^{-1}$$

A reaction with a negative ΔH is termed exothermic and a reaction with a positive ΔH is called endothermic. ΔH_f° for glucose is negative, indicating that energy is released to the surroundings on forming glucose from its elements.

COMMENT: It is tempting to relate these ΔH° values to the utilisation of glucose as an energy source by living systems. Breaking glucose down to its elements requires a large input of energy from the surroundings, $\Delta H^\circ = +1250{\cdot}1\ kJ\ mol^{-1}$, suggesting that it would not be possible to use the reaction as a source of energy in a biochemical system. In contrast, the oxidation of glucose releases considerable energy to the surroundings, $\Delta H^\circ = -2821{\cdot}5\ kJ\ mol^{-1}$, and in fact the final end-products of aerobic glucose metabolism are CO_2 and H_2O. However, while this argument appears to provide a full explanation for the observed facts, it is based on the erroneous assumption that it is ΔH that determines whether a reaction is favourable or not. As we shall see, a complete analysis involves more than just considerations of ΔH.

The first law and the direction of a reaction

The first law of thermodynamics only deals with the energy balance during a process. It does not give us any information about whether that process will occur or not, just that we cannot create or destroy energy if it does. Similarly, the enthalpy change, ΔH, during a reaction does not give us any information on whether the reaction is likely to occur. For example

- Dissolution of $NaNO_3$ in water spontaneous ΔH = +ve
- Dissolution of NaOH in water spontaneous ΔH = −ve
- Diffusion of Na^+ ions down a concentration gradient spontaneous ΔH = 0

Thus, in order to predict whether a reaction will occur we need to consider something other than the total energy change in the system.

Available energy, work and change

To change a system from one state to a second state is an orderly process. We want to make a system change but we also want to make it change in a specific way. To do this, we must control the energy that we use. For example, we cannot just heat a pile of bricks and hope that the energy we put in will rearrange them into a house. We have already made the distinction between work, the energy involved in orderly movement, and heat, the energy involved in disorderly movement. In order to make a system change, work energy must be used.

> Work energy is the driving force for change.

The work energy can come from within the system itself, in which case the system is unstable and can change spontaneously. Alternatively, the work energy can come from the surroundings, for example a stable system can be forced to change by the input of energy. As the first law of thermodynamics only deals with the sum of heat and work and not work on its own, it is not surprising that it cannot predict the direction of a reaction.

Statement of the second law of thermodynamics

The second law of thermodynamics relates work and change.

> SECOND LAW
> Spontaneous changes are those that can be made to do work.

This is just a formal statement of the ideas in the previous section. A system that is unstable and can change spontaneously must have available work energy. During a spontaneous change work energy is released and dissipated as heat. However, this work energy could be harnessed and used. A system that is stable and cannot change has no available work energy. In this case, change can only occur by driving the process using an external source of work energy. In order to predict whether a system will change spontaneously, we need to know whether the system can do work, i.e. whether there is work energy available in the system which can be released during the change.

The concept of entropy

In order to calculate whether there is any available work energy in a system, we have to introduce another thermodynamic quantity, the *entropy*, S. The simplest way to visualise S is

> Entropy is a measure of the disorder of a system.

For instance, the entropy of liquid water at 0°C is much greater than that of ice at 0°C, because the molecules are held in fixed positions in ice, whereas in water they are free to move. For liquid water, as we increase the temperature the molecules move faster and so the disorder, or entropy, of the system increases. Similarly, the entropy of water vapour is much greater than that of liquid water because of the even greater freedom of movement of the molecules in the vapour. To increase the temperature of a liquid we have to put in heat. To go from the solid to liquid or liquid to gas we also have to put in heat, the latent heat of fusion or vaporisation. Thus, the input of heat energy leads to an increase in disorder, whether or not the temperature changes.

This leads us to a second way of looking at entropy.

> Entropy (S) multiplied by temperature, TS, is a measure of the amount of heat energy, q, present.

If we know how much heat energy is present and what the total internal energy is, then the difference between the two must be the work energy in the system.

Alternative expression of the second law of thermodynamics

As TS is a measure of the heat energy, $T\Delta S$ gives the change in heat energy in a system. Δq is the heat energy transferred to the system from the surroundings. For a system that is not changing, called a system at equilibrium, no work energy is being used and so the change in heat energy in the system equals the amount of heat energy added to the system.

At equilibrium

$$T\Delta S = \Delta q$$

During a spontaneous change, work energy is being released and converted to heat energy. Thus, the change in heat energy of the system, $T\Delta S$, must be greater than the heat absorbed from the surroundings, Δq.

Spontaneous change

$$T\Delta S > \Delta q$$

The second law can then be expressed by the equation $T\Delta S \geq \Delta q$. The difference between $T\Delta S$ and Δq gives the amount of work energy that has been released during a change.

The concept of free energy

Rather than use the change in heat energy as an indirect measure of the change in work energy, it would be more convenient to deal with work energy directly because it is this that can be used to drive change. For convenience, we can define a property called the *Gibbs free energy*, G, as

$$G = H - TS$$

For a reaction occurring at constant pressure and temperature, the change in G is given by

$$\Delta G = \Delta H - T\Delta S$$

This can be rewritten as

Change in free energy	=	Change in total internal energy allowing for expansion	−	Change in heat energy

Thus,

> G is a measure of the total work energy of a system and ΔG is a measure of the change in work energy of a system during a reaction.

For a reaction to proceed spontaneously, work energy in the system is being released and so G must decrease, $\Delta G < 0$. At equilibrium there is no change in work energy, $\Delta G = 0$.

For an isolated system, $\Delta H = 0$ and so the criterion for a spontaneous change is that $\Delta S > 0$. The most common isolated system is the universe itself and so the entropy, or randomness, of the universe must always increase during any change. The entropy in a small part of the universe can decrease, as long as there is an equal or larger increase elsewhere to balance it.

Like enthalpy, entropy and free energy are extensive state functions. Thus we can add and subtract entropy and free energy changes in the same way that we did for changes in enthalpy.

WORKED EXAMPLE

Given that $\Delta G^{o\prime}$ for the hydrolysis of ATP to ADP and P_i (phosphate) is $-30{\cdot}5$ kJ mol^{-1} and $\Delta G^{o\prime}$ for the hydrolysis of ADP to AMP and P_i is $-31{\cdot}1$ kJ mol^{-1}, what is $\Delta G^{o\prime}$ for the following process?

$$\text{ATP} + \text{AMP} \rightleftharpoons 2\text{ADP}$$

SOLUTION: As G is an *extensive* state function, ΔG values can be added and subtracted directly. Thus, the above reaction can be rewritten as

$$\begin{array}{ll} \text{ATP} + H_2O \rightleftharpoons \text{ADP} + P_i & \Delta G^{o\prime} = -30{\cdot}5\ \text{kJ mol}^{-1} \\ + & + \\ \text{AMP} + P_i \rightleftharpoons \text{ADP} + H_2O & \Delta G^{o\prime} = +31{\cdot}1\ \text{kJ mol}^{-1} \\ = & = \\ \text{ATP} + \text{AMP} \rightleftharpoons 2\text{ADP} & \Delta G^{o\prime} = 0{\cdot}6\ \text{kJ mol}^{-1} \end{array}$$

COMMENT: The reaction will not occur spontaneously under standard conditions because $\Delta G^{o\prime}$ is positive. $\Delta G^{o\prime}$ for the reverse reaction, 2ADP $\rightleftharpoons$ ATP + AMP, is $-0{\cdot}6$ kJ mol^{-1} and so this reaction could occur spontaneously in the biochemical standard state. In practice, a solution of ADP is stable because the reaction is too slow to be observed.

WORKED EXAMPLE

Calorimetric measurements give $\Delta H^{o\prime} = -20{\cdot}1$ kJ mol^{-1} for the hydrolysis of ATP to ADP at 310 K. What is the value of $\Delta S^{o\prime}$ for this process?

SOLUTION: From the worked example above, $\Delta G^{o\prime} = -30{\cdot}5$ kJ mol^{-1}

$$\Delta G^{o\prime} = \Delta H^{o\prime} - T\Delta S^{o\prime}$$

$$(-30{\cdot}5 \times 10^3) = (-20{\cdot}1 \times 10^3) - (310 \times \Delta S^{o\prime})$$

Thus

$$\Delta S^{o\prime} = 33{\cdot}5\ \text{JK}^{-1}\text{mol}^{-1}$$

COMMENT: As $\Delta H^{o\prime}$ is negative and $\Delta S^{o\prime}$ is positive, the hydrolysis of ATP is both enthalpically and entropically favourable. It is tempting to try to interpret the $\Delta H^{o\prime}$ and $\Delta S^{o\prime}$ values in terms of the phosphate bond in ATP being broken and this bond is sometimes referred to as a 'high-energy' phosphate bond. However, the values of $\Delta H^{o\prime}$ and $\Delta S^{o\prime}$ refer to changes in the whole system and so depend on many other factors (see the worked example in the last section of this chapter).

TABLE 2.1

Type of system	Exchange with surroundings	Criterion for equilibrium	Criterion for spontaneous change
Isolated system	None	$\Delta S = 0$	$\Delta S > 0$
Closed system	Energy	$\Delta G = 0$	$\Delta G < 0$

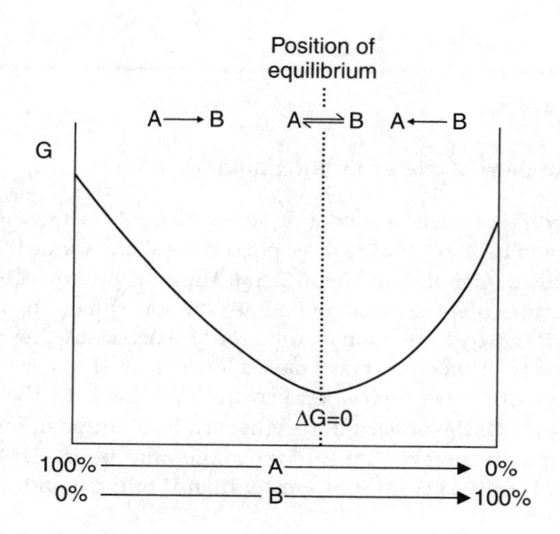

Fig. 2.4 A plot of free energy, G, versus composition for A interconverting with B. The system will change, either A will convert to B or B will convert to A, in the direction of minimising G until the point at which $\Delta G = 0$. At this point, A is in equilibrium with B and no further change will occur.

Free energy and equilibrium

Taken together, the first and second laws of thermodynamics provide criteria for predicting whether a system is at equilibrium or whether it can undergo a spontaneous change (summarised in Table 2.1).

These criteria also allow us to predict the eventual position of equilibrium for a spontaneous reaction, the point at which the system will stop changing. Consider the simplest possible chemical reaction, a species A interconverting with a species B. We shall see many real examples of such systems in subsequent chapters. If G can be determined for any given composition of A and B, G can be plotted versus composition (figure 2.4). At whatever composition we start from, the reaction will proceed in the direction of reducing G until a minimum is reached at which point $\Delta G = 0$. Thus, the same position of equilibrium will be reached whether we start from pure A or pure B or any mixture of A and B.

Thermodynamics applied to real systems

The thermodynamic principles we have discussed so far are completely independent of any knowledge about the system we are studying. No assumptions have been made about the nature of the system. In order to relate thermodynamic quantities to things that can be measured, such as pressures or concentrations, a model of the system is needed, for instance we need to know about atoms and molecules and bonds to relate changes in H and S to the details of a chemical reaction.

WORKED EXAMPLE

Is a compound more stable as a solid, liquid or gas?

SOLUTION: *Step 1*—We need a model of our system that will let us predict the values of H, S, and hence G, for the three phases. We shall assume that our system is made of totally rigid molecules (figure 2.5a). The entropy term comes from how ordered these molecules are in each phase. In the solid phase the molecules are highly ordered and so have a small value of S, whereas in the gas phase the molecules are very disordered and so have a large value of S. The enthalpy term comes from the attractive forces between the molecules. In the solid, strong attractive forces hold the molecules together. To break these to form the liquid requires the input of energy. Thus H of the liquid must be greater than H of the solid. Similarly H of the gas must be greater than H of the liquid.

Phase	Model	H	S
Solid	Molecules held in place by strong intermolecular attractions	Small and +ve	Small and +ve
Liquid	Molecules held together by weaker intermolecular attractions	Medium and +ve	Medium and +ve
Gas	Molecules are not held together at all and are completely free to move	Large and +ve	Large and +ve

Step 2—We use $G = H - TS$ to plot G versus temperature for each phase. For each phase, this gives a straight line with a y-intercept of H and a slope of $-S$ (figure 2.5b).

Step 3—At any given temperature, the phase with the lowest value of G is the most stable. When two phases have equal values of G then they are in equilibrium at that temperature since ΔG to convert one to the other is zero.

COMMENT: The model predicts that the solid form is the most stable at low temperatures and the gaseous form is most stable at high temperatures. As this corresponds to what we find experimentally, our model is satisfactory but, as we shall see below, it needs to be extended to explain the phase changes in any more detail.

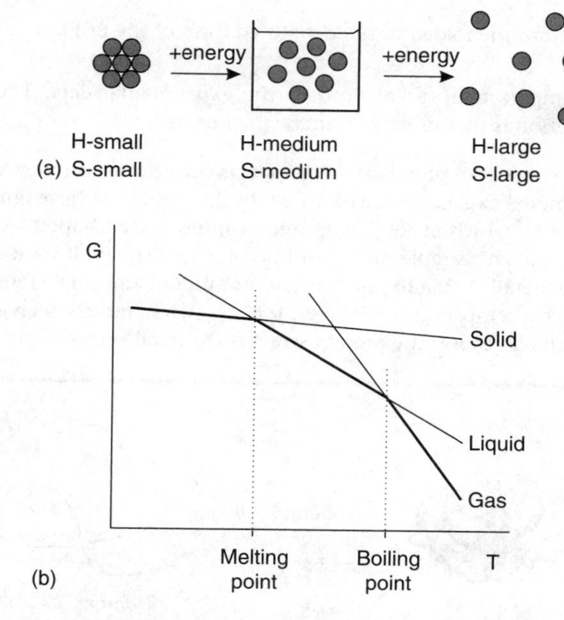

Fig. 2.5 (a) A schematic representation of the solid to liquid and liquid to gas phase changes. The molecules are most ordered in the solid phase and least ordered in the gas phase, and hence the gas has the highest value of S. Energy is required to break the intermolecular forces holding the molecules in the solid together to give the liquid and gas phases, hence the gas has the highest value of H. (b) A plot of free energy, G, versus temperature, T, for a solid, liquid and gas based on the simplified model of these three phases given in the worked example. For each line, the intercept on the y-axis equals H and the gradient equals –S. At any given temperature the phase with the lowest value of G will be the most stable.

WORKED EXAMPLE

Why do proteins denature when they are heated?

SOLUTION: *Step 1*—We propose a simple model of the system (figure 2.6a).

Phase	Model	H	S
Folded protein	There is a single conformation that maximises intramolecular attractions (H-bonds, van der Waals forces, etc.)	Small and +ve	Small and +ve
Denatured protein	Can adopt many different conformations, each characterised by weak intermolecular interactions	Large and +ve	Large and +ve

Step 2—We make predictions based on the model, using $G = H - TS$ to plot G versus temperature for each phase, as before (figure 2.6b). The model predicts a

phase change from the folded to the denatured form of the protein as the temperature rises.

Step 3—We compare these predictions to the experimental data. The experimental observation is that proteins unfold when heated.

COMMENT: This does not prove that the model is correct, you can never do this, only that the model explains the data so far. In this model we have ignored the role of the solvent, which makes things more complex (see Chapter 18), and so we can expect that this simple model will become inadequate if we look at the system in more detail. In reality, proteins do not unfold at a single temperature but unfold over a temperature range, and at low temperatures increasing the temperature actually causes the protein structure to stabilise.

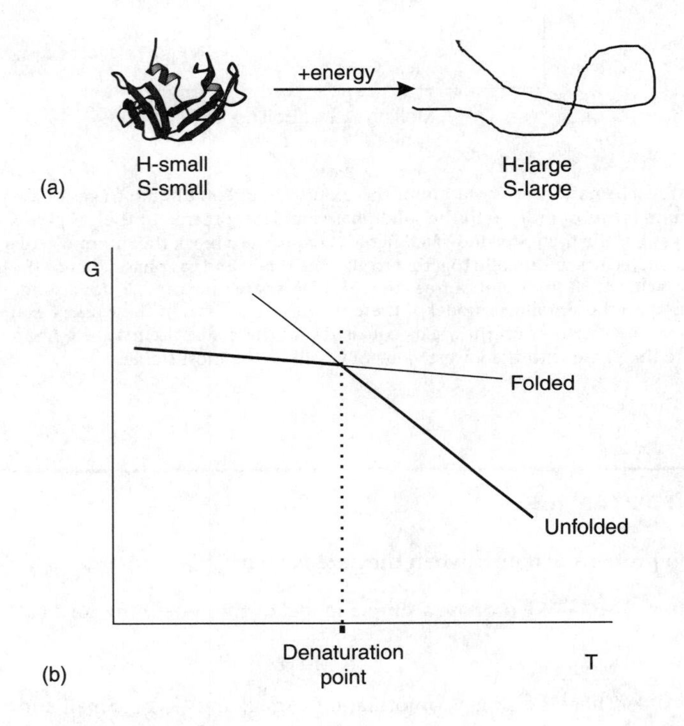

Fig. 2.6 (a) A schematic representation of protein unfolding. The folded form is the more ordered, and hence the unfolded form has the higher value of S. Energy is required to break the intramolecular forces in the folded form, and hence the unfolded form has the higher value of H. (b) A plot of free energy, G, versus temperature, T, for a protein in the folded and unfolded states based on the simplified model given in the worked example. At any given temperature the phase with the lower value of G will be the more stable.

The worked examples above use very simplified physical models to explain the phenomena of melting/boiling and protein denaturation. In order to make a better thermodynamic model of a system, and hence to be able to explain behaviour more accurately or to make more accurate predictions, we need a more realistic physical model.

To improve the model for melting/boiling, we shall start by considering the possible physical models for a gas. The simplest model of a gas is of non-interacting molecules that move totally randomly. Such a gas is called an *ideal gas* and its behaviour is described mathematically by an equation called the *ideal gas law*

$$PV = nRT$$

where P is the pressure, V is the volume, n is the number of moles of gas present, R is the gas constant and T is the temperature. We can calculate the thermodynamic behaviour of an ideal gas by combining the thermodynamic equations derived from the first and second laws with the equation for an ideal gas (figure 2.7). This gives the equation

$$G = G^{\circ} + nRT \log_e\left(\frac{P}{P^{\circ}}\right)$$

where G is the free energy of the gas at pressure P and G° is the free energy of the gas in the standard state, defined as $P^{\circ} = 1$ atm. [*Note: $\log_e(x)$ can also be written as ln(x), see Appendix 2.*] This equation describes how the free energy of an ideal gas varies with pressure. If a gas does not obey this equation, it is because the physical model of an ideal gas does not apply in that case.

We will deal with the properties of ideal gases in more detail in Chapter 8, as well as giving the formal derivation for this equation. At this stage we will simply use the result.

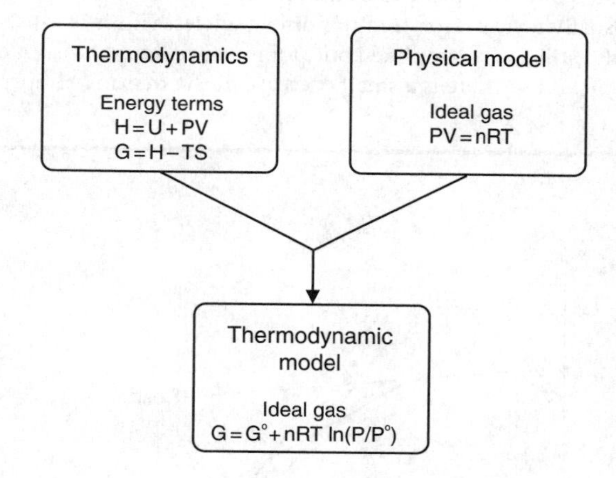

Fig. 2.7 A thermodynamic model of a system is obtained by combining the equations based on the laws of thermodynamics with equations that describe the physical model of the system. If the thermodynamic model does not match the experimental data, it is the physical model of the system that is incorrect.

WORKED EXAMPLE

How are the relative stabilities of the solid, liquid and gas phases affected by pressure?

SOLUTION: *Step 1*—The basic model of solids, liquids and gases above needs to be extended to deal with changes in pressure. In particular, gases can be compressed which involves doing work on the system, thus increasing the internal energy.

Phase	Model	H	S
Solid	Incompressible, not affected by pressure	No change	No change
Liquid	Incompressible, not affected by pressure	No change	No change
Gas	Compressible, PV = nRT for an ideal gas	May change	May change

Step 2—For solids and liquids, the plot of G versus T is the same. As we have seen, for an ideal gas G varies with pressure according to the equation

$$G = G^{\circ} + nRT \log_e \left(\frac{P}{P^{\circ}} \right)$$

Step 3—As P decreases, $\log_e(P/P^{\circ})$ becomes smaller and the free energy of an ideal gas decreases. This makes the gas phase more stable relative to the liquid phase and reduces the boiling point (figure 2.8).

A full phase diagram can be constructed by plotting the melting and boiling points on a P versus T graph (figure 2.9). The melting point is independent of pressure whereas the boiling point drops as the pressure decreases. Below a critical pressure, the liquid phase is no longer stable and a direct solid to gas transition takes place, called sublimation.

COMMENT: The experimental phase diagrams for many liquids, such as water, have the same qualitative features as that in figure 2.9. However, more detailed inspection usually reveals a lot of minor discrepancies. This is because the model we have used is still a very simplified one, for instance liquids are not completely incompressible and so there is a small decrease in the freezing temperature with pressure.

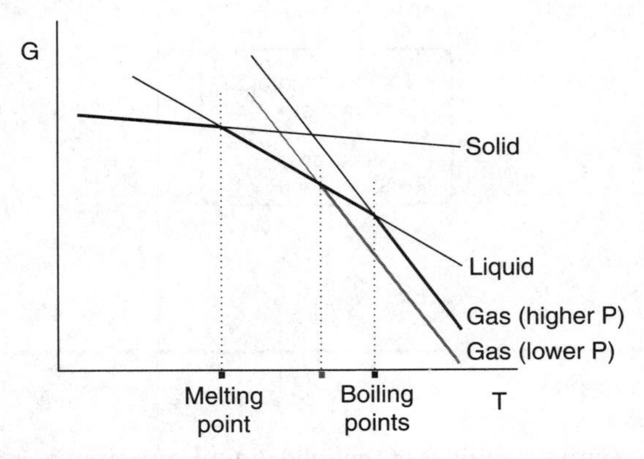

Fig. 2.8 A plot of free energy, G, versus temperature, T, for a solid, a liquid and an ideal gas showing how the boiling point varies with pressure. As the pressure is decreased, the free energy of the gas decreases (grey line) and so the boiling point decreases. If the pressure is low enough, the boiling point becomes less than the melting point and the liquid phase is not observed.

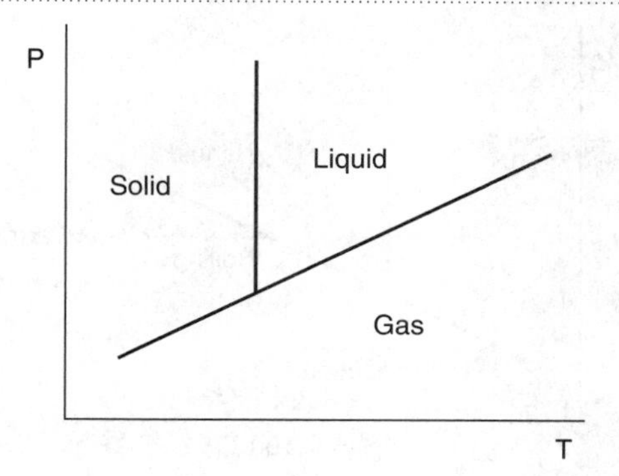

Fig. 2.9 A phase diagram based on the free energy plots in figure 2.8. The melting point is independent of pressure and the boiling point decreases as the pressure is decreased. This plot has all the main features of the phase diagram for water, although some of the details are incorrect because the physical model used is not sufficiently realistic.

Finally, once the model explains all the required experimental data, we can use it to make predictions and design further experiments.

WORKED EXAMPLE

How can you dry biological samples, such as proteins, without destroying activity?

SOLUTION: It is essential to find a method that avoids protein denaturation, since protein denaturation leads to loss of activity. However, if we heat a protein solution to boil off the water, then the heat denatures the protein. Similarly, if we reduce the pressure to reduce the boiling point, the protein concentrates at the surfaces of the vapour bubbles as they form in the liquid and this denatures the protein.

The alternative is to work at very low temperatures and to avoid the liquid to gas transition completely. Inspection of the phase diagram for water shows that we can lower the temperature, usually to below −70°C, to form ice and then lower the pressure, usually to below 0·001 atm, until the ice converts directly to the vapour (figure 2.10). This procedure is called lyophilisation, or freeze-drying, and it is used in the manufacture of instant coffee and rations for astronauts as well as to dry biological samples.

COMMENT: Although we know that the phase diagram for water in figure 2.10 is an oversimplification, it is perfectly adequate for this analysis. We do not need a more complicated model to be able to explain lyophilisation.

Molecular basis of enthalpy and entropy

Enthalpy and entropy are macroscopic quantities. They are used to describe an entire system and so far we have not been concerned with the properties of the molecules that make up the system. However, we know that molecules do exist

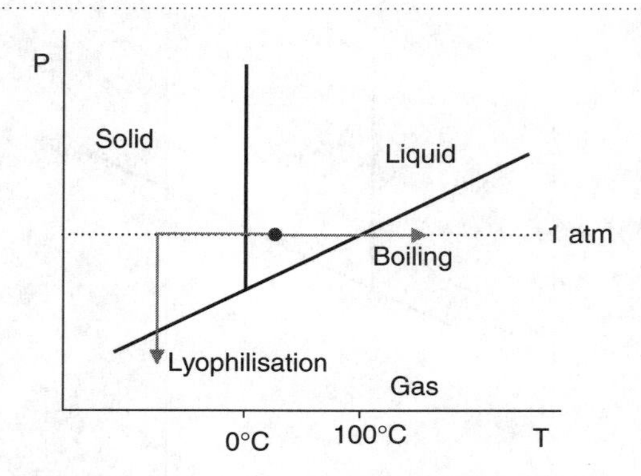

Fig. 2.10 A simplified phase diagram for water showing the process of boiling and freeze-drying. The filled circle represents room temperature and pressure. To boil a protein solution, the temperature is increased until the water converts to steam. This denatures the protein. To freeze-dry a protein solution, the temperature is reduced until the water freezes and then the pressure is reduced until the ice converts directly to the gas. This dries the protein sample without causing denaturation.

and that macroscopic behaviour is simply a consequence of atomic and molecular interactions, so the values of enthalpy and entropy for a given macroscopic system must be determined by the energies of the component molecules. The calculation of thermodynamic values from molecular energies is called *statistical thermodynamics*. This is beyond the scope of this book and we will only give the basic concepts here.

The molecular basis of enthalpy or internal energy is relatively straightforward. A single molecule has a total energy that is the sum of all the forms of energy it possesses, e.g. translational energy, rotational energy, vibrational energy, chemical bond energy, etc. In addition there is energy associated with intermolecular interactions, e.g. van der Waals forces, hydrogen bonds, etc. The internal energy of the system is simply the sum of the energies of all the molecules and all the intermolecular interactions.

The molecular basis of entropy is a little more difficult to understand. A single molecule cannot have entropy. Entropy is a measure of randomness and a single molecule cannot be random. The concept of entropy arises as a probability function when lots of molecules are taken together. Entropy is a measure of the number of ways that the internal energy of the system can be distributed between all the molecules. This can be described mathematically by

$$S = k \log_e(W)$$

where k is the Boltzmann constant and W is the number of ways of distributing energy in the system. The chemical bond energies, etc., of each molecule are usually fixed and so the only significant contributions to entropy come from the distribution of translational, rotational and vibrational energies. For example, a solid has low entropy because all the molecules are fixed in space and so have no translational or rotational energy. The only variations are in the distribution of vibrational energy between the different molecules. A gas has very high entropy because each molecule can have translational, rotational and vibrational energy and there are lots of possible combinations of these that can give the same total energy.

WORKED EXAMPLE

For the hydrolysis of ATP to ADP in aqueous solution at 310 K, $\Delta H^{o\prime} = -20{\cdot}1$ kJ mol^{-1} and $\Delta S^{o\prime} = +33{\cdot}5$ J K^{-1} mol^{-1}. Comment on the molecular factors that contribute to these thermodynamic values.

SOLUTION: Suppose that the reaction at pH = 7 is

$$ATP^{4-} + H_2O \rightleftharpoons ADP^{3-} + P_i^{2-} + H^+$$

where P_i is phosphate.

$\Delta H^{o\prime}$ is the change in total energy of the system, after allowing for expansion. For chemical reactions, $\Delta H^{o\prime}$ is usually dominated by the energy involved in making and breaking bonds. This reaction involves breaking an O–P bond in ATP and an O–H bond in water and making an O–H and an O–P bond to give ADP and P_i (figure 2.11), thus any differences in these bond energies will contribute to $\Delta H^{o\prime}$. There is also a reduction in intramolecular electrostatic repulsion, ATP having four negative charges and ADP only three. In addition, the reactants and products are hydrated, forming hydrogen bonds with water, and any difference in hydration will contribute to $\Delta H^{o\prime}$. In this case, the products are more charged than the reactants and so will former stronger bonds with the solvent leading to a large negative contribution to $\Delta H^{o\prime}$.

$\Delta S^{o\prime}$ depends on the change in the ways of distributing energy between the reactants and products. In this case, two reactant molecules are forming three product molecules. There will be more ways of distributing translational and rotational energy among three molecules than two and this will make a positive contribution to $\Delta S^{o\prime}$. However, the products are more hydrated than the reactants and this will have an ordering effect on the solvent leading to a negative contribution to $\Delta S^{o\prime}$.

COMMENT: Although the phosphate bond in ATP is sometimes referred to as a 'high-energy' bond, the breaking of the O–P bond is offset by the making of another O–P bond. In fact, the enthalpic driving force for this reaction comes from changes in intramolecular electrostatic repulsion and in solvation and there is also a considerable entropic driving force provided by splitting one molecule into two.

Considerable free energy is released on the hydrolysis of ATP. However, what makes ATP a useful store of energy in biochemical systems is that, although thermodynamically favourable, hydrolysis is normally very slow (see Chapter 9 for chemical kinetics) unless catalysed by an enzyme (see Chapter 11 for enzyme catalysis). This allows separate control of the synthesis and breakdown of ATP and thus the concentrations of ATP, ADP and P_i within a cell can be controlled. As we shall see in the next chapter, this is critical in enabling free energy to be obtained from ATP hydrolysis. Free energy is used in the cell to drive the synthesis of ATP.

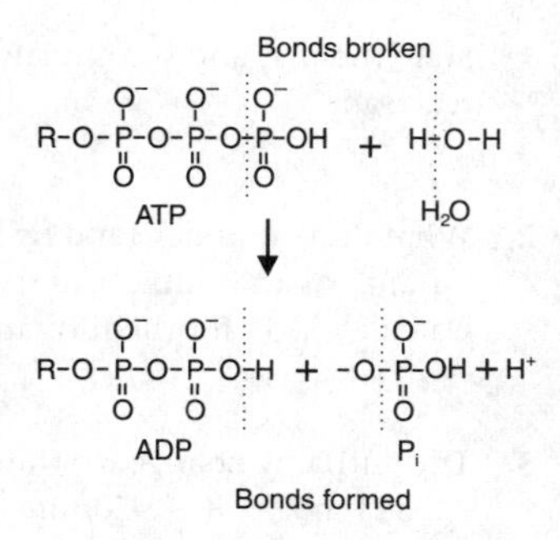

Fig. 2.11 Schematic diagram of the hydrolysis of ATP showing which bonds are broken and formed. This does not represent the actual mechanism for the reaction. [Note: the charges are shown as −3 for ATP, −2 for ADP and −2 for inorganic phosphate (P_i). The actual charges at pH = 7·0 will be non-integral because all of these species have pK_a values close to 7·0 and so will be in equilibrium with their conjugate acids or bases (Chapter 5).]

ATP stores the free energy in a kinetically stable form. When this energy is required, an enzyme is used to catalyse the hydrolysis of ATP. This releases the stored free energy and the enzyme uses this free energy to drive the required process (Chapter 13). Thus ATP acts as a 'molecular battery'.

There are usually many different factors at the molecular level that contribute to the observed values of ΔH and ΔS for a reaction. Some of these factors may be favourable and some unfavourable. Indeed, a small value of ΔH or ΔS often occurs as the result of taking the difference between two or more large numbers. For example, the overall $\Delta H^{o\prime}$ value of $-20{\cdot}1\ \text{kJ mol}^{-1}$ for the hydrolysis of ATP includes a term $\approx 540\ \text{kJ mol}^{-1}$ for the breaking of an O–P bond and a negative term of similar magnitude for making a different O–P bond. Thus, great care has to be taken in trying to explain ΔH and ΔS values.

FURTHER READING

P.W. Atkins, 1998, 'Physical chemistry', Oxford University Press—*a well written physical chemistry textbook that provides a useful introduction to the laws of thermodynamics from a chemical perspective.*

E.B. Smith, 1990, 'Basic chemical thermodynamics', Oxford University Press—*an excellent introduction to thermodynamics, again written from a chemical perspective.*

PROBLEMS

1. State the first and second laws of thermodynamics and discuss the implications of these laws for chemical reactions.

2. What do you understand by the terms enthalpy, entropy and free energy of a reaction? $\Delta G^{o\prime}$ and $\Delta H^{o\prime}$ for the binding of an inhibitor to trypsin at 298 K were found to be $-20{\cdot}9\ \text{kJ mol}^{-1}$ and $0\ \text{kJ mol}^{-1}$ respectively. Calculate $\Delta S^{o\prime}$ for the formation of the complex and comment on the result.

3. The enthalpy change for the folding of ribonuclease at pH 6 is $-209\ \text{kJ mol}^{-1}$ and the entropy change is $-554\ \text{J mol}^{-1}\ \text{K}^{-1}$. Comment on the magnitude of these numbers. Will the protein spontaneously fold at 298 K? Assuming that ΔH and ΔS do not change with temperature, at what temperatures will the protein not fold spontaneously?

4. Calculate the bond dissociation enthalpy for the C–H bond in methane, CH_4.

$$C(s) + 2H_2(g) \rightleftharpoons CH_4(g) \qquad \Delta H^\circ = -74{\cdot}8\ \text{kJ mol}^{-1}$$

$$H_2(g) \rightleftharpoons 2H(g) \qquad \Delta H^\circ = +434{\cdot}7\ \text{kJ mol}^{-1}$$

$$C(s) \rightleftharpoons C(g) \qquad \Delta H^\circ = +719{\cdot}0\ \text{kJ mol}^{-1}$$

[Hint: The bond dissociation enthalpy is a quarter of the enthalpy change for $CH_4(g) \rightleftharpoons C(g) + 4H(g)$.] Assuming that the C–H bond dissociation enthalpy is the same in methane and ethene and that $\Delta H^\circ_{formation}$ for ethene is $+54{\cdot}3\ \text{kJ mol}^{-1}$, calculate the C=C bond dissociation enthalpy in ethene.

5. Calculate the resonance stabilisation energy of benzene from the values for $\Delta H_{hydrogenation}$ of ethene, C_2H_4, and benzene, C_6H_6, at 298 K.

$$C_2H_4(g) + H_2(g) \rightleftharpoons C_2H_6(g) \qquad \Delta H^\circ = -136{\cdot}3\ kJ\ mol^{-1}$$
$$C_6H_6(g) + 3H_2(g) \rightleftharpoons C_6H_{12}(g) \qquad \Delta H^\circ = -208{\cdot}2\ kJ\ mol^{-1}$$

[Hint: The resonance stabilisation energy of benzene is the additional stability of benzene compared to three isolated double bonds.]

6. Use the enthalpies of formation at pH 7 given below to calculate ΔH° for the hydrolysis of the amide bond in the dipeptide Gly–Gly.

$$NH_3^+CH_2CONHCH_2COO^-(aq) + H_2O \rightleftharpoons 2NH_3^+CH_2COO^-(aq)$$

Compound	Gly–Gly(aq)	Gly(aq)	H_2O
$\Delta H^\circ_{formation}$	$-747.7\ kJ\ mol^{-1}$	$-527.5\ kJ\ mol^{-1}$	$-285.8\ kJ\ mol^{-1}$

Comment on the value of $\Delta H^\circ_{hydrolysis}$(Gly–Gly) that you have calculated in the light of the following average single bond dissociation enthalpies.

Bond	Average dissociation enthalpy
C–N	$+275\ kJ\ mol^{-1}$
C–O	$+330\ kJ\ mol^{-1}$
H–N	$+385\ kJ\ mol^{-1}$
H–O	$+455\ kJ\ mol^{-1}$

7. Explain the following observations

Dissolution of $NaNO_3$ in water	spontaneous	$\Delta H = +ve$
Dissolution of NaOH in water	spontaneous	$\Delta H = -ve$
Diffusion of Na^+ ions down a concentration gradient	spontaneous	$\Delta H = 0$

3 Chemical potential and multiple component systems

KEY POINTS:

- The *chemical potential* (μ) of a compound in a particular system is a measure of the free energy of one mole of the compound in that system. It is independent of the size of the system.
- In an ideal solution, μ varies with concentration, $\mu_X = \mu_X^\circ + RT\log_e[X]$.
- The total free energy, G, of a system can be obtained by adding together the chemical potentials multiplied by the amount present of all the components of the system, $G = \sum_X \mu_X n_X$.
- Combining these equations gives the variation of ΔG with concentration for a reaction, $\Delta G = \Delta G^\circ + RT\log_e \Gamma$, where Γ is the *mass action ratio*.
- Measurement of Γ at equilibrium gives the *equilibrium constant*, K, and allows ΔG° to be determined.
- Assuming that ΔH° and ΔS° are independent of temperature, K varies with temperature according to the *van't Hoff isochore*. Measuring K as a function of temperature allows ΔH° to be determined.
- In some cases ΔH° and ΔS° cannot be assumed to be independent of temperature. This is especially important for protein denaturation.

In Chapter 2, we saw how the relative stability of different states of the same system could be assessed in terms of the Gibbs free energy, G. G varies with the temperature and pressure of the system and we also stated that G may vary with composition (see figure 2.4). Thus, the next step is to extend the quantitative analysis to include situations in which chemical reactions and/or transport processes alter the composition of the system. This requires a consideration of the relationship between the composition of a system and its free energy.

The concept of chemical potential

In the last chapter, we only considered the free energy, G, of an entire system. We did not consider the energy associated with the separate molecules in the system. If we double the size of a system, for instance go from 1 litre of 1 mM sucrose to 2 litres of 1 mM sucrose, the free energy of the system must also double. G doubles because the number of molecules we are considering doubles. The average energy of each molecule, and thus its chemical properties, remains the same. This shows that minimum G is not a useful criterion for thermodynamic stability when considering systems in which the amount of matter present can change, for instance 1 litre of 1 mM sucrose is not more stable than 2 litres of 1 mM sucrose even though G is twice as large for the latter. The concept of chemical potential is introduced to allow us to deal with the thermodynamics of such open systems.

The *chemical potential* of a compound, μ, is defined as the free energy per mole of that compound under a given set of conditions. μ is also referred to as the *partial molar Gibbs free energy*. For a one-component system, μ is given by dividing the free energy of the system by the number of moles present, n.

$$\mu = \frac{G}{n}$$

The chemical potential is an intensive state function. If we double the size of the system, G doubles and n doubles but μ does not. μ is therefore a useful measure of the average free energy of a molecule in a system as it only depends on the environment of the molecule and is independent of the size of the system.

For a multicomponent system, G is given by the summation of the chemical potentials of all the species present.

$$G = \sum_x \mu_x n_x$$

Alternatively, the chemical potential of x can be defined as the rate of change of free energy of the whole system as the amount of x present changes at constant temperature, pressure and concentration of all other species, i.e.

$$\mu_x = \left(\frac{\delta G}{\delta n_x}\right)_{T,P,n_j} \quad \text{or} \quad \delta G = \mu_x \delta n_x$$

where δn_x is the change in x measured in moles.

Chemical potential and change

From a thermodynamic point of view, any chemical process can be thought of as the removal of the reactants from the system followed by the addition of the products (figure 3.1).

$$\text{Chemical process} = \text{Removal of reactants} + \text{Addition of products}$$

The change in G for removal of the reactants is given by

$$\Delta G = \sum_{\text{reactants}} \mu_{\text{reactants}} \Delta n_{\text{reactants}}$$

where Δn is negative, i.e. the number of moles present decreases. The change in G for addition of the products is given by

$$\Delta G = \sum_{\text{products}} \mu_{\text{products}} \Delta n_{\text{products}}$$

where Δn is positive. The overall change in G for the reaction is given by the sum of these, i.e.

$$\Delta G = \sum_{x} \mu_x \Delta n_x$$

Removal of reactants can occur by converting them into other chemicals or transporting them out of the system. Similarly, addition of products can occur by generating them from other chemicals or transporting them into the system. Thus, by using chemical potentials, chemical reactions and transport processes can be treated in exactly the same way.

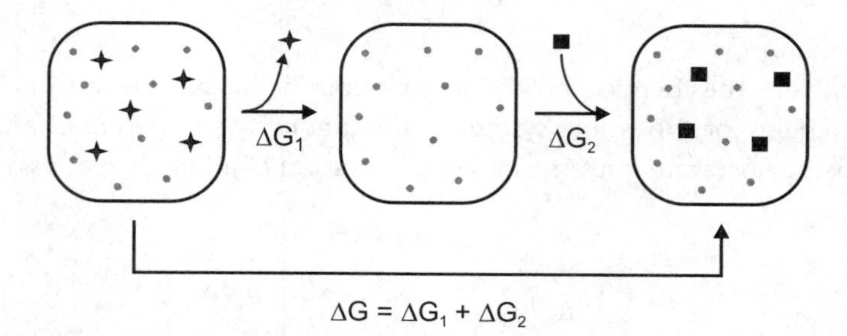

Fig. 3.1 Any chemical process can be thought of in thermodynamic terms as removal of reactants (✦) followed by addition of products (■), ΔG for the overall process being the sum of the ΔG values for the two separate steps. Removal of reactants may be due to them undergoing a chemical reaction or due to them being transported out of the system. Similarly, addition of products may be due to them being formed during a chemical reaction or being transported into the system (• represents solvent.)

WORKED EXAMPLE

Given the chemical potentials of ATP, AMP, P_i and H_2O, what is the free energy change for the hydrolysis of one mole of ATP to AMP?

SOLUTION: In order to calculate ΔG we must consider *all* the species that are changing during the reaction. The overall *balanced* reaction is

$$ATP + 2H_2O \rightleftharpoons AMP + 2P_i$$

This reaction is equivalent to removing 1 mole of ATP and 2 moles of H_2O and adding 1 mole of AMP and 2 moles of P_i. Thus

$$\Delta G = \sum_x \mu_x \Delta n_x = (\mu_{ATP} \times -1) + (\mu_{H_2O} \times -2) + (\mu_{AMP} \times +1) + (\mu_{P_i} \times +2)$$

$$\Delta G = \mu_{AMP} + 2\mu_{P_i} - \mu_{ATP} - 2\mu_{H_2O}$$

Spontaneous reactions and equilibria

The next step is to determine the criteria for spontaneous change and equilibrium in multiple component systems. We shall start with the simple reaction where n moles of a compound A are converted into n moles of a compound B. We can treat this as the loss of n moles of A ($\mu = \mu_A$, $\Delta n = -n$) and the gain of n moles of compound B ($\mu = \mu_B$, $\Delta n = +n$). Overall, ΔG for the conversion of A into B is

$$\Delta G = \mu_B n - \mu_A n$$

If μ_B is less than μ_A, then ΔG for the reaction is negative and it can occur spontaneously. At equilibrium $\Delta G = 0$ and so $\mu_A = \mu_B$. In this case, the criterion for spontaneous change is to lower the chemical potential and equilibrium occurs when the chemical potentials are equal.

For a more complex reaction, such as

$$aA + bB \rightleftharpoons cC + dD$$

where a moles of A and b moles of B are converted into c moles of C and d moles of D, the free energy change is given by

$$\Delta G = \sum_x \mu_x \Delta n_x = c\mu_C + d\mu_D - a\mu_A - b\mu_B$$

For a spontaneous reaction $\Delta G < 0$ and so $c\mu_C + d\mu_D$ must be less than $a\mu_A + b\mu_B$. At equilibrium $\Delta G = 0$ and so $c\mu_C + d\mu_D = a\mu_A + b\mu_B$.

The criterion for spontaneous change is that the weighted sum of the chemical potentials can be lowered, and the criterion for equilibrium is that the weighted sums of the chemical potentials for the reactants and products are equal.

TABLE 3.1

Type of system	Exchange with surroundings	Criterion for equilibrium	Criterion for spontaneous change
Isolated system	None	$\Delta S = 0$	$\Delta S > 0$
Closed system	Energy	$\Delta G = 0$	$\Delta G < 0$
Open system	Energy + matter	$\sum_{\text{reactants}} \mu n = \sum_{\text{products}} \mu n$	$\sum_{\text{reactants}} \mu n > \sum_{\text{products}} \mu n$

This is a more general condition of equilibrium than that based on ΔG, as we can now deal with the transfer of matter, as well as energy, into and out of the system (Table 3.1).

Variation of chemical potential with concentration

The last step is to determine how the chemical potential varies with concentration. This will be dealt with more rigorously in Chapter 8. The starting point is the equation relating G for an ideal gas to the pressure (see previous chapter).

$$G = G^\circ + nRT\log_e\left(\frac{P}{P^\circ}\right)$$

As $\mu = G/n$ for a pure compound, direct substitution gives

$$\mu = \mu^\circ + RT\log_e\left(\frac{P}{P^\circ}\right)$$

where μ° is the chemical potential of the gas in the standard state. Unlike G, which applies to the whole system, the equation for μ can be extended to one component of a mixture of ideal gases to give

$$\mu_x = \mu_x^\circ + RT\log_e\left(\frac{P_x}{P_x^\circ}\right)$$

where P_x is the partial pressure of x, i.e. the pressure exerted by x, the total pressure being given by the sum of all the partial pressures for all the different gases present. If the proportion of x present changes at constant total pressure, then P_x will change. For example, if there is less x present it will make a smaller contribution to the total pressure. Thus the partial pressure of x depends on the number of moles of x present. This can be represented by the formula

$$\frac{P_x}{P_x^\circ} = \frac{\text{no. of moles of x present}}{\text{no. of moles of x present in the standard state}}$$

which is effectively a ratio of concentrations and so

$$\mu_x = \mu_x^\circ + RT\log_e\left(\frac{\text{'concentration of x'}}{\text{'concentration of x in the standard state'}}\right)$$

This approach can be generalised to any type of compound. We shall do this more rigorously in Chapter 8 but for now we shall simply state the results, which are

STATE	EQUATION FOR μ	DEFINITION
Gas	$\mu_x = \mu_x^\circ + RT\log_e\left(\frac{P_x}{P_x^\circ}\right)$	P_x is the partial pressure of x and P_x° is the pressure in the standard state, i.e. $P_x^\circ = 1$ atm.
Liquid/ solvent	$\mu_x = \mu_x^\circ + RT\log_e\left(\frac{[x]}{[x]^\circ}\right)$	[x] is the concentration of x and $[x]^\circ$ is the concentration in the standard state, i.e. the concentration of the pure liquid.
Solute	$\mu_x = \mu_x^\circ + RT\log_e\left(\frac{[x]}{[x]^\circ}\right)$	[x] is the concentration of x and $[x]^\circ$ is the concentration in the standard state, i.e. $[x]^\circ = 1$ M.
Solid	$\mu_x = \mu_x^\circ$	All pure solids are in their standard state by definition.

WORKED EXAMPLE

What are the chemical potentials at 300 K of:

(1) $O_2(g)$ at a partial pressure of 0·5 atm
(2) water in a 90% water:10% methanol mix
(3) 1 M $Na^+(aq)$
(4) 2 M $Na^+(aq)$

(1) SOLUTION: For a gas, the chemical potential is given by

$$\mu_{O_2} = \mu_{O_2}^\circ + RT\log_e\left(\frac{P_{O_2}}{P_{O_2}^\circ}\right)$$

The standard state of a gas is the pure gas at 1 atm pressure, so $P_{O_2}^\circ = 1$ atm. Thus

$$\mu_{O_2} = \mu_{O_2}^\circ + RT\log_e\left(\frac{0{\cdot}5}{1{\cdot}0}\right) = \mu_{O_2}^\circ - 1728{\cdot}8\,\text{J mol}^{-1}$$

(2) SOLUTION: For a liquid, the chemical potential is given by

$$\mu_{H_2O} = \mu_{H_2O}^\circ + RT\log_e\left(\frac{[H_2O]}{[H_2O]^\circ}\right)$$

The standard state of a liquid is the pure liquid. The concentration of pure water is approximately 55 M, so $[H_2O]^\circ = 55$ M. The concentration of 90% water is approximately $55 \times 0{\cdot}9$ M. Thus

$$\mu_{H_2O} = \mu^\circ_{H_2O} + RT\log_e\left(\frac{49{\cdot}5}{55}\right) = \mu^\circ_{H_2O} - 262{\cdot}8\,J\,mol^{-1}$$

(3) SOLUTION: For a solute, the chemical potential is given by

$$\mu_{Na^+} = \mu^\circ_{Na^+} + RT\log_e\left(\frac{[Na^+]}{[Na^+]^\circ}\right)$$

The standard state for a solute in water is 1 M, so $[Na^+]^\circ = 1{\cdot}0$ M. In this case, $[Na^+] = [Na^+]^\circ$ and so

$$\mu_{Na^+} = \mu^\circ_{Na^+}$$

(4) SOLUTION: The same approach is used as for (3) except that $[Na^+] \neq [Na^+]^\circ$, thus

$$\mu_{Na^+} = \mu^\circ_{Na^+} + RT\log_e\left(\frac{2{\cdot}0}{1{\cdot}0}\right) = \mu^\circ_{Na^+} + 1728{\cdot}8\,J\,mol^{-1}$$

COMMENT: μ can be larger or smaller than μ° depending on the concentration. μ increases as the concentration increases. Thus, chemicals are more thermodynamically stable at lower concentrations. This is because the entropy of a component is larger at lower concentrations than it is at higher concentrations.

The most important equation for chemical potential, and the one that we will use extensively, is the version that applies to solvents and solutes. This is usually written as

$$\mu_x = \mu^\circ_x + RT\log_e[x]$$

but it must be emphasised that [x] is expressed relative to the concentration in the standard state.

WORKED EXAMPLE

What is the chemical potential for 0·1 M H^+(aq) relative to μ° and $\mu^{\circ\prime}$ at 300 K?

SOLUTION: The chemical potential for H^+(aq) relative to the standard state is given by

$$\mu_{H^+} = \mu^\circ_{H^+} + RT\log_e\left(\frac{[H^+]}{[H^+]^\circ}\right)$$

where $[H^+]^\circ = 1{\cdot}0\,M$. Thus

$$\mu_{H^+} = \mu^\circ_{H^+} + RT\log_e\left(\frac{0{\cdot}1}{1{\cdot}0}\right) = \mu^\circ_{H^+} - 5743{\cdot}1\,J\,mol^{-1}$$

The chemical potential for $H^+(aq)$ relative to the biochemical standard state is given by

$$\mu_{H^+} = \mu^{\circ\prime}_{H^+} + RT\log_e\left(\frac{[H^+]}{[H^+]^{\circ\prime}}\right)$$

where $[H^+]^{\circ\prime} = 10^{-7}M$. Thus

$$\mu_{H^+} = \mu^{\circ\prime}_{H^+} + RT\log_e\left(\frac{0{\cdot}1}{10^{-7}}\right) = \mu^{\circ\prime}_{H^+} + 34458{\cdot}6\,J\,mol^{-1}$$

COMMENT: These two equations must give the same value of μ_{H^+} as a species can only ever have one chemical potential under one set of conditions. The differences in the $\log_e$ terms reflect the different reference points used for the calculation, i.e. the difference in value between $\mu^\circ_{H^+}$ and $\mu^{\circ\prime}_{H^+}$.

Dependence of ΔG on concentration

We are now in a position to deal with the thermodynamics of chemical reactions. Consider the reaction

$$aA + bB \rightleftharpoons cC + dD$$

where [A], [B], [C] and [D] are the concentrations of the reactants and the products expressed relative to the standard states and a, b, c and d are the stoichiometry coefficients. At constant temperature and pressure

$$\Delta G^\circ = \sum_x \mu^\circ_x \Delta n_x = c\mu^\circ_C + d\mu^\circ_D - a\mu^\circ_A - b\mu^\circ_B$$

and

$$\Delta G = \sum_x \mu_x \Delta n_x = c\mu_C + d\mu_D - a\mu_A - b\mu_B$$

Substituting for μ in the equation for ΔG gives

$$\Delta G = c\{\mu^\circ_C + RT\log_e[C]\} + d\{\mu^\circ_D + RT\log_e[D]\} - a\{\mu^\circ_A + RT\log_e[A]\} - b\{\mu^\circ_B + RT\log_e[B]\}$$

This can be rearranged to give

$$\Delta G = \{c\mu^\circ_C + d\mu^\circ_D - a\mu^\circ_A - b\mu^\circ_B\} + RT\{\log_e[C]^c + \log_e[D]^d - \log_e[A]^a - \log_e[B]^b\}$$

The first term is just $\sum_x \mu^\circ_x \Delta n_x$ which is the standard free energy change for 1 mole of reactants giving 1 mole of products, all in their standard states (ΔG°), and the second term gives the change in ΔG as the concentrations change from the standard state. The equation can be rewritten as

$$\Delta G = \Delta G^\circ + RT\log_e\left(\frac{[C]^c[D]^d}{[A]^a[B]^b}\right)$$

or in the more general form, applicable to any reaction,

$$\Delta G = \Delta G^\circ + RT\log_e\left(\frac{[\text{product 1}]^{m_1}[\text{product 2}]^{m_2}[\text{product 3}]^{m_3}\ldots}{[\text{reactant 1}]^{m_1}[\text{reactant 2}]^{m_2}[\text{reactant 3}]^{m_3}\ldots}\right)$$

This can also be written more compactly as

$$\Delta G = \Delta G^\circ + RT\log_e\left(\frac{\prod_{\text{products}}[x]^{m_x}}{\prod_{\text{reactants}}[x]^{m_x}}\right)$$

where m_x is the stoichiometry coefficient of x. [Note: the Π sign means multiply terms together, rather than Σ which means add terms together, see Appendix 2.]

WORKED EXAMPLE

The cellular concentrations of ATP, ADP and P_i were found to be 0·01 M, 0·003 M and 0·001 M respectively. Given that $\Delta G^{o\prime}$ for the hydrolysis of ATP to ADP is $-30{\cdot}5\ \text{kJ mol}^{-1}$, how much free energy would be produced by this reaction in a cell at pH = 7·0 and 37°C?

SOLUTION: The balanced reaction is

$$ATP + H_2O \rightleftharpoons ADP + P_i + H^+$$

ΔG is given by

$$\Delta G = \Delta G^{o\prime} + RT\log_e\left(\frac{[ADP][P_i][H^+]}{[ATP][H_2O]}\right)$$

H_2O is approximately in its standard state, i.e. a pure liquid, as all other components are at very low concentrations. Thus, $[H_2O] \approx [H_2O]^{o\prime} = 55\ M$ and so $[H_2O]$ expressed relative to the standard state is $[H_2O]/[H_2O]^{o\prime} = 1$. H^+ is in its biochemical standard state, pH = 7, so $[H^+] = [H^+]^{o\prime} = 10^{-7}\ M$ and $[H^+]$ expressed relative to the standard state is $[H^+]/[H^+]^{o\prime} = 1$. For all the other components, the standard state is 1 M and so $[ATP]^{o\prime} = [ADP]^{o\prime} = [P_i]^{o\prime} = 1{\cdot}0\ M$.

The above equation then simplifies to

$$\Delta G = \Delta G^{o\prime} + RT\log_e\left(\frac{[ADP][P_i]}{[ATP]}\right)$$

Thus

$$\Delta G = -30\,500 + RT \log_e\left(\frac{0{\cdot}003 \times 0{\cdot}001}{0{\cdot}01}\right) = -51{\cdot}4\,\text{kJ mol}^{-1}$$

COMMENT: In this case, ΔG is negative and so work energy can be obtained from the hydrolysis of ATP and this can be used to drive other reactions. It is important to note that the value of ΔG depends on the concentration of species in solution. Any chemical reaction can be set up to have a negative value of ΔG as long as the reactants are in sufficient excess relative to products.

Mass action ratios and equilibrium constants

For convenience, the term

$$\left(\frac{\prod\limits_{\text{products}} [x]^{m_x}}{\prod\limits_{\text{reactants}} [x]^{m_x}}\right)$$

is often referred to as the *mass action ratio*, Γ, and so

$$\Delta G = \Delta G^\circ + RT \log_e \Gamma$$

At equilibrium, $\Delta G = 0$. The mass action ratio at equilibrium is called the *equilibrium constant*, K, given by

$$K = \left(\frac{\prod\limits_{\text{products}} [x]_{eq}^{m_x}}{\prod\limits_{\text{reactants}} [x]_{eq}^{m_x}}\right)$$

where subscript $_{eq}$ indicates the concentration at equilibrium. Thus

$$\Delta G^\circ = -RT \log_e K$$

Note that both Γ and K have no dimensions because all the concentration terms are expressed relative to the standard state concentrations, i.e. $[x]/[x]^\circ$. Dimensions are added frequently, as we will do in brackets, but should be used only to indicate the dimensions of the quantities used to calculate Γ or K.

Combining the equations for ΔG and ΔG° gives

$$\Delta G = RT \log_e\left(\frac{\Gamma}{K}\right)$$

The ratio of Γ to K gives a measure of ΔG for the reaction. If $\Gamma = K$, then $\Delta G = 0$ and the system is at equilibrium, if $\Gamma < K$ then $\Delta G < 0$ and the reaction can proceed spontaneously.

WORKED EXAMPLE

After adding triosephosphate isomerase to a 0·05 M solution of glyceraldehyde 3-phosphate at 298 K, the concentration fell to 0·002 M. What is ΔG° for the reaction catalysed by the enzyme?

SOLUTION: The reaction is

$$\text{glyceraldehyde 3-phosphate} \rightleftharpoons \text{dihydroxyacetone phosphate}$$

At equilibrium, [glyceraldehyde 3-phosphate] = 0·002 M and so [dihydroxyacetone phosphate] = 0·05 − 0·002 = 0·048 M. Thus the equilibrium constant, K, is given by

$$K = \frac{[\text{dihydroxyacetone phosphate}]}{[\text{glyceraldehyde 3-phosphate}]} = 24\ (\text{M})$$

The standard free energy is given by

$$\Delta G^\circ = -RT \log_e K = -7{\cdot}87\ \text{kJmol}^{-1}$$

COMMENT: Measuring the concentrations of reactants and products at equilibrium is an easy and often convenient method of determining ΔG° for a reaction.

WORKED EXAMPLE

The conversion of glucose 6-phosphate to fructose 6-phosphate is catalysed by the enzyme phosphoglucoisomerase. $\Delta G^{\circ\prime}$ for this reaction is 2·1 kJ mol^{-1}. If we start with 0·1 M glucose 6-phosphate and add phosphoglucoisomerase, what is the final composition of the solution at 298 K?

SOLUTION: The reaction being catalysed is

$$\text{glucose 6-phosphate} \overset{K}{\rightleftharpoons} \text{fructose 6-phosphate} \qquad K = \exp\left(\frac{-\Delta G^{\circ\prime}}{RT}\right) = 0{\cdot}43$$

If x M glucose 6-phosphate is converted to fructose 6-phosphate at equilibrium, then

$$[\text{glucose 6-phosphate}] = (0.1 - x)\ \text{M} \qquad \text{and} \qquad [\text{fructose 6-phosphate}] = x\,\text{M}$$

and so

$$K = \frac{[\text{fructose 6-phosphate}]}{[\text{glucose 6-phosphate}]} = \frac{x}{0.1 - x} = 0.43$$

Thus, x = 0·03 and so at equilibrium [glucose 6-phosphate] = 0·07 M and [fructose 6-phosphate] = 0·03 M.

WORKED EXAMPLE

$\Delta G^{o\prime}$ for the hydrolysis of phosphocreatine is $-37{\cdot}6\ \text{kJ mol}^{-1}$ at 310 K. A solution contains phosphocreatine at 0·0001 mM, creatine at 0·01 mM and phosphate at 0·5 mM. Is this solution at equilibrium? If not, in which direction will the reaction proceed and what is the value of ΔG?

SOLUTION: The reaction is

$$\text{phosphocreatine} + H_2O \rightleftharpoons \text{creatine} + P_i$$

From $\Delta G^{o\prime} = -RT \log_e K'$

$$K' = 2{\cdot}2 \times 10^6\ (\text{M})$$

Assuming H_2O is in the standard state, Γ' is given by

$$\Gamma' = \frac{[\text{creatine}][P_i]}{[\text{phosphocreatine}]} = 5 \times 10^{-2}\ (\text{M})$$

As $\Gamma' \neq K'$ the reaction is not at equilibrium. Moreover, as Γ' is less than K' the reaction will proceed to the right, forming more creatine. ΔG is given by

$$\Delta G = RT \log_e \left(\frac{\Gamma'}{K'}\right) = -45{\cdot}3\ \text{kJ mol}^{-1}$$

COMMENT: The ratio Γ/K is frequently used to determine whether reactions in metabolic pathways are at equilibrium. If $\Gamma/K \neq 1$ the reaction is not at equilibrium; however, values over a range near to one (for instance 0·1 to 10) are often taken to indicate that the reaction is close enough to equilibrium to be treated as such. This can be a useful simplification for qualitative discussion but more care needs to be taken for quantitative analysis, as we shall see in Chapter 13.

Alternative view of equilibrium constants

A chemical system is at equilibrium if the concentrations of the species present do not change with time. The system appears to be static at a macroscopic level. However, this does not mean that changes are not occurring at the microscopic level. In any chemical system at a microscopic level reactants are continually being converted to products and products are continually being converted to reactants. If the former reaction occurs faster than the latter, then the net result at the macroscopic level is the conversion of reactants to products. However, if these two reactions occur at the same rate then the net result is no change.

A chemical equilibrium is a dynamic process in which the forward and backward reactions occur at the same rate.

Consider the single-step chemical reaction

$$aA + bB \underset{k_{-1}}{\overset{k_1}{\rightleftharpoons}} cC + dD$$

As we shall see in Chapter 9, the rate of the forward reaction converting A and B into C and D is

$$\text{forward rate} = k_1 [A]^a [B]^b$$

where k_1 is the rate constant for the forward reaction. Similarly, the rate of the back reaction is given by

$$\text{backward rate} = k_{-1} [C]^c [D]^d$$

where k_{-1} is the rate constant for the backward reaction. At equilibrium, the forward rate equals the backward rate and so

$$k_1 [A]^a [B]^b = k_{-1} [C]^c [D]^d$$

This can be rearranged to give

$$\frac{k_1}{k_{-1}} = \frac{[C]^c [D]^d}{[A]^a [B]^b} = K$$

The equilibrium constant for a single-step reaction can also be thought of as the ratio of the forward and backward rate constants for the reaction. The equilibrium constant for a multistep reaction is not the ratio of the forward and backward rate constants for the reaction, although the equilibrium for each separate step is given by the ratio of the rate constants for that step.

Variation of equilibrium constant with temperature

The position of equilibrium for a reaction, and hence its equilibrium constant, varies with temperature. For a chemical reaction at equilibrium

$$\Delta G^\circ = -RT \log_e K$$

and

$$\Delta G^\circ = \Delta H^\circ - T\Delta S^\circ$$

Thus

$$\log_e K = \frac{\Delta S^\circ}{R} - \frac{\Delta H^\circ}{RT}$$

Assuming that ΔH° and ΔS° are independent of temperature, a plot of $\log_e K$ versus $1/T$ gives a straight line of slope $-\Delta H^\circ/R$. Thus, the temperature dependence of K allows ΔH° to be determined. Differentiation of the above equation with respect to T gives

$$\frac{d(\log_e K)}{dT} = \frac{\Delta H^\circ}{RT^2}$$

This equation is known as the *van't Hoff isochore*. Integration between temperatures T_1 and T_2, with values for the equilibrium constant of K_{T_1} and K_{T_2} respectively, gives

$$\log_e\left(\frac{K_{T_2}}{K_{T_1}}\right) = -\frac{\Delta H^\circ}{R}\left(\frac{1}{T_2} - \frac{1}{T_1}\right)$$

WORKED EXAMPLE

The equilibrium constant for a reaction doubled on raising the temperature from 298 K to 308 K. What is the value of ΔH°?

SOLUTION: $K_{T_2}/K_{T_1} = 2$, $T_2 = 308$ K and $T_1 = 298$ K. Substituting into the equation

$$\log_e\left(\frac{K_{T_2}}{K_{T_1}}\right) = -\frac{\Delta H^\circ}{R}\left(\frac{1}{T_2} - \frac{1}{T_1}\right)$$

gives

$$\Delta H^\circ = 52{\cdot}9\ \mathrm{kJ\ mol^{-1}}$$

WORKED EXAMPLE

The value of the equilibrium constant for the binding of phosphate to aldolase is 540 (M) at 298 K. Direct measurements show that the enthalpy change is $-87{\cdot}8$ kJ mol^{-1}. Predict the value of the equilibrium constant at 310 K.

SOLUTION: $K_{T_1} = 540$, $\Delta H^\circ = -87\,800\ \mathrm{J\ mol^{-1}}$, $T_1 = 298$ K, $T_2 = 310$ K. Substituting into the same equation gives

$$K_{T_2} = 137\ (\mathrm{M})$$

which is the equilibrium constant at 310 K.

COMMENT: Both these calculations use the implicit assumption that ΔH° and ΔS° are independent of temperature. The only way to determine whether or not this assumption is valid is to measure K as a function of T over the temperature range of interest, plot $\log_e K$ versus $1/T$ and check to see if the plot is linear.

This leads to a more general point. It is important to understand the derivation of any equation that you use, either for prediction or data analysis, so that you are aware of any implicit assumptions that you are making.

Dependence of enthalpy and entropy on temperature

The assumption that ΔH° and ΔS° are both independent of temperature is usually satisfactory if the temperature range considered is fairly small. However, the assumption is usually invalid over large temperature ranges, and for some processes, such as protein folding, ΔH and ΔS are strongly temperature dependent over quite small temperature ranges.

The variation of the enthalpy with temperature at constant pressure for a single compound is given by

$$\left(\frac{dH}{dT}\right)_p = C_p$$

where C_p is the molar specific heat capacity, defined as the amount of heat that one mole must absorb to raise its temperature by 1 K. For a reaction

$$\Delta H = H_{final} - H_{initial}$$

and so at constant pressure

$$\frac{d(\Delta H)}{dT} = \frac{d(H_{final})}{dT} - \frac{d(H_{initial})}{dT} = (C_p)_{final} - (C_p)_{initial}$$

or

$$\frac{d(\Delta H)}{dT} = \Delta C_p$$

where ΔC_p is the change in heat capacity between the reactants and products. This is known as Kirchoff's law. If ΔC_p is independent of temperature, this can be integrated to give

$$\Delta H_{T_2} - \Delta H_{T_1} = \Delta C_p(T_2 - T_1)$$

The temperature dependence of the entropy change during a reaction also depends on C_p. At equilibrium, $\delta H = T\delta S$ and so

$$\frac{dH}{dT} = \frac{TdS}{dT} = C_p \quad \text{or} \quad \frac{dS}{dT} = \frac{C_p}{T}$$

For a reaction

$$\frac{d(\Delta S)}{dT} = \frac{\Delta C_p}{T}$$

Assuming that C_p is independent of temperature, this can be integrated to give

$$\Delta S_{T_2} - \Delta S_{T_1} = \Delta C_p \log_e\left(\frac{T_2}{T_1}\right)$$

In cases where ΔC_p is large, such as the denaturation of proteins, the temperature dependence of ΔH and ΔS is also large. For instance, ΔH° for the unfolding of lysozyme by guanidinium hydrochloride is 62·7 kJ mol^{-1} at 293 K and 117·9 kJ mol^{-1} at 303 K. It is tempting to interpret the large heat capacity change that occurs on protein denaturation as resulting from the large conformational change of the protein. However, the major contribution to ΔC_p comes from the difference in the structure of the water surrounding the folded and unfolded protein. This emphasises again the importance of taking a system-based view of thermodynamic parameters.

Measurement of the thermodynamic quantities of reactions

We can now summarise some of the methods for measuring the thermodynamic quantities for a reaction.

- ΔG° — This is usually determined by measuring the concentrations of the reactants and products at equilibrium, i.e. by determining the equilibrium constant, and using

$$\Delta G^\circ = -RT \log_e K$$

If this is not possible, either because the equilibrium lies too far to one side or because the reaction itself cannot be done, then ΔG° can be calculated from an appropriate thermodynamic cycle of reactions for which the ΔG° values are known.

- ΔG — This is determined from ΔG° by measuring the concentrations of the reactants and products under the desired conditions and using

$$\Delta G = \Delta G^\circ + RT \log_e \Gamma$$

- ΔH° — This can be determined by measuring the equilibrium constant as a function of temperature. A plot of $\log_e K$ versus 1/T gives a straight line of slope $-\Delta H^\circ/R$. If this is not possible, either because the equilibrium lies too far to one side or because the reaction itself cannot be done, then ΔH° can be calculated from an appropriate thermodynamic cycle of reactions for which the ΔH° values are known.

- ΔH — This can be determined by direct calorimetric measurements made at constant pressure, microcalorimetry being the most useful technique for biochemical reactions.

- ΔS° and ΔS — These are usually obtained by knowing ΔG° or ΔG and ΔH° or ΔH at a given temperature and substituting in the equations

$$\Delta G^\circ = \Delta H^\circ - T\Delta S^\circ \quad \text{or} \quad \Delta G = \Delta H - T\Delta S$$

As we shall see in Chapter 6, thermodynamic quantities for redox reactions can also be obtained with great precision from measurement of cell e.m.f. values.

WORKED EXAMPLE

The following data were obtained for the temperature variation of the equilibrium constant of an inhibitor binding to the enzyme carbonic anhydrase.

T/K	289·0	294·2	298·0	304·9	310·5
$K \times 10^{-7}/(M)$	7·25	5·25	4·17	2·66	2·0

What are the values of ΔG°, ΔH° and ΔS° for this process at 298 K?

SOLUTION: At 298 K, $K = 4{\cdot}17 \times 10^7$. Using the equation $\Delta G^\circ = -RT \log_e K$ gives

$$\Delta G^\circ = -43{\cdot}45\ \text{kJ mol}^{-1}$$

The plot of $\log_e K$ versus 1/T (figure 3.2) has a slope of $-\Delta H^\circ/R$, assuming that ΔH° and ΔS° are independent of temperature. This gives

$$\Delta H^\circ = -45{\cdot}1\ \text{kJ mol}^{-1}$$

Using the equation $\Delta G^\circ = \Delta H^\circ - T\Delta S^\circ$ and the values of ΔG° and ΔH° already determined gives

$$\Delta S^\circ = -5{\cdot}53\ \text{J mol}^{-1}\ \text{K}^{-1}$$

COMMENT: In this case, a plot of $\log_e K$ versus 1/T is a straight line and so the assumption that ΔH° and ΔS° are independent of temperature over this temperature range is valid.

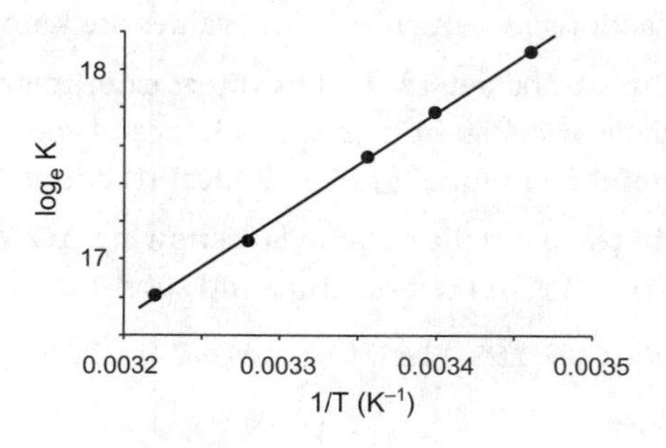

Fig. 3.2 A plot of $\log_e K$ versus 1/T. The slope of the straight line equals $-\Delta H^\circ/R$.

FURTHER READING

R.A. Alberty, 2000, 'Calculating apparent equilibrium constants of enzyme-catalysed reactions at pH 7', in Biochemical Education, vol. 28, pp. 12–17—*a useful source of primary data and worked examples.*

P.W. Atkins, 1998, 'Physical chemistry', Oxford University Press—*excellent coverage of all aspects of chemical thermodynamics, including many topics of only marginal interest to biochemists.*

E.B. Smith, 1990, 'Basic chemical thermodynamics', Oxford University Press—*a less comprehensive and more user-friendly treatment of chemical thermodynamics.*

PROBLEMS

1. The standard free energies of formation at 298 K for citrate, *cis*-aconitate and water are – 1167·1, – 921·7 and – 237·0 kJ mol^{-1} respectively. What is the equilibrium constant for the reaction

$$\text{citrate}^{3-} \rightleftharpoons cis\text{-aconitate}^{3-} + H_2O$$

and what concentration of citrate will be in equilibrium with 0·4 mM *cis*-aconitate?

2. $\Delta G^{o\prime}$ for the reaction

$$\text{glycerol} + P_i \rightleftharpoons \text{glycerol 1-phosphate} + H_2O$$

is 11·1 kJ mol^{-1} at 298 K. What concentration of glycerol 1-phosphate will be present at equilibrium if we start with 1 M glycerol and 0·5 M phosphate?

3. Glycogen phosphorylase catalyses the reaction

$$(\text{glycogen})_n + P_i \rightleftharpoons (\text{glycogen})_{n-1} + \text{glucose 1-phosphate} \quad \Delta G^{o\prime} = +3{\cdot}05\ \text{kJ mol}^{-1}$$

Assuming that the concentrations of P_i and glucose 1-phosphate are equal, does the equilibrium lie in favour of glycogen synthesis or degradation at 310 K?
In muscle, the concentrations of phosphate and glucose 1-phosphate are 10 mM and 30 μM respectively. Does glycogen phosphorylase cause synthesis or degradation of glycogen in muscle?

4. The following metabolite concentrations were measured in perfused rat heart at 308 K

Metabolite	Concentration
fructose 6-phosphate	60 μM
fructose bisphosphate	9 μM
ATP	5·3 mM
ADP	1·1 mM
AMP	95 μM

The enzyme phosphofructokinase catalyses the reaction

$$\text{fructose 6-phosphate} + \text{ATP} \rightleftharpoons \text{fructose bisphosphate} + \text{ADP} \quad \Delta G^{o\prime} = -17{\cdot}7\,\text{kJ mol}^{-1}$$

and the enzyme adenylate kinase catalyses the reaction

$$2\text{ADP} \rightleftharpoons \text{ATP} + \text{AMP} \quad \Delta G^{o\prime} = +2{\cdot}1\,\text{kJ mol}^{-1}$$

Are these enzyme catalysed reactions at equilibrium in perfused rat heart? If not, what are the values of $\Delta G^{o\prime}$ for the reactions? Comment briefly on your results.

5. In most organisms, pyruvate kinase catalyses the conversion of phosphoenolpyruvate to pyruvate

$$\text{phosphoenolpyruvate} + \text{ADP} \rightleftharpoons \text{pyruvate} + \text{ATP}$$

It has been proposed that when growing on pyruvate a novel micro-organism is able to convert pyruvate to phosphoenolpyruvate via the pyruvate kinase reaction. To test this hypothesis a suspension of growing cells at 30°C, shown to contain $5 \times 10^{-5}\,\text{dm}^3$ intracellular water, was disrupted and the following amounts of metabolites found.

Phosphoenolpyruvate	5 nmol
Pyruvate	7 nmol
ADP	6 nmol
ATP	500 nmol
P_i	50 nmol

Given the following $\Delta G^{o\prime}$ values, evaluate the validity of the proposal for the route of phosphoenolpyruvate synthesis.

$$\text{phosphoenolpyruvate} + H_2O \rightleftharpoons \text{pyruvate} + P_i \quad \Delta G^{o\prime} = -62\,\text{kJ mol}^{-1}$$

$$\text{ATP} + H_2O \rightleftharpoons \text{ADP} + P_i \quad \Delta G^{o\prime} = -30\,\text{kJ mol}^{-1}$$

An alternative hypothesis suggests that a novel enzyme catalysing the reaction

$$\text{pyruvate} + 2\text{ATP} + H_2O \rightleftharpoons \text{phosphoenolpyruvate} + 2\text{ADP} + P_i$$

is responsible. Evaluate by calculation whether this reaction is more or less likely than the pyruvate kinase reaction to account for the synthesis of phosphoenolpyruvate.

6. The enzyme phosphorylase b can be activated by the addition of AMP. The equilibrium constant for the reaction

$$\text{enzyme : AMP} \overset{K}{\rightleftharpoons} \text{enzyme} + \text{AMP}$$

is given as a function of temperature below

Temperature/K	285·5	289	300	312·5
$K \times 10^5/(\text{M})$	2·75	3·1	4·2	5·9

Calculate ΔH°, ΔS° and ΔG° for this reaction at 303 K, stating any assumptions that you make.

7. In the biosynthesis of proteins it is important that the correct tRNA binds to its corresponding amino acid tRNA ligase enzyme. The following data were obtained for the binding of two tRNAs to isoleucine tRNA ligase from *E. coli*

Substrate	$\Delta H°$/kJ mol^{-1}	$\Delta S°$/J mol^{-1} K^{-1}
Isoleucine tRNA	0	+142·0
Valine tRNA	+33·4	+225·7

By what factor is the binding of the correct tRNA favoured over the incorrect tRNA at 293 K and 313 K?

4 Binding of ligands to macromolecules

KEY POINTS:

- The binding of a ligand to a protein is quantified by a *dissociation constant*, K_d. Knowledge of K_d and the concentration of free ligand allows the degree of saturation of the protein to be calculated.
- For a protein with more than one ligand site, independent binding occurs when the affinity of one site for a ligand is unaffected by binding at the other site.
- *Cooperativity* occurs when the binding of a ligand to one site changes the affinity of a second site on the protein for its ligand.
- *Positive* and *negative* cooperativity alter the response of a protein to changes in ligand concentration.
- *Multivalent interactions* enhance binding, thus multiple weak interactions can lead to high affinity binding.

It is convenient to discuss the analysis of ligand binding data in a separate section, as several additional points are brought out even though no new fundamental principles are involved. We consider the binding of ligands to proteins, but the equations apply for the binding of any species to any type of macromolecule.

Ligand binding to a single site on a protein

We start by considering the simplest case of a single ligand, L, binding to a protein P.

$$P + L \overset{K}{\rightleftharpoons} PL$$

The equilibrium constant, K, for this reaction is given by

$$K = \frac{[PL]}{[P][L]}$$

However, it is standard practice for biochemists to refer to a *dissociation constant*, K_d, which is the equilibrium constant for the reverse reaction.

$$PL \overset{K_d}{\rightleftharpoons} P + L \qquad K_d = \frac{[P][L]}{[PL]} = \frac{1}{K}$$

When more than one ligand is binding to a protein, it can be confusing to determine which ligand dissociation is being referred to and so we shall use the nomenclature $K_d^{L/PL}$ to represent the K_d where L is dissociating from PL.

The next quantity that it is useful to define is the average number of ligand molecules bound to each protein, $\bar{n}$.

$$\bar{n} = \frac{\text{concentration of L bound to P}}{\text{total concentration of P}}$$

For a single ligand binding to a protein, $\bar{n}$ is given by

$$\bar{n} = \frac{\text{concentration of L bound to P}}{\text{total concentration of P}} = \frac{[PL]}{[P] + [PL]}$$

and since $[PL] = [P][L]/K_d^{L/PL}$

$$\bar{n} = \frac{\dfrac{[P][L]}{K_d^{L/PL}}}{[P] + \dfrac{[P][L]}{K_d^{L/PL}}}$$

Dividing both top and bottom by [P] and multiplying by $K_d^{L/PL}$ gives

$$\boxed{\bar{n} = \frac{[L]}{K_d^{L/PL} + [L]}}$$

where [L] is the concentration of free ligand. In this case, $\bar{n} \leq 1$ and is the same as the *fractional saturation* of the protein, i.e. the fraction of the protein sites that are occupied by ligand. In general, the fractional saturation, θ, is given by

$$\theta = \frac{\bar{n}}{n}$$

where n is the total number of binding sites for the ligand.

WORKED EXAMPLE

Mg^{2+} and ADP form a 1:1 complex. In an experiment, the concentration of ADP was kept constant at 80 μM and the concentration of Mg^{2+} was varied. The following results were obtained.

Total Mg^{2+}/μM	**20**	**50**	**100**	**150**	**200**	**400**
Mg^{2+} bound to ADP/μM	**11·6**	**26·0**	**42·7**	**52·8**	**59·0**	**69·5**

Determine the dissociation constant for Mg:ADP under these conditions.

SOLUTION: At each value of the total Mg^{2+} concentration, the free Mg^{2+} concentration ([L] in the equations) can be evaluated by the difference between total and bound. The value of θ is found by dividing the concentration of Mg:ADP complex (equal to the concentration of bound Mg^{2+}) by the total ADP concentration (i.e. 80 μM).

Total $Mg^{2+}/\mu M$	20	50	100	150	200	400
Bound $Mg^{2+}/\mu M$	11·6	26·0	42·7	52·8	59·0	69·5
Free $Mg^{2+}/\mu M$	8·4	24·0	57·3	97·2	141·0	330·5
θ	0·145	0·325	0·534	0·660	0·738	0·869

As the total number of sites is one, $\theta = \bar{n}$ and so θ is given by

$$\theta = \frac{[L]}{K_d + [L]}$$

This equation can be rearranged in several ways to give a straight line. The two most common are

$$\frac{1}{\theta} = 1 + \frac{K_d}{[L]}$$ Plot of $1/\theta$ versus $1/[L]$ gives a straight line of slope K_d

$$\frac{\theta}{[L]} = \frac{1}{K_d} - \frac{\theta}{K_d}$$ Plot of $\theta/[L]$ versus θ gives a straight line of slope $-1/K_d$

Plots of θ versus $[Mg^{2+}]$, $1/\theta$ versus $1/[Mg^{2+}]$ and $\theta/[Mg^{2+}]$ versus θ are shown in figure 4.1. K_d can be determined from the slope of both straight line plots, giving

$$K_d = 50\ (\mu M)$$

COMMENT: We would not normally do both straight line plots. It is worth noting that the distribution of points is very different between the two plots. In general, it is easier to determine the best straight line when the data points are distributed evenly. Thus, the plot of $\theta/[L]$ versus θ is more likely to give an accurate result.

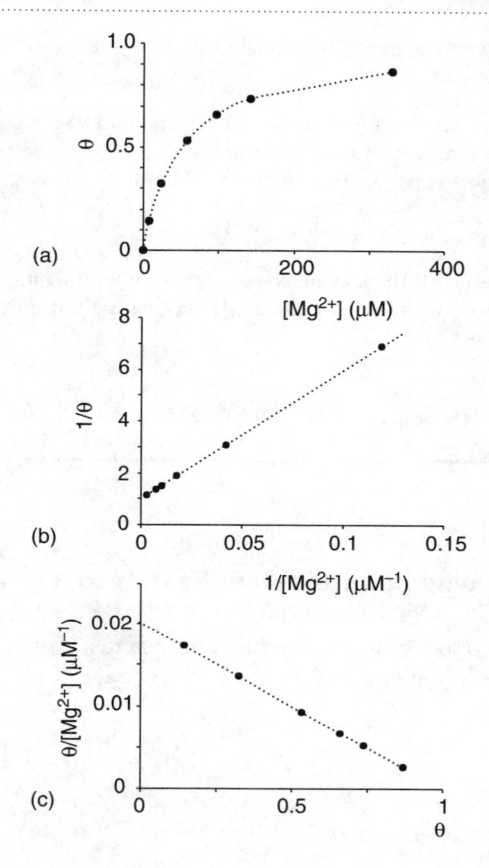

Fig. 4.1 Plots of (a) θ versus $[Mg^{2+}]$, (b) $1/\theta$ versus $1/[Mg^{2+}]$ (Hughes–Klotz plot), and (c) $\theta/[Mg^{2+}]$ versus θ (Scatchard plot), for the 1:1 complex formation between Mg^{2+} and ADP.

WORKED EXAMPLE

What are the fractional saturations of a single-site protein, $K_d = 10^{-6}$ (M), at a total protein concentration of 1 μM and total ligand concentrations of (a) 2 μM and (b) 200 μM?

SOLUTION:

(a) If the concentration of complex formed is x μM, then

$$\underset{(1\text{-}x)}{P} + \underset{(2\text{-}x)}{L} \rightleftharpoons \underset{x}{PL}$$

$$K_d = 1 \times 10^{-6} = \frac{[P][L]}{[PL]} = \frac{(1-x) \times 10^{-6} \times (2-x) \times 10^{-6}}{x \times 10^{-6}}$$

Rearranging this gives

$$x^2 - 4x + 2 = 0$$

Solving the quadratic equation gives x = 0·59 μM. As the total concentration of protein is 1 μM, the fractional saturation is 59%.

(b) The same approach gives a fractional saturation of 99·5% with a ligand concentration of 200 μM.

COMMENT: It is worth noting that when the ligand is in large excess compared to the enzyme, we can make the simplifying assumption that the concentration of free ligand equals the concentration of total ligand.

$$[\mathrm{L}] = [\mathrm{L}]_{\mathrm{total}} - [\mathrm{PL}] \approx [\mathrm{L}]_{\mathrm{total}}$$

In this case, even if all the protein were complexed with ligand, the free ligand concentration would still be 199 μM. The fractional saturation can then be calculated directly.

$$\theta \approx \frac{[\mathrm{L}]_{\mathrm{total}}}{\mathrm{K_d} + [\mathrm{L}]_{\mathrm{total}}} = \frac{200 \times 10^{-6}}{(1 \times 10^{-6}) + (200 \times 10^{-6})} = 99.5\%$$

Simultaneous binding of different ligands to a protein

The next situation to consider is two different ligands, L and L′, binding to two different sites on a protein.

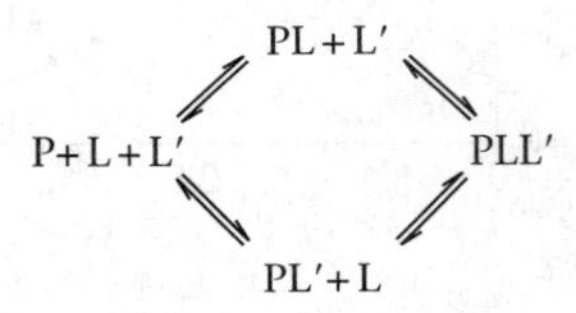

$\mathrm{K_d^{L/PL}}$ is the dissociation constant for L leaving PL and $\mathrm{K_d^{L/PLL'}}$ is the dissociation constant for L leaving PLL′

$$\mathrm{K_d^{L/PL}} = \frac{[\mathrm{P}][\mathrm{L}]}{[\mathrm{PL}]} \quad \text{and} \quad \mathrm{K_d^{L/PLL'}} = \frac{[\mathrm{PL'}][\mathrm{L}]}{[\mathrm{PLL'}]}$$

and similarly $\mathrm{K_d^{L'/PL'}}$ and $\mathrm{K_d^{L'/PLL'}}$ are the dissociation constants for L′ from PL′ and PLL′ respectively.

Binding of L and L′ independent

If the binding of L does not affect the binding of L′ and vice versa, then $\mathrm{K_d^{L/PL}} = \mathrm{K_d^{L/PLL'}}$ and $\mathrm{K_d^{L'/PL'}} = \mathrm{K_d^{L'/PLL'}}$. In this case, we can treat each site separately and calculate their fractional saturations

$$\theta_{\mathrm{L}} = \frac{[\mathrm{L}]}{\mathrm{K_d^{P/PL}} + [\mathrm{L}]} \quad \text{and} \quad \theta_{\mathrm{L'}} = \frac{[\mathrm{L'}]}{\mathrm{K_d^{P/PL'}} + [\mathrm{L'}]}$$

where [L] and [L′] are the concentrations of the unbound ligands. The fraction of the protein that has both sites occupied is given by $\theta_{\mathrm{LL'}} = \theta_{\mathrm{L}} \times \theta_{\mathrm{L'}}$.

Binding of L and L′ not independent

In this case, the dissociation constant for L depends on whether the site for L′ is occupied or not and the dissociation constant for L′ depends on whether the site for L is occupied or not, $K_d^{L/PL} \neq K_d^{L/PLL'}$ and $K_d^{L'/PL'} \neq K_d^{L'/PLL'}$. This leads us to the concept of *cooperative binding*.

Positive cooperativity	The binding of L to the protein makes the binding of L′ stronger and the binding of L′ to the protein makes the binding of L stronger. $K_d^{L/PL} > K_d^{L/PLL'}$ and $K_d^{L'/PL'} > K_d^{L'/PLL'}$
Negative cooperativity	The binding of L to the protein makes the binding of L′ weaker and the binding of L′ to the protein makes the binding of L weaker. $K_d^{L/PL} < K_d^{L/PLL'}$ and $K_d^{L'/PL'} < K_d^{L'/PLL'}$

The fractional saturation of the site for L on the protein at a given value of [L] now depends on the concentration of L′. For positive cooperativity, an increase in [L′] will increase θ_L whereas for negative cooperativity an increase in [L′] will decrease θ_L. We can no longer use independent equations for θ_L. and $\theta_{L'}$, but have to solve the four equations for the equilibrium constants simultaneously.

The occurrence of cooperativity is taken as evidence for conformational changes in the protein induced by ligand binding (figure 4.2). For positive cooperativity, the binding of one ligand induces a conformational change in the other binding site that increases its affinity for the other ligand.

A useful measure of the degree of cooperativity is the change in $\Delta G°$ of binding caused by saturating levels of the other ligand, $\Delta\Delta G°$.

$$\Delta\Delta G°(L) = \Delta G°(L \text{ binding to } PL') - \Delta G°(L \text{ binding to } P)$$

For independent binding sites $\Delta\Delta G° = 0$ whereas for positive cooperativity $\Delta\Delta G° < 0$. An example of this is given for haemoglobin below.

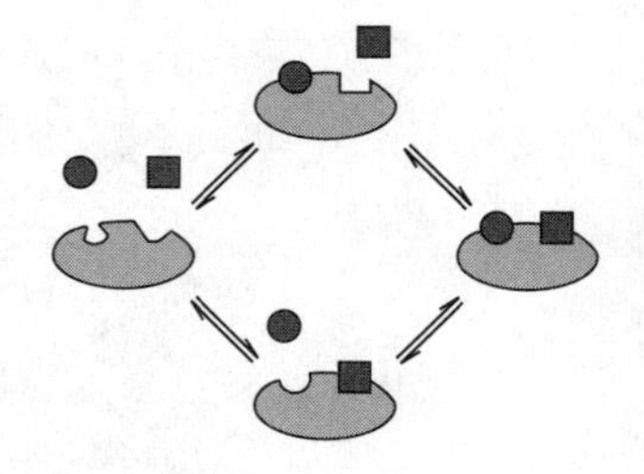

Fig. 4.2 Schematic diagram showing the induced conformational changes on ligand binding that result in positive cooperativity.

A single ligand binding to multiple sites on a protein

The situation becomes a little more complex when one type of ligand can bind to more than one site on the protein. For a protein with n binding sites,

$$P + nL \rightleftharpoons PL + (n-1)L \rightleftharpoons PL_2 + (n-2)L \rightleftharpoons \cdots \rightleftharpoons PL_n$$

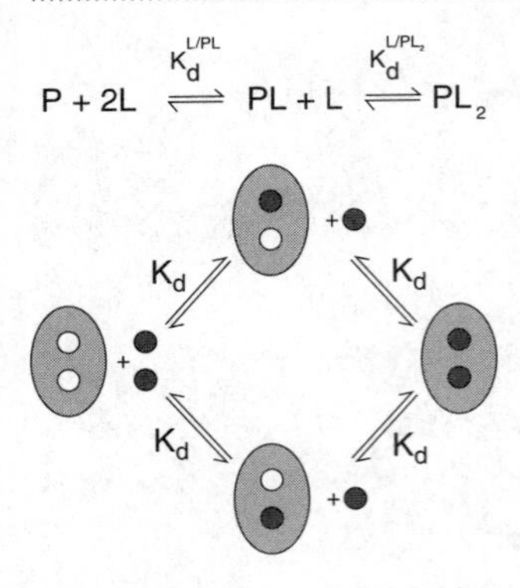

Fig. 4.3 $K_d^{L/PL}$ is a macroscopic dissociation constant that includes the probability of a ligand binding at one of a number of sites. κ_d is a microscopic dissociation constant for a ligand binding at a specific site.

the average number of ligands bound per protein is given by

$$\bar{n} = \frac{\text{concentration of L bound to P}}{\text{total concentration of P}} = \frac{[PL] + 2[PL_2] + \cdots + n[PL_n]}{[P] + [PL] + [PL_2] + \cdots + [PL_n]}$$

The experimentally determined dissociation constants are

$$K_d^{L/PL} = \frac{[P][L]}{[PL]} \qquad K_d^{L/PL_2} = \frac{[PL][L]}{[PL_2]} \qquad K_d^{L/PL_n} = \frac{[PL_{n-1}][L]}{[PL_n]}$$

These are termed *macroscopic dissociation constants*. We can define a *microscopic dissociation constant*, κ_d, as the dissociation constant for a ligand at a single site (figure 4.3). So, for example, for the equilibrium

$$P + nL \rightleftharpoons PL + (n-1)L$$

κ_d is the dissociation constant for the binding of the ligand to a specific site on the protein

$K_d^{L/PL}$ is the dissociation constant for the binding of the first ligand to any of the sites on the protein

These two values are not the same. The microscopic and macroscopic dissociation constants can be expressed in terms of ratios of rate constants.

$$\kappa_d = \frac{\text{rate constant for dissociation of L from a single site}}{\text{rate constant for binding of L to a single site}} = \frac{k_d}{k_a}$$

$$K_d^{L/PL} = \frac{\text{rate constant for dissociation of L from PL}}{\text{rate constant for binding of L to P}}$$

Assume that all of the sites are identical and so have identical values of the microscopic dissociation constant, κ_d. There are n sites that the ligand can bind to on the protein, so the rate of binding of L to P is n times the rate of binding of L to a single site. There is only one site occupied at a time on PL, so the rate of dissociation of L from PL equals the rate of dissociation of L from a single site.

$$K_d^{L/PL} = \frac{\text{rate constant for dissociation of L from PL}}{\text{rate constant for binding of L to P}} = \frac{k_d}{n \times k_a} = \frac{1}{n}\kappa_d$$

All the other macroscopic dissociation constants can be calculated in a similar manner. However, it is worth stressing that even if all the sites are identical with identical values of κ_d, none of the macroscopic dissociation constants is the same.

$$K_d^{L/PL} \neq K_d^{L/PL_2} \neq \cdots \neq K_d^{L/PL_n}$$

The microscopic dissociation constants give more insight into the binding properties of the protein because they relate directly to how strongly a binding site interacts with its ligand. The macroscopic dissociation constants include an additional probability term due to the possibility of any given ligand molecule binding to one of several sites.

Independent binding to identical sites

In this case, the microscopic dissociation constants for the different sites are all the same and do not change as more sites are occupied. Each site behaves independently and has a fractional saturation given by

$$\theta = \frac{[L]}{K_d + [L]}$$

where K_d is the microscopic dissociation constant. The average number of ligands bound per protein is the fractional saturation at each site multiplied by the number of sites, n.

$$\bar{n} = \frac{n[L]}{K_d + [L]}$$

$\bar{n}$ has values between 0 and n. This equation can be rearranged in several ways to give straight line plots (Table 4.1).

Table 4.1

Equation	Plot	K_d	n
Hughes–Klotz plot			
$\frac{1}{\bar{n}} = \frac{K_d}{n}\frac{1}{[L]} + \frac{1}{n}$	$\frac{1}{\bar{n}}$ versus $\frac{1}{[L]}$	x-intercept $= -\frac{1}{K_d}$	y-intercept $= \frac{1}{n}$
Scatchard plot			
$\frac{\bar{n}}{[L]} = \frac{n}{K_d} - \frac{\bar{n}}{K_d}$	$\frac{\bar{n}}{[L]}$ versus $\bar{n}$	slope $= -\frac{1}{K_d}$	x-intercept $= n$

WORKED EXAMPLE

In an experiment the concentration of an enzyme was kept constant at 11 μM, and the concentration of inhibitor [I] varied. The following results were obtained.

$[I]_{total}$/μM	5·2	10·4	15·6	20·8	31·2	41·6	62·4
$[I]_{free}$/μM	2·3	4·8	7·95	11·3	18·9	27·4	45·8

Determine the dissociation constant for the enzyme–inhibitor complex and the number of inhibitor binding sites on the enzyme.

SOLUTION: At each value of $[I]_{total}$ we can evaluate $[I]_{bound}$ by subtraction. $\bar{n}$ is obtained by dividing $[I]_{bound}$ by the concentration of enzyme (i.e. 11 μM).

$[I]_{bound}/\mu M$	2·9	5·6	7·65	9·5	12·3	14·2	16·6
$\bar{n}$	0·264	0·510	0·695	0·864	1·118	1·291	1·510
$1/\bar{n}$	3·793	1·964	1·438	1·158	0·894	0·775	0·663
$\bar{n}/[I]_{free}/\mu M^{-1}$	0·115	0·106	0·087	0·076	0·059	0·047	0·033
$1/[I]_{free}/\mu M^{-1}$	0·435	0·208	0·126	0·088	0·053	0·036	0·022

The Hughes–Klotz and Scatchard plots are shown in figure 4.4. Both of these give $n = 2$ and $K_d = 15{\cdot}2 \times 10^{-6}$(M).

COMMENT: Both these plots give straight lines, proving that the binding of the inhibitor at the two sites is independent.

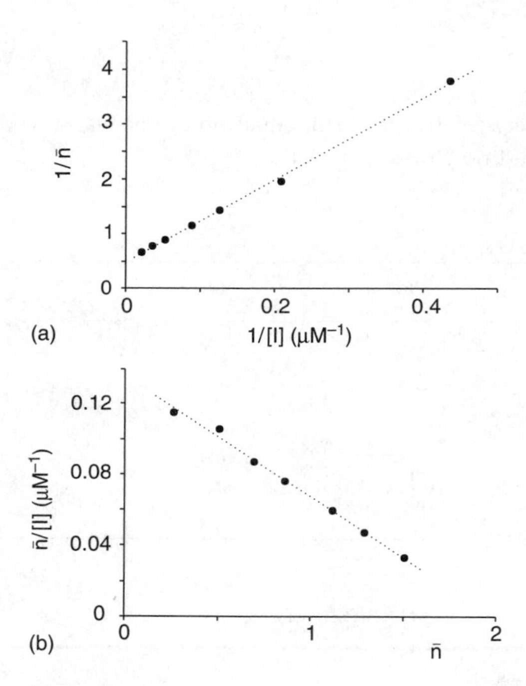

Fig. 4.4 (a) Hughes–Klotz plot, $1/\bar{n}$ versus $1/[I]$, and (b) Scatchard plot, $\bar{n}/[I]$ versus $\bar{n}$, for an inhibitor binding to an enzyme.

Non-independent binding to identical sites

In this case, the microscopic dissociation constants for the different sites are all the same but are now dependent on the level of occupancy of the other sites. The origin of this behaviour is the same as before, ligand binding at one site inducing conformational changes at the other sites. For positive cooperativity, as more sites become occupied, the affinity of the remaining sites for ligand increases. This can easily be identified by non-linear behaviour in Hughes–Klotz or Scatchard plots (figure 4.5).

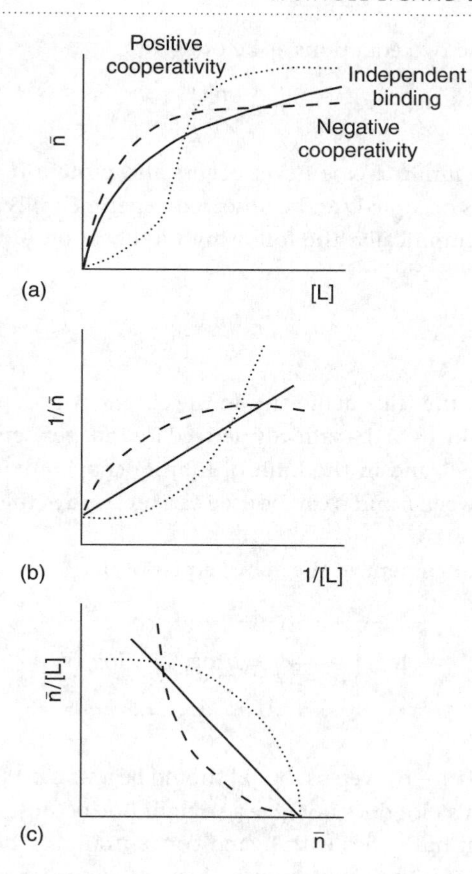

Fig. 4.5 Plots of (a) $\bar{n}$ versus [L], (b) $1/\bar{n}$ versus 1/[L] and (c) $\bar{n}$/[L] versus $\bar{n}$ for independent binding, positive cooperativity and negative cooperativity. These plots are only intended to show the different shapes of the curves.

The limiting case of positive cooperativity occurs when the only species present are P and PL_n, species such as PL or PL_{n-1} being present at very low concentration. The binding of the first ligand increases the affinity at the other sites so much that they immediately become saturated and the only equilibrium that occurs is

$$P + nL \rightleftharpoons PL_n$$

All the ligands have the same microscopic dissociation constant. This is the same as the macroscopic equilibrium constant for dissociation of all ligands because dissociation of one ligand causes dissociation of all the rest.

$$K_d = \frac{[P][L]^n}{[PL_n]}$$

and the average number of ligands bound per protein is

$$\bar{n} = \frac{n[PL_n]}{[P] + [PL_n]}$$

Combining these two equations as before gives

$$\bar{n} = \frac{n[L]^n}{K_d + [L]^n}$$

In practice, this limiting case never occurs and protein molecules with only some of the sites occupied can be observed experimentally. To try and model this behaviour empirically, the following modification of the equation for $\bar{n}$ can be used

$$\bar{n} = \frac{n[L]^h}{K_d + [L]^h}$$

where h is called the Hill coefficient. In the absence of cooperativity $h = 1$ and this equation reduces to that already derived for independent sites. For positive cooperativity $h > 1$ and in the limit of infinite cooperativity $h = n$. Thus, the comparison between h and n can be used as a measure of the degree of cooperativity, as well as $\Delta\Delta G°$.

A useful rearrangement of the above equation is

$$\log_e\left(\frac{\bar{n}}{n - \bar{n}}\right) = h \log_e[L] - \log_e(K_d)$$

A plot of $\log_e(\bar{n}/(n - \bar{n}))$ versus $\log_e[L]$ should be a straight line with a gradient h. In practice, this plot does not give a straight line because the Hill coefficient is not a constant but varies with ligand concentration. The value of h at any particular value of [L] can be determined from the gradient of the line. At very low and very high values of [L], h always approaches one. This is because the induced protein conformational changes that lead to cooperativity cannot occur until some ligand has bound, hence h must be one at very low $\bar{n}$. Similarly, once only one binding site remains unfilled, further ligand binding cannot result in any further cooperativity.

WORKED EXAMPLE

Haemoglobin is a tetrameric oxygen-binding protein with four identical binding sites. The following data give the fractional saturation, θ, of haemoglobin, concentration $1{\cdot}55 \times 10^{-5}$ μM, 300 K, at various partial pressures of oxygen. Calculate the Hill coefficients at 0%, 50% and 100% saturation.

$pO_2 \times 10^3$/atm	0·3	0·5	1·1	1·7	2·8	3·8
θ	0·007	0·013	0·030	0·066	0·136	0·273
$pO_2 \times 10^3$/atm	5·7	10·1	15·8	20·4	36·6	109·6
θ	0·500	0·864	0·953	0·978	0·991	0·997

SOLUTION: $\bar{n}$ is given by θ times the number of binding sites, n. For haemoglobin, n = 4.

$\bar{n}$	0·028	0·052	0·120	0·264	0·544	1·092
$\log_e(\bar{n}/(n-\bar{n}))$	−4·95	−4·33	−3·48	−2·65	−1·85	−0·98
$\log_e(pO_2)$	−8·24	−7·55	−6·86	−6·37	−5·89	−5·57
$\bar{n}$	2·000	3·456	3·812	3·912	3·964	3·988
$\log_e(\bar{n}/(n-\bar{n}))$	0·00	1·85	3·01	3·79	4·70	5·81
$\log_e(pO_2)$	−5·17	−4·59	−4·15	−3·89	−3·31	−2·21

The plot of $\log_e(\bar{n}/(n-\bar{n}))$ versus $\log_e(pO_2)$ is shown in figure 4.6. Measuring the gradients of the plot at appropriate points gives

at 0% saturation, $h = 0{\cdot}9$

at 50% saturation, $h = 2{\cdot}9$

at 100% saturation, $h = 0{\cdot}9$

COMMENT: As expected, the Hill coefficient is approximately 1 at 0% and 100% saturation. These correspond to the states where very little ligand is bound and no induced conformational changes have occurred (called the tense state, T, of haemoglobin) and where the protein is almost saturated and no further induced conformational changes can occur (called the relaxed state, R). The value of the Hill coefficient increases with increasing saturation until it peaks at a value of 2·9 at 50% saturation and then decreases again. The maximum theoretical value of the Hill coefficient is 4 in this case and haemoglobin is classified as highly cooperative. The triggering mechanism for the conformational changes caused by binding of O_2 is discussed in Chapter 16.

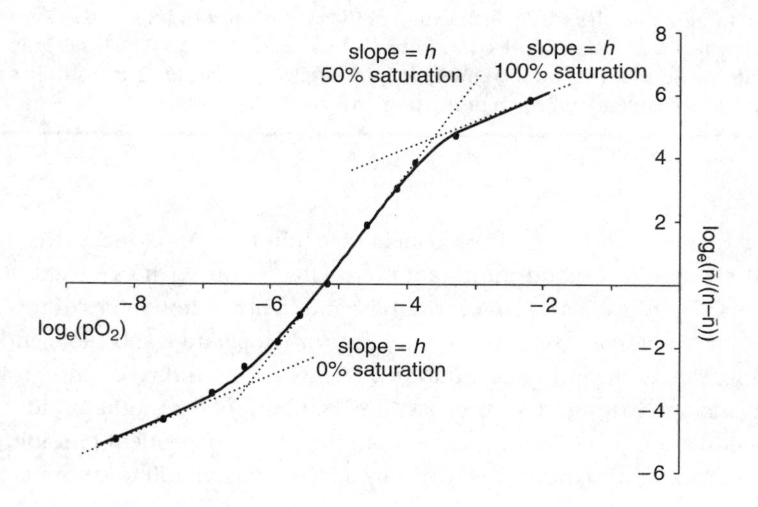

Fig. 4.6 Plot of $\log_e(\bar{n}/(n-\bar{n}))$ versus $\log_e(pO_2)$ for haemoglobin. The slope of this line gives the Hill coefficient, *h*. At 0% and 100% saturation, *h* is necessarily one. The value for *h* at 50% saturation is taken as a measure of the degree of cooperativity.

WORKED EXAMPLE

Using the data in the previous worked example, calculate the microscopic dissociation constants for the tense and relaxed forms of haemoglobin and calculate the difference in ΔG° binding between the two ($\Delta\Delta G^\circ$).

SOLUTION: The linear regions of the $\log_e(\bar{n}/(n-\bar{n}))$ versus $\log_e[L]$ curve are given by the equation

$$\log_e\left(\frac{\bar{n}}{n-\bar{n}}\right) = h\log_e[L] - \log_e(K_d)$$

The y-intercept of this line equals $-\log_e(K_d)$. Haemoglobin is entirely in the tense form at 0% saturation and in the relaxed form at 100% saturation. Measuring the y-intercepts of the linear regions at 0% saturation and 100% saturation (figure 4.7) gives

at 0% saturation (tense form), $K_d = 0{\cdot}08$ (atm)
at 100% saturation (relaxed form), $K_d = 0{\cdot}0003$ (atm)

Thus, the dissociation constant for O_2 binding to the relaxed form is approximately 250 times less than for the tense form. ΔG° binding is given by

$$\Delta G^\circ = -RT\log_e(K) = -RT\log_e\left(\frac{1}{K_d}\right)$$

$\Delta\Delta G^\circ$ is given by

$$\Delta\Delta G^\circ = \Delta G^\circ(\text{relaxed form}) - \Delta G^\circ(\text{tense form})$$

$$\Delta\Delta G^\circ = \left\{-RT\log_e\left(\frac{1}{K_d}\right)\right\}_{\text{relaxed}} - \left\{-RT\log_e\left(\frac{1}{K_d}\right)\right\}_{\text{tense}}$$

$$= RT\log_e\left(\frac{K_d(\text{relaxed})}{K_d(\text{tense})}\right) = -13{\cdot}9\ \text{kJ mol}^{-1}$$

COMMENT: This value of $\Delta\Delta G^\circ$ corresponds to the effect that the binding of the first three molecules of O_2 to haemoglobin has on the binding of the fourth molecule. Each occupied site contributes $\sim 4{\cdot}5$ kJ mol^{-1} to $\Delta\Delta G^\circ$. This is only a small fraction of the binding energy for any single O_2 molecule but it still has a considerable impact on the binding properties of haemoglobin.

The major effect of positive cooperative binding is to make the fractional saturation respond more rapidly to changes in ligand concentration (figure 4.8). At low ligand concentrations, and hence at low values of $\bar{n}$ when $h \approx 1$, the fractional saturation is the same for cooperative and independent binding. As the ligand concentration increases, the affinity of the protein for ligand increases and so the fractional saturation rises more rapidly for cooperative compared to independent binding. For independent binding, the ligand concentration needs to change by a factor of over 100 to go from close

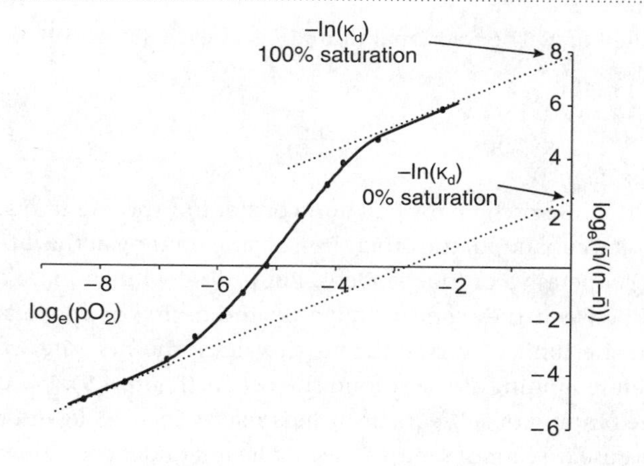

Fig. 4.7 Plot of $\log_e(\bar{n}/(n-\bar{n}))$ versus $\log_e(pO_2)$ for haemoglobin. The y-intercepts of the linear regions at 0% and 100% saturation give the K_d values for the two limiting conformations of haemoglobin.

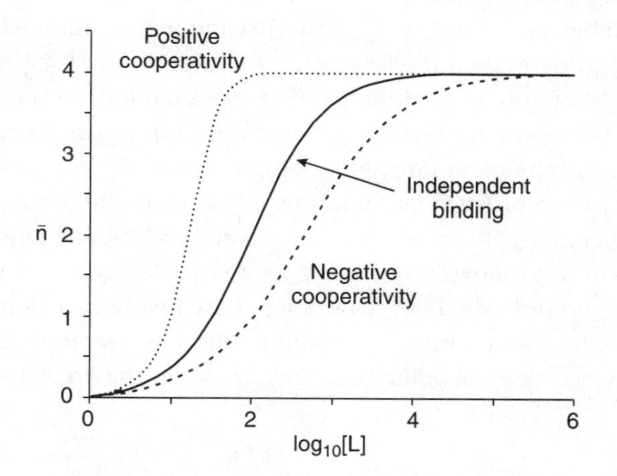

Fig. 4.8 A plot of $\bar{n}$ versus $\log_{10}[L]$ is a convenient way of showing the change in ligand concentration necessary to go from ~0% to ~100% saturation. Plots of $\bar{n}$ versus $\log_{10}[L]$ are shown for a protein with four identical binding sites showing either independent (solid line), positive cooperative (dotted line) or negative cooperative (dashed line) binding. K_d is the same in all cases, $h = 2$ for the dotted line and $h = 0{\cdot}75$ for the dashed line. The slopes of the cooperative curves are different from that for independent binding because K_d is either increasing or decreasing during the binding titration.

to 0% saturation to close to 100% saturation. For cooperative binding, a change in ligand concentration by a factor of 10 may be sufficient. In the case of haemoglobin, the change in partial pressure of O_2 between lungs and muscle is only a factor of 3 and so cooperativity in O_2 binding is essential if haemoglobin is to be saturated with O_2 in the lungs and yet release O_2 to muscle. In contrast, negative cooperativity results in the fractional saturation being less sensitive to changes in ligand concentration, a larger change in ligand concentration being required to go from ≈0% to ≈100% saturation (figure 4.8).

Binding to non-identical sites

In this case, the K_d values are not the same for all the ligand sites. There will not be random filling of the sites on the protein, but those with the lowest values of

K_d will be filled first. The fractional saturation at each site is still given by the equation

$$\theta = \frac{[L]}{K_d + [L]}$$

where K_d is the microscopic dissociation constant of the site in question at a given level of occupancy of the other sites. Ligand binding at the different sites can still be cooperative or independent, but it is now much more difficult to distinguish between these and to quantify the degree of cooperativity. For example, in the limit of very different K_d values, the first site will be fully occupied before binding at the second site occurs (figure 4.9). We cannot tell whether the binding of subsequent ligands makes the first ligand bind more strongly because we cannot remove the first ligand under conditions in which the other ligands remain bound. Similarly, we cannot measure the dissociation of the second ligand in the absence of the first and so we cannot tell whether it would bind more weakly if the first ligand was absent.

An example of this is calcium binding to calmodulin. Calmodulin is a monomeric two-domain protein, one domain binding two calcium ions at high affinity and the other domain binding two calcium ions at low affinity. The binding of the first two calcium ions to the high-affinity domain shows positive cooperativity, the binding of the second two calcium ions to the low-affinity domain also shows positive cooperativity. However, it is exceedingly difficult to determine whether the binding of the first two calcium ions alters the affinity for the second two calcium ions, i.e. whether there is any cooperativity between the domains as well as within the domains.

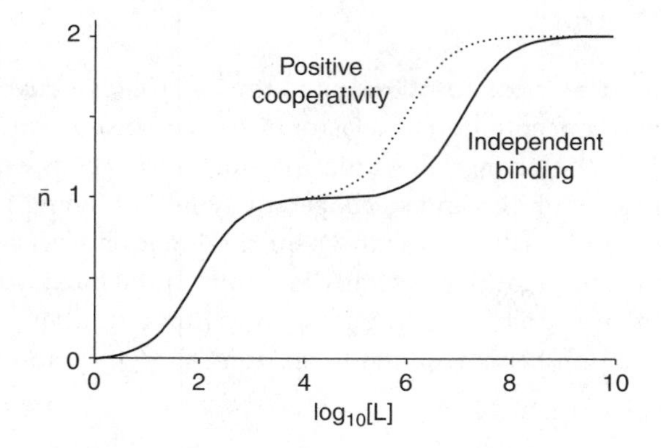

Fig. 4.9 Plot of $\bar{n}$ versus $\log_{10}[L]$ for a protein with two binding sites with different ligand affinities. The solid line corresponds to a case with independent binding, the dotted line to the same case with cooperative binding. The slope of the cooperative curve is the same as that for independent binding because K_d is not altering during the binding titration of either site. In practice, only one curve can be measured and you cannot tell from the shape whether the binding is cooperative or not.

The binding of multivalent ligands to multivalent proteins

So far we have only considered cases where a monovalent ligand binds to one site on a protein. However, many biological binding events involve multivalent ligands binding to multivalent receptors. Multivalent ligands may include ligands with several different functional groups that bind to different sites on a protein, or ligands that are multiply presented on a cell surface and so can bind simultaneously to multidomain proteins.

We shall start with a simplified example of this type of interaction. Consider a protein, P, that has two binding sites for different ligands, A and B. The binding of A to P results in a free energy change ΔG_a° and the binding of B to P results in a free energy change ΔG_b° (figure 4.10a and b). These two monovalent ligands are then joined by a long flexible linker to give the divalent ligand AB (figure 4.10c). Assume that the binding of AB to site A on P is independent of the binding of AB to site B on P, i.e. the binding of the A part of AB is unaffected by whether the B part is bound or not. The free energy change for the binding of AB to P (figure 4.10c) is

$$\Delta G_{ab}^\circ = \Delta G_a^\circ + \Delta G_b^\circ$$

Substituting for ΔG° using the equation $\Delta G^\circ = -RT\log_e(K) = -RT\log_e(K_d^{-1})$ gives

$$RT\log_e K_d^{AB/PAB} = RT\log_e K_d^{A/PA} + RT\log_e K_d^{B/PB}$$

or

$$K_d^{AB/PAB} = K_d^{A/PA} \times K_d^{B/PB}$$

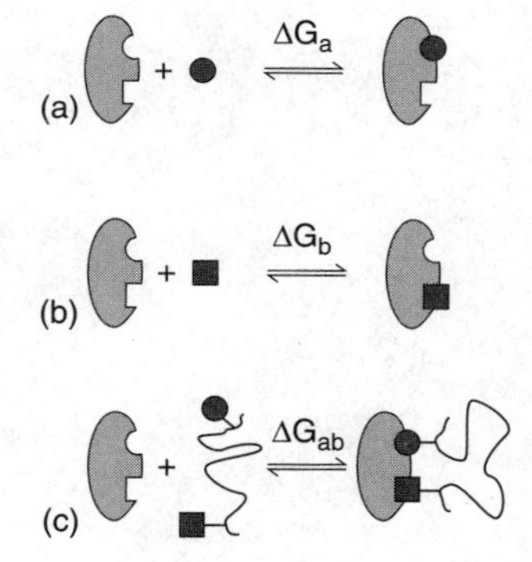

Fig. 4.10 (a) and (b) The binding of two ligands in solution, A (circle) and B (square), to two sites on a protein are characterised by free energy changes ΔG_a and ΔG_b. (c) If these two ligands are joined together by a highly flexible linker, the free energy change on binding A–B to the protein is given by $\Delta G_{ab} = \Delta G_a + \Delta G_b$.

In general for independent binding, the dissociation constant for a multivalent ligand equals the product of the dissociation constants for each individual interaction. Thus, a protein with three binding sites that each bind weakly to a monovalent ligand will bind very strongly to a trivalent ligand. If each binding site has a millimolar affinity for the monovalent ligand, i.e.

$$K_d^{A/PA} = K_d^{B/PB} = K_d^{C/PC} = 10^{-3}\ (M)$$

the protein will have a nanomolar affinity for the trivalent ligand, i.e.

$$K_d^{ABC/PABC} = K_d^{A/PA} \times K_d^{B/PB} \times K_d^{C/PC} = 10^{-9}\ (M)$$

This increase in affinity on binding to a multivalent ligand rather than a monovalent ligand is called avidity.

WORKED EXAMPLE

Immunoglobulin G, IgG, is a soluble protein that binds with high affinity to small regions of molecules, called antigens, on the surface of foreign cells. This initiates an immune system response, called the complement pathway, that results in the foreign cells being lysed. The first step of complement activation is the binding of C1q, a multidomain protein, to IgG in a region of IgG remote from the antigen binding site. Soluble IgG does not bind to C1q and neither can IgG that has bound small, soluble antigenic molecules. Explain these observations and comment on the mechanism of triggering complement.

SOLUTION: C1q is a multidomain protein that binds very weakly to soluble IgG. The binding of antigen to IgG does not affect the binding of C1q to a single IgG molecule. However, if multiple antigens are present on a cell, lots of IgG molecules will bind to the cell surface, i.e. the IgG molecules are effectively cross-linked by the cell surface. C1q can bind with high affinity to the immobilised IgG, several domains binding to several different IgG molecules on the cell surface. Having bound tightly to the cell surface, C1q can then activate the rest of the complement pathway to cause lysis of the cell.

The mechanism of triggering the complement pathway by IgG is not a conformational change in IgG induced by antigen binding that is relayed to the C1q binding site. The mechanism is the aggregation of IgG caused by binding to multiple antigens on a surface, which converts the low-affinity binding of a single C1q domain to a single IgG molecule into high-affinity binding of several C1q domains to the aggregated IgG.

COMMENT: It is important that the immune system is not triggered inadvertently. If a single IgG:antigen complex could bind with high affinity to C1q and activate complement, then small fragments of foreign cells would trigger cell lysis when only host cells are present to be lysed. By ensuring that C1q only binds to aggregated IgG, the complement pathway will only be triggered by multiple presentation of antigens, i.e. by the presence of a foreign cell rather than its components. At present it is thought that at least four domains of C1q need to bind to aggregated IgG to activate the rest of the complement pathway.

The analysis of the binding of multivalent ligands to multivalent proteins is usually more complex than above because the binding to the different sites is not independent. The most important additional factor to consider is the loss of entropy of the ligands on binding to the protein. In the simplified example above, the binding of A to P results in the loss of the entropy associated with the translational and rotational motion of A and the binding of B to P results in a similar loss of entropy for B. These entropy terms are included in the values of ΔG°_a and ΔG°_b and so the value of $\Delta G^\circ_a + \Delta G^\circ_b$ includes the sum of these two entropy terms. The binding of AB to P also results in a loss of entropy associated with translational and rotational motion, but the ligand AB has less translational and rotational entropy than A plus B and so has less to lose. In the limit of AB being completely rigid (figure 4.11a), after AB has bound to site A on the protein there is no remaining translational and rotational entropy to lose on binding to site B and so the binding to site B is much stronger than predicted by ΔG^o_b. This results in the binding of AB being stronger than predicted on the basis of the binding of A and B.

$$\Delta G^o_{ab} < \Delta G^o_a + \Delta G^o_b \quad \text{and} \quad K_d^{AB/PAB} < K_d^{A/PA} \times K_d^{B/PB}$$

This situation is found for the binding of desthiobiotin to avidin where the dissociation constant for the binding of the multivalent ligand is 10^4 smaller, i.e. the binding is 10^4 stronger, than predicted by the simple model (see Jencks, 1981, in further reading below).

An opposite effect is seen where AB is rigid but A and B are held in the wrong orientation to bind to P (figure 4.11b). In this case, the ligand needs to be distorted for both interactions to occur, which results in an additional unfavourable ΔH term and hence weaker binding. This effect can be much larger than the favourable entropy contribution, in which case

$$\Delta G^o_{ab} > \Delta G^o_a + \Delta G^o_b \quad \text{and} \quad K_d^{AB/PAB} > K_d^{A/PA} \times K_d^{B/PB}$$

This situation is found with the binding of meromyosin to actin where the dissociation constant for the binding of the multivalent ligand is 10^4 larger, i.e. the binding is 10^4 weaker, than predicted by the simple model.

Cell recognition and cell–cell interactions are very important in multicellular organisms, as we have seen above in the immune system. If these relied on the high-affinity interaction between a single protein and its ligand, then biological responses could be triggered by soluble ligand as well as by the presence of a cell. Low-affinity interactions provide a method for distinguishing between isolated recognition events involving single molecules and the multiple recognition events that can occur at cell surfaces.

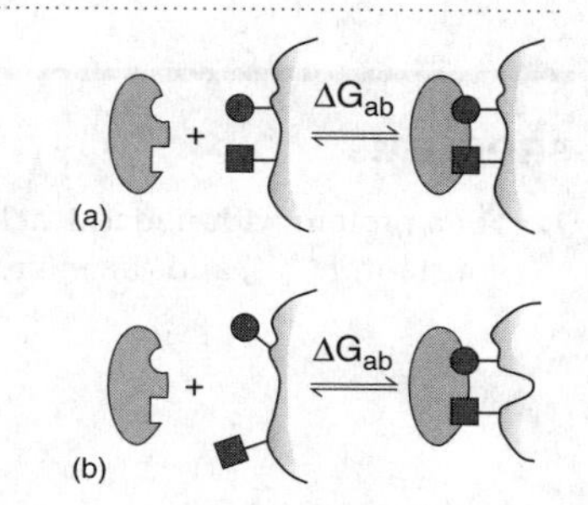

Fig. 4.11 (a) For two ligands, A (circle) and B (square), joined by a rigid linker in the correct orientation to bind to the protein, the loss in entropy of binding A–B to the protein is less than the loss in entropy of binding A and B to the protein and so $\Delta G_{ab} < \Delta G_a + \Delta G_b$. (b) For two ligands, A and B, joined by a rigid linker in an incorrect orientation to bind to the protein, binding of A–B to the protein requires distortion of the linker and so $\Delta G_{ab} > \Delta G_a + \Delta G_b$.

FURTHER READING

T.E. Creighton, 1993, 'Proteins: structure and molecular properties', Freeman—*includes an informative chapter on the binding properties of proteins.*

W.P. Jencks, 1981, 'On the attribution and additivity of binding energies', in Proceedings of the National Academy of Sciences USA, vol. 78, pp. 4046–50—*a useful discussion of the concept of avidity with biological examples.*

M.F. Perutz, 1990, 'Mechanisms of cooperativity and allosteric regulation in proteins', Cambridge University Press—*detailed accounts of the structural basis of numerous allosteric effects.*

PROBLEMS

1. For a protein with n identical binding sites, show that the general relationship between the macroscopic dissociation constant, K_d^{L/PL_i}, and the microscopic dissociation constant, κ_d, for the equilibrium

$$PL_{i-1} + L \rightleftharpoons PL_i$$

is

$$K_d^{L/PL_i} = \frac{i}{(n-i)+1}\kappa_d$$

and hence show that κ_d is the geometric mean of all the K_d values, i.e $\kappa_d = (K_d^{L/PL} \times K_d^{L/PL_2} \times \cdots \times K_d^{L/PL_n})^{1/n}$.

2. Consider the binding of substrate (S) to enzyme (E). If an inhibitor, I, that can bind to the enzyme (but not to the ES complex) is added to the system we have two simultaneous equilibria to be considered

$$E + S \rightleftharpoons ES \qquad K_d^{S/ES} = \frac{[E][S]}{[ES]}$$

and

$$E + I \rightleftharpoons EI \qquad K_d^{I/EI} = \frac{[E][I]}{[EI]}$$

Discuss (i.e. examine the algebra for) the evaluation of [E], [EI], and [ES] as fractions of $[E]_{total}$ for
(a) the case where

$[E]_{total} = 1\ \mu M$ $\quad$ $[S]_{total} = 1\ \mu M$ $\quad$ $[I]_{total} = 1.5\ \mu M$

$K_d^{S/ES} = 50 \times 10^{-6}(M)$ $\quad$ $K_d^{I/EI} = 75 \times 10^{-6}(M)$

(b) as above except that $[S]_{total}$ and $[I]_{total}$ are 150 μM and 150 μM respectively.

3. The number of ligand binding sites on a macromolecule can be determined by direct titration if the binding of the ligand to the macromolecule is very tight. An example is the binding of NADPH to isocitrate dehydrogenase, which can be monitored by the increase in fluorescence of NADPH when it is bound to the enzyme. The following data were obtained when NADPH was added to a 0·0135 g dm^{-3} solution of the enzyme (a small correction has been made for the fluorescence of free NADPH). The molecular weight of the enzyme is 90 000 Da.

NADPH added/μM	0·07	0·14	0·21	0·28	0·35	0·50	0·75
Fluorescence increase	2·3	4·7	7·0	9·3	9·9	10	10

Determine the number of sites for NADPH on the enzyme.

4. The following data were obtained for the binding of Mn^{2+} to the enzyme phosphorylase a

$[Mn^{2+}]_{total}$/μM	110	190	290	500	750
$[Mn^{2+}]_{bound}$/μM	75	125	175	250	300

The enzyme concentration is 100 μM. By means of a suitable plot calculate n and K.

5. The binding of a substrate (ADP) to the enzyme pyruvate kinase was studied by measuring the quenching of the protein fluorescence. The following results were obtained.

Total concentration of ADP/mM	0·29	0·36	0·42	0·53	0·84	1·18	1·89
mol ADP bound/mol protein	1·25	1·35	1·62	1·82	2·26	2·62	3·03

The enzyme concentration was 4 μM throughout the titration. Calculate the number of substrate binding sites on the enzyme and the dissociation constant, from these data.

6. The binding of NAD^+ to yeast glyceraldehyde 3-phosphate dehydrogenase was studied by equilibrium dialysis with the following results

$[NAD^+]_{total}$/μM	41	78	132	187	230	285	374	474
$[NAD^+]_{free}$/μM	13	21	30	39	48	68	125	211

The enzyme concentration is 71 μM. What type of binding is being observed in this case? Determine the Hill coefficients at 50% and 100% saturation.

7. The enzyme aspartate transcarbamoylase is inhibited by CTP. Measurements of the binding of CTP to the enzyme were made by equilibrium dialysis.

Experiment A					
$[CTP]_{total}$/μM	0·78	1·29	2·35	4·63	11·65
$[CTP]_{free}$/μM	0·24	0·43	1·00	2·56	8·86
Experiment B					
$[CTP]_{total}$/μM	144·9	172·0	203·7	269·5	416·0
$[CTP]_{free}$/μM	27·9	43·0	65·7	122·5	260·0

The concentrations of enzyme in experiments A and B were 0·27 and 9 g dm^{-3} respectively. Aspartate transcarbamoylase has been shown to consist of six catalytic and six regulatory subunits with a total molecular weight of 300 000 Da. What can you determine from these data?

8. A binding assay involves linking either a protein or its ligand to a solid support and monitoring the binding of the other component that is in solution in contact with the support. When this assay is used to measure the binding constant for an antibody to an antigen, different results are obtained when the antibody is linked to the support and the antigen is in solution compared to when the antigen is linked to the support and the antibody is in solution. Explain this observation and suggest which of the two assays would give the smaller measured dissociation constant.

9. For a protein that has two ligand binding sites with similar dissociation constants showing cooperativity, estimate the maximum useful value of $\Delta\Delta G°$. [Hint: The maximum useful value of $\Delta\Delta G°$ is that which results in binding at the second site going from $< 10\%$ to $> 90\%$ saturation on addition of the first ligand, i.e. increases the binding of the second ligand by about two orders of magnitude. Any larger value than this will result in free energy being wasted.]

5 Acids, bases and pH regulation

KEY POINTS:

- pH, defined as $-\log_{10}[H^+]$, is a convenient scale for measuring H^+ concentrations. The pH of pure water is approximately 7.
- Compounds can be classified according to their ability to reduce (acidic compound) or increase (basic compound) the pH of a solution.
- The strength of an acid or base can be quantified by comparing its pK_a value to the pH of the solution.
- The pK_a value of a group varies with environment. This is particularly important for amino acid side chains in proteins.
- The pH of a solution can be controlled in two ways: (i) by the presence of buffers in the solution; and (ii) by changing the concentrations of ions present. Both of these methods are found in biological systems.

Water is one of the most important molecules in biology. It acts as a solvent for many molecules of biological interest and thus provides a medium in which reactions can occur. Moreover, most biological molecules are solvated to some degree and their interactions with water modify their physical and chemical properties. Thus the balance between the hydrophilic interactions of biological macromolecules and the hydrophobic effect is important in determining macromolecular structures (see Chapter 18), and similar considerations underpin the assembly of membranes. The ionisation state of molecules that dissociate is strongly dependent on the presence of water, and this can have a major influence on chemical reactivity, with some reactions being favoured in aqueous environments and others in hydrophobic, i.e. non-aqueous, regions of the cell. For example, one of the major roles of enzymes in phosphate transfer reactions is to exclude water from the reactive site in order to prevent the hydrolysis of the phosphate group. In fact water is frequently involved directly in the reactions that occur in cells, serving as reactant in hydrolytic processes and as a product in condensation reactions. Water is also produced when oxygen is reduced, and the photolysis of water that occurs in photosynthetic organisms was simultaneously responsible for the appearance of oxygen in the earth's atmosphere, the evolution of aerobic organisms and the subsequent development of the eukaryotes. Thus the chemical properties of water are of great biological importance.

The ionic dissociation of water

Water dissociates according to the following equation

$$2H_2O \rightleftharpoons H_3O^+ + OH^-$$

where H_3O^+ represents a solvated H^+ ion. The equilibrium constant for this process is given by

$$K = \frac{[H_3O^+][OH^-]}{[H_2O]^2}$$

As this equilibrium lies mostly to the left, H_2O is approximately in its standard state and so $[H_2O]/[H_2O]^\circ = 1{\cdot}0$. Thus

$$K_w = [H_3O^+][OH^-]$$

K_w is the ionic dissociation constant for water, called the *ionic product*. It has a value of 10^{-14} at 298 K and increases with temperature. For pure water, the concentration of H_3O^+ must equal the concentration of OH^-, and so

$$[H_3O^+] = [OH^-] = 10^{-7}\,M$$

The Arrhenius definition of acids and bases

We can classify other compounds in terms of their effects on the concentrations of H_3O^+ and OH^-. This is the *Arrhenius* definition of acids and bases.

- Acid—increases the concentration of H_3O^+ ions in solution
- Base—increases the concentration of OH^- ions in solution

If we increase the concentration of H_3O^+ ions in solution, we also decrease the concentration of OH^- ions because $[H_3O^+] \times [OH^-]$ must remain constant. We can then define acidic and basic solutions as

- Acidic solution
  $[H_3O^+] > [OH^-]$ (i.e. $[H_3O^+] > 10^{-7}\,M$ and $[OH^-] < 10^{-7}\,M$)
- Basic solution
  $[H_3O^+] < [OH^-]$ (i.e. $[H_3O^+] < 10^{-7}\,M$ and $[OH^-] > 10^{-7}\,M$)

WORKED EXAMPLE

1 ml of 0·1 M HCl is added to a litre of water. What is the ratio of H_3O^+ (aq) ions to OH^-(aq) ions in the final solution?

SOLUTION: HCl is an acid that will dissociate fully in water to give 1 mole of H_3O^+ (aq) ions per mole of acid. The initial HCl solution is diluted about a thousand times (1/1001 to be accurate) so the final concentration of H_3O^+(aq) ions in the solution will be

$$0{\cdot}1/1001 \approx 10^{-4}$$

As

$$[H_3O^+] \times [OH^-] = 10^{-14}$$

$$[OH^-] = 10^{-10}\,M$$

and

$$[H_3O^+]/[OH^-] = 10^{-4}/10^{-10} = 10^6$$

COMMENT: Even at relatively low acid concentrations, the amount of OH^- is insignificant. Similarly, at relatively low base concentrations, the amount of H_3O^+ is insignificant.

An acid can work either by donating a proton, H^+, to the solution or by removing OH^- from a water molecule. Similarly, a base can work either by donating OH^- to the solution, for instance

$$NaOH(aq) \rightleftharpoons Na^+(aq) + OH^-(aq)$$

or by removing a proton from a water molecule, for instance

$$NH_3(aq) + H_2O \rightleftharpoons NH_4^+(aq) + OH^-(aq)$$

This definition of acids and bases is based on the equilibrium for ionic dissociation of water. It is thus a thermodynamic definition rather than a molecular definition.

The concept of pH

pH is a convenient way of expressing the concentration of H_3O^+ ions in solution, avoiding the use of large negative powers of 10. The pH of a solution is defined as

$$pH = -\log_{10}[H_3O^+]$$

From the equation for the ionic product of water, this can also be written as

$$pH = 14 + \log_{10}[OH^-]$$

For pure water at 298 K, $[H_3O^+] = 10^{-7}$ and so $pH = 7{\cdot}0$. We can then define acidic and basic solutions by their pH.

- Acidic solution $[H_3O^+] > 10^{-7}$ M $pH < 7$
- Basic solution $[H_3O^+] < 10^{-7}$ M $pH > 7$

The pH range for most experiments is from 0, corresponding to a 1 M solution of H_3O^+ ions, to 14, corresponding to a 1 M solution of OH^- ions. In contrast, the biologically relevant pH range is smaller than this with the majority of biological systems having pH values between 6 and 8 (see below).

Conjugate acids and bases

Consider an acid, HA, that can donate a proton to water.

$$HA(aq) + H_2O \rightleftharpoons A^-(aq) + H_3O^+(aq)$$

The A^- species formed in this reaction is also capable of removing a proton from water and thus acting as a base.

$$A^-(aq) + H_2O \rightleftharpoons HA(aq) + OH^-(aq)$$

A^- is called the *conjugate base* of the acid HA. Any acid will have a conjugate base and similarly any base will have a conjugate acid.

WORKED EXAMPLE

Identify which of the following species are capable of acting as acids or bases and give their conjugate acids or bases: (a) CH_3COOH; (b) NH_3; (c) HSO_4^-.

SOLUTION: For the first two species there is a single equilibrium with water.

(a) $\underset{\text{Acid}}{CH_3COOH} + H_2O \rightleftharpoons \underset{\text{Conjugate base}}{CH_3COO^-} + H_3O^+$

(b) $\underset{\text{Base}}{NH_3} + H_2O \rightleftharpoons \underset{\text{Conjugate acid}}{NH_4^+} + OH^-$

HSO_4^- is capable of acting as either an acid or a base, i.e. by donating or receiving a proton.

(c) $\underset{\text{Acid}}{HSO_4^-} + H_2O \rightleftharpoons \underset{\text{Conjugate base}}{SO_4^{2-}} + H_3O^+$

$\underset{\text{Base}}{HSO_4^-} + H_2O \rightleftharpoons \underset{\text{Conjugate acid}}{H_2SO_4} + OH^-$

COMMENT: Although CH_3COOH, NH_3 and HSO_4^- are potential acids and bases, whether they actually behave as acids or bases depends on the concentrations of OH^- and H_3O^+ in solution. At very high H_3O^+ concentrations, the equilibria

$$CH_3COOH + H_2O \rightleftharpoons CH_3COO^- + H_3O^+$$

and

$$HSO_4^- + H_2O \rightleftharpoons SO_4^{2-} + H_3O^+$$

lie a long way to the left and so addition of CH_3COOH or HSO_4^- will not increase the H_3O^+ concentration significantly. Thus, CH_3COOH and HSO_4^- will not act as acids at very low pH. Similarly, NH_3 and HSO_4^- will not act as bases if the concentration of OH^- is too high, i.e. at very high pH.

Quantifying acid and base strengths

The Arrhenius acid definition is based on a chemical equilibrium, such as

$$HA(aq) + H_2O \rightleftharpoons A^-(aq) + H_3O^+(aq)$$

We can use the position of this equilibrium as a quantitative measure of the strength of an acid. The equilibrium constant for this reaction, called the acid dissociation constant K_a, is

$$K_a = \frac{[H_3O^+][A^-]}{[HA][H_2O]}$$

$[H_2O]$ is usually omitted because $[H_2O]/[H_2O]^\circ \approx 1$. In the same way that we defined pH, we can define a pK_a value

$$pK_a = -\log_{10} K_a$$

If the equilibrium lies to the right, addition of HA leads to an increase in the concentration of H_3O^+ and HA is called a strong acid. If the equilibrium lies to the left, addition of HA is ineffective at increasing the concentration of H_3O^+ and HA is called a weak acid. A low value of the pK_a indicates that the acid is largely dissociated in water and is thus a strong acid.

Similarly, we can consider the ionic dissociation of a base using the equilibrium,

$$BOH(aq) \rightleftharpoons B^+(aq) + OH^-(aq)$$

The base dissociation constant, K_b, is defined as

$$K_b = \frac{[OH^-][B^+]}{[BOH]}$$

and pK_b is given by

$$pK_b = -\log_{10} K_b$$

This gives us separate scales for the strengths of acids and bases, pK_a and pK_b. We can put both acids and bases on the same scale by using the strength of the conjugate acid B^+ as a measure of the strength of the base BOH.

$$B^+(aq) + 2H_2O \rightleftharpoons BOH(aq) + H_3O^+(aq)$$

For this equilibrium, K_a is given by

$$K_a = \frac{[BOH][H_3O^+]}{[B^+][H_2O]^2}$$

Multiplying K_a and K_b gives, assuming that $[H_2O]/[H_2O]^\circ = 1$

$$K_a \times K_b = \frac{[BOH][H_3O^+]}{[B^+]} \times \frac{[B^+][OH^-]}{[BOH]} = [H_3O^+][OH^-]$$

which is simply the ionic product of water. Thus

$$K_a \times K_b = 10^{-14} \qquad \text{or} \qquad pK_a + pK_b = 14$$

Therefore, a strong acid, $pK_a < 7$, will have a weak conjugate base, $pK_b > 7$, and similarly a strong base will have a weak conjugate acid. The strength of a base is usually given as a pK_a value, i.e. as the strength of its conjugate acid.

WORKED EXAMPLE

The pK_a of formic (methanoic) acid is 3·77 at 298 K. What is the pH of a 0·01 M solution of formic acid in water?

SOLUTION: Formic acid dissociates in water to give the following equilibrium.

$$HCOOH + H_2O \rightleftharpoons HCOO^- + H_3O^+$$

Let the concentration of H_3O^+ at equilibrium be x, then the concentration of $HCOO^-$ is also x and the concentration of HCOOH is $0{\cdot}01 - x$. As

$$pK_a = -\log_{10} K_a$$

$$K_a = \frac{[HCOO^-][H_3O^+]}{[HCOOH]} = 1{\cdot}695 \times 10^{-4}\ (M)$$

so

$$\frac{x^2}{(0{\cdot}01 - x)} = 1{\cdot}695 \times 10^{-4}$$

This rearranges to

$$x^2 + (1{\cdot}695 \times 10^{-4})x - (1{\cdot}695 \times 10^{-6}) = 0$$

Taking the positive root gives $x = 1{\cdot}22 \times 10^{-3}$ M. This is the concentration of H_3O^+ and so from the formula $pH = -\log_{10}[H_3O^+]$, the solution pH is 2·91.

COMMENT: If formic acid were fully dissociated in water, the concentration of H_3O^+ would be 0·01 M, corresponding to a pH of 2·0. Formic acid is a weak acid in water.

WORKED EXAMPLE

The pK_a of methylamine is 10·6 at 298 K. What is the pH of a 0·01 M solution in water?

SOLUTION: Methylamine is a base, giving the following equilibrium in water,

$$CH_3NH_2(aq) \rightleftharpoons CH_3NH_3^+(aq) + OH^-(aq)$$

where

$$K_b = \frac{[CH_3NH_3^+][OH^-]}{[CH_3NH_2]}$$

The pK_a value given is for the dissociation of the conjugate acid, $CH_3NH_3^+$.

$$CH_3NH_3^+(aq) + H_2O \rightleftharpoons CH_3NH_2(aq) + H_3O^+(aq)$$

This can be converted into the pK_b for the base, CH_3NH_2, by using the equation $pK_a + pK_b = 14$. Thus, pK_b for methylamine is 3·4.

The problem can then be solved in an identical fashion to the previous worked example to give the concentration of OH^- as $1{\cdot}8 \times 10^{-3}$ M at equilibrium. This can be converted into pH by using the equation $pH = 14 + \log_{10}[OH^-]$ to give a final pH of 11·26.

Relative and absolute acid and base strengths

The relative strengths of two acids or bases can be determined by comparing their pK_a values. The acid with the lower pK_a will be more dissociated and so will be the stronger. Similarly, the base with the higher pK_a will be the stronger.

The absolute strength of an acid depends on the position of an equilibrium of the form

$$HA(aq) + H_2O \rightleftharpoons A^-(aq) + H_3O^+(aq)$$

The position of this equilibrium also depends on the concentration of H_3O^+, i.e. the pH of the solution. At lower pH the equilibrium will lie further to the left and so the acid will behave as a weaker acid, at higher pH the equilibrium will lie further to the right and so the acid will behave as a stronger acid. We can determine the absolute strength of an acid or base in any solution by comparing the pK_a value with the pH of the solution. The greater the difference between the pK_a and the pH, the stronger is the acid or base.

WORKED EXAMPLE

Calculate the degree of dissociation of a 0·01 M solution of formic acid, pK_a of 3·77, at pH 4 and pH 10.

SOLUTION: From the previous worked example, K_a for formic acid is given by

$$K_a = \frac{[HCOO^-][H_3O^+]}{[HCOOH]} = 1{\cdot}695 \times 10^{-4}(M)$$

At pH 4, $[H_3O^+] = 10^{-4}$ M. Thus

$$\frac{[HCOO^-]}{[HCOOH]} = 1{\cdot}695$$

As $[HCOOH] + [HCOO^-] = 0{\cdot}01$ M

$$[HCOO^-] = 0{\cdot}0063\,M \quad \text{and} \quad [HCOOH] = 0{\cdot}0037\,M$$

or 63% of the acid is dissociated.

At pH 10, $[H_3O^+] = 10^{-10}$ M. A similar calculation gives $[HCOO^-] = 0{\cdot}01$ M and $[HCOOH] = 5{\cdot}8 \times 10^{-9}$ M, or ~ 100% of the acid is dissociated.

COMMENT: At pH 4 formic acid is a very weak acid, with only slightly more than 50% dissociation. At pH 10 formic acid is a very strong acid, effectively completely dissociated.

Variation of pK_a with environment

The dissociation constants of acidic and basic groups are not constant, but vary with temperature and chemical environment. Consider the dissociation of a carboxylic acid

$$RCOOH + H_2O \rightleftharpoons RCOO^- + H_3O^+$$

The position of the equilibrium depends on several factors, including the pK_a of the carboxylic acid, the temperature, since pK_a values are related to ΔG° values and these are usually temperature dependent, and the ionic strength of the solution, since pK_a values are ultimately defined in terms of activities rather than concentrations (see Chapter 8). Ignoring the effects of temperature and ionic strength, the most important factor is the nature and identity of the group R. For example, formic, acetic and propanoic acids, corresponding to $R = H-$, CH_3- and CH_3CH_2- respectively, have different pK_a values because the nature of the R group influences the stability of the dissociated acid. Stabilisation of $RCOO^-$ results in the equilibrium lying further to the right and hence a lower value for the pK_a. Similarly, the pK_a of a carboxyl group in the side chain of an amino acid will depend on the chemical environment in its vicinity when the amino acid is incorporated into a polypeptide.

WORKED EXAMPLE

How will the pK_a of a glutamic acid side chain vary if it is: (a) on the protein surface; (b) in a positively charged region; (c) in a negatively charged region; (d) in a non-polar region; or (e) buried within the centre of the protein structure?

SOLUTION: The glutamic acid side chain contains a carboxyl group with a pK_a of 4·4.

$$-CH_2CH_2COOH + H_2O \rightleftharpoons -CH_2CH_2COO^- + H_3O^+$$

Consider the effect of the environment on the stability of the $-COOH$ and $-COO^-$ groups. If the latter is stabilised relative to the former, the equilibrium will shift to the right and the pK_a will go down and vice versa.

	Environment	Effect on side chain	Effect on pK_a
(a)	Protein surface	Side chain is solvated by water, similar to the free amino acid in solution	No change
(b)	Positively charged region	$-COO^-$ will be stabilised by favourable electrostatic interactions	Lowered
(c)	Negatively charged region	$-COO^-$ will be destabilised by unfavourable electrostatic interactions	Raised
(d)	Non-polar region	$-COO^-$ will be destabilised because it will be less hydrated in a non-polar region than it is in water	Raised
(e)	Buried in the protein	Water cannot get access to the side chain, it is in a non-aqueous environment	pK_a is undefined

COMMENT: The pK_a value of an ionisable amino acid side chain in a protein is influenced by its immediate chemical environment and hence by the three-dimensional structure of the protein. This is particularly important for catalytic residues in the active sites of enzymes, where a specific acid or base strength may be required for optimum activity. For example, the side chain of histidine has a pK_a of 6·9 in free solution, whereas a histidine residue in an enzyme can have a pK_a value ranging from below 6 to above 8, and is thus capable of acting as an acid or a base depending on its microenvironment. For example, ribonuclease has two histidines in its active site: (i) His 12, which acts as a base early in the reaction mechanism and as an acid in a later step; and (ii) His 119, which acts as an acid early in the reaction mechanism and as a base in a later step.

The neutralisation of acids and bases

The change in pH on adding a strong base, NaOH, to a solution of a strong acid, HCl, or a weak acid, CH_3COOH, is shown in figure 5.1. These curves are called titration curves and to explain their shape, we need to consider the equilibria that have been established and which of these is significant in determining the pH at each stage of the titration. Before the addition of any base, the only equilibrium that has been established is

$$HA(aq) + H_2O \overset{K_a}{\rightleftharpoons} A^-(aq) + H_3O^+(aq)$$

The pH depends on the value of the pK_a of the acid. HCl is a strong acid and so this equilibrium lies almost entirely to the right. CH_3COOH is a weak acid and so less dissociated, leading to a higher pH.

After NaOH has been added, a second equilibrium is established in the solution

$$NaOH \overset{K_b}{\rightleftharpoons} Na^+(aq) + OH^-(aq)$$

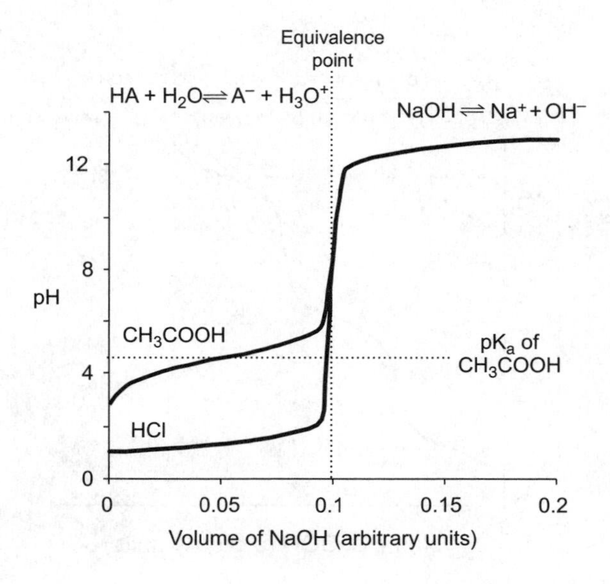

Fig. 5.1 Titration curves for 0·1 M solutions of HCl, pKa ≈ − 1, and CH_3COOH, pKa = 4·8, on addition of NaOH. Before the equivalence point, the pH of the solution is calculated using the equilibrium for dissociation of the acid. After the equivalence point, when all the acid has been neutralised, the pH of the solution is calculated using the equilibrium for dissociation of the base. Note that both these equilibria are present at all pH values.

This equilibrium lies almost entirely to the right and the production of OH^- results in neutralisation of some H_3O^+.

$$OH^- + H_3O^+ \rightleftharpoons 2H_2O$$

For HCl, which is fully dissociated already, this has no effect on the position of the equilibrium for the acid dissociation and so there is a steady rise in pH as more base is added. For CH_3COOH, the reduction in the concentration of H_3O^+ causes the equilibrium between HA and A^- to shift further to the right.

As long as the acid is in excess compared to the NaOH, the concentration of OH^- remains insignificant and so the equilibrium for the dissociation of the base can be ignored when calculating the pH (approximately [NaOH] $< 0{\cdot}09$ M in figure 5.1). The shape of the titration curve can be calculated by using the equation for the dissociation of the acid. Once all the acid present has been neutralised and the base is in excess, the concentration of H_3O^+ becomes insignificant compared to OH^- and the equilibrium between HA and A^- can be ignored when calculating the pH (approximately [NaOH] $> 0{\cdot}11$ M in figure 5.1). The pH is calculated using the equation for the dissociation of NaOH and thus both the HCl and CH_3COOH solutions give the same values. When similar amounts of the acid and base are present, the equilibria for both the acid and base dissociations need to be considered when calculating the pH (corresponding to the steeply sloping sections of the titration curves in figure 5.1).

For polyprotic acids, i.e. acids that can lose more than one proton, a series of proton dissociations occurs as we add a strong base. Between each equivalence point a different equilibrium is significant when calculating the pH. The equivalence points are where the solution pH swaps from being determined by one equilibrium to being determined by the next. The titration curve for H_3PO_4 is shown in figure 5.2. In this case, the pK_a values for the three

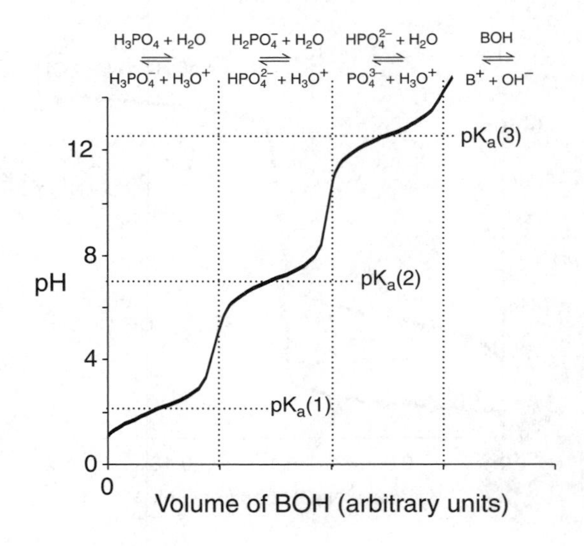

Fig. 5.2 H_3PO_4 has three acidic protons with pK_a values of 2·1, 7·2 and 12·4. A well-defined titration curve can be observed for each proton. At each stage during the titration, the equilibrium that is important for determining the pH is given. The equivalence points correspond to the changeovers between the different equilibria.

acid dissociations are sufficiently distinct to allow each separate step in the titration curve to be observed. However, if the pK_a values were similar, then the separate curves would lie on top of one another and a complex curve would be observed. This is the situation for most proteins, which contain a large number of titratable groups from the amino acid side chains with similar pK_a values.

For a molecule with two or more separate titratable groups, each group will have a pK_a value and each ionisation can be considered independently.

WORKED EXAMPLE

Describe how the charge on a free amino acid changes with pH.

SOLUTION: Free amino acids, of general formula $H_2N-CHR-COOH$ where R is the variable side chain, have an amino group ($-NH_2$) with a pK_a of $\approx 9{\cdot}5$ and a carboxyl group ($-COOH$) with a pK_a of $\approx 2{\cdot}5$. In addition, certain amino acids have acidic or basic groups in their side chains.

Amino acid	Side-chain equilibrium	pK_a
Asp and Glu	$-COOH + H_2O \rightleftharpoons -COO^- + H_3O^+$	4·4
Histidine	$=NH^+- + H_2O \rightleftharpoons =N- + H_3O^+$	6·9
Cysteine	$-SH + H_2O \rightleftharpoons -S^- + H_3O^+$	8·5
Tyrosine	$-OH + H_2O \rightleftharpoons -O^- + H_3O^+$	10·0
Lysine	$-NH_3^+ + H_2O \rightleftharpoons -NH_2 + H_3O^+$	10·0
Arginine	$= NH_2^+ + H_2O \rightleftharpoons = NH + H_3O^+$	12·0

First consider an amino acid without a titratable group in the side chain. Below pH 2·5, both the carboxyl group and the amino group will be protonated, giving a net charge of +1. Above pH 2·5, the carboxyl group will lose its proton and above pH 9·5 the amino group will lose its proton.

$$NH_3^+-CHR-COOH \rightleftharpoons NH_3^+ - CHR-COO^- \rightleftharpoons NH_2-CHR-COO^-$$

pH	0	7	14
Net charge (approx.)	+1	0	−1

Near neutral pH, although there are charged groups present, the overall charge on the molecule will be close to zero. This type of species is called a *zwitterion*.

The presence of ionisable side chains alters this analysis.

Glutamic and aspartic acid have an acidic group on the side chain. At pH 0 this will be protonated and so the net charge will still be 1. At pH 7 the second carboxyl group will be deprotonated and so the net charge will be −1. At pH 14 the net charge will be −2.

Lysine and arginine have a basic group on the side chain. At pH 0 and 7 this will be protonated giving net charges of +2 and +1 respectively. At pH 14 this will be deprotonated and so the net charge will be −1.

COMMENT: The concentrations of positive and negative charges will exactly balance, and the overall charge on the molecule will be zero, at a specific pH value. This pH value is known as the *isoelectric point*, or pI. The pI is equal to the average of the pK_a values for the groups that contribute to the zwitterion. Thus, for an amino acid without a titratable group in the side chain the pI is $(9{\cdot}5 + 2{\cdot}5)/2 = 6{\cdot}0$.

The amino and carboxyl groups of all amino acids in proteins, except the first and last of the chain, are used to form the peptide backbone. Thus, the pH behaviour of proteins depends almost entirely on the amino acid side chains. The pI value for the protein will then be the average of the pK_a values for all the ionisable side chains. However, as we have seen, the pK_a value for a side chain depends on its environment. Thus, two proteins with similar compositions of titratable amino acids can have very different pI values. The pH dependence of the total charge of a protein can be exploited as a means of protein separation using a technique called isoelectric focusing (figure 5.3).

pH in biological systems

The properties of many biological molecules are pH dependent, and as a result many cellular processes are also sensitive to pH. It follows that uncontrolled variations in pH are potentially disruptive and so pH is tightly controlled in

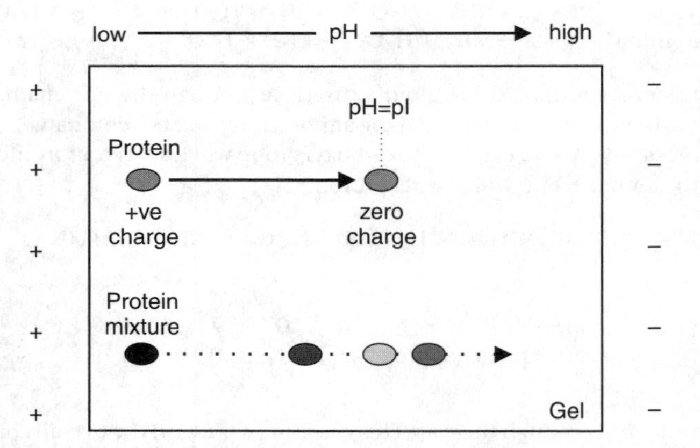

Fig. 5.3 Isoelectric focusing involves loading a protein solution on a gel with a pH gradient and then generating an electric field across the gel. At very low pH, all the acidic and basic side chains of the amino acids will be protonated giving the protein a positive charge. In the presence of the electric field the protein will move towards the negative pole. As the pH increases, the acidic and basic side chains will start to become deprotonated and so the charge on the protein will decrease. When the pH equals the pI, the protein will have no net charge, will no longer be affected by the electric field and so will stop moving. If a mixture of proteins is loaded onto the gel each protein will continue migrating until it reaches the point where the pH equals its pI, enabling proteins with different pI values to be separated.

TABLE 5.1

System		Typical pH	Comment
Extra-cellular	Blood	7·3	
	Stomach	1·5	Low pH favours hydrolysis.
Intra-cellular	Lungs	7·6	Decreased pH reduces the affinity of haemoglobin for O_2 and so promotes its release
	Muscle (resting)	7·2	
	Muscle (active)	6·0	
	Intermembrane space	7·5	pH difference important in ATP synthesis
	Mitochondrial matrix	8·0	
	Endoplasmic reticulum	7·2	pH gradient is thought to be involved in protein trafficking and localisation
	Golgi	6·5	
	Trans Golgi network	5·9	
	Lysosomes	5·0	Optimum pH for hydrolytic enzymes
	Plant cell vacuoles	2–6·5	Typically 5–5·5, but falling to less than 2 overnight in succulent plants using Crassulacean acid metabolism
	Plant cell cytosol	7·5	
	Chloroplast stroma (dark)	7·5	
	Chloroplast stroma (light)	8·0	pH difference important in ATP synthesis
	Intrathylakoid space (light)	5·0	

most biological systems. However, there is no requirement for the pH in different organisms or in different tissues and compartments of a particular organism to be the same and indeed the pH differences between different regions are often important for biological function (Table 5.1).

Thus pH regulation in biological systems is very important. This regulation is achieved by a combination of two methods

- Buffering. Buffers are substances that, when added to a solution, enable the solution to resist changes in pH.
- Active pumping of ions across membranes. The relative pH values of two different compartments can by altered by pumping ions, such as H^+ and Cl^-, between two compartments.

The pH of an intracellular compartment depends on its buffering capacity (see below) and its ionic composition. The buffering capacity is determined by the concentration of weak acids and bases in the compartment and depends on the metabolism of the cell, while the ionic composition depends on the ion transport processes that occur across the membranes that define the compartment. Changes in pH can be caused by changes in metabolism, which can alter the buffering capacity, and by changes in ion transport. These changes can be resisted by compensating metabolic and transport processes.

The buffering capacity is an important feature of pH homeostasis in large extracellular compartments, for example in the stomach and the blood, whereas intracellular pH regulation is often strongly influenced by ion transport. ATP-driven cation pumps, such as the plasma membrane H^+-ATPases found in plants and bacteria, make an important contribution to ion transport but the contribution of these electrogenic pumps to pH regulation is constrained by the need to maintain electroneutrality. Thus pH regulation cannot be explained in terms of H^+-pumping alone and it is necessary to consider the potential contribution of all the ion transport processes that alter the composition of the compartment of interest.

Buffer solutions

Buffers are substances that increase the amount of acid or alkali that must be added to a solution to cause a unit change in pH. Thus the presence of a buffer enables the solution to resist changes in pH. Buffering is an essentially passive mechanism of pH regulation in cells, since it is always possible to overwhelm the buffering capacity by increasing the pH stress.

A buffer consists simply of an equilibrium mixture of an acid and its conjugate base, or a base and its conjugate acid. As can be seen from the titration curves in figure 5.1, until the equivalence point is reached the pH of such a solution remains relatively constant as base is added. Buffering can be understood in terms of the following equilibrium.

$$\mathrm{HA(aq) + H_2O \rightleftharpoons A^-(aq) + H_3O^+(aq)}$$

Addition of H^+ will result in the equilibrium shifting to the left to produce more HA, thus eliminating most of the additional H^+. Addition of OH^- will result in some of the H_3O^+ reacting with the OH^- to form H_2O. The equilibrium will then shift to the right, partly restoring the H_3O^+ concentration.

To calculate the pH of a buffer mixture of an acid and its conjugate base, we can rearrange the equation

$$K_a = \frac{[H_3O^+][A^-]}{[HA]}$$

Taking negative logarithms on both sides gives

$$-\log_{10}(K_a) = pK_a = -\log_{10}\left(\frac{[H_3O^+][A^-]}{[HA]}\right)$$

which can be rearranged to give

$$pK_a = -\log_{10}[H_3O^+] - \log_{10}\left(\frac{[A^-]}{[HA]}\right)$$

As $-\log_{10}[H_3O^+]$ is just the pH of the solution, this equation becomes

$$pH = pK_a + \log_{10}\left(\frac{[A^-]}{[HA]}\right)$$

This equation, which relates the composition of a solution to its pH, is called the *Henderson–Hasselbalch* equation. When the concentration of the acid and that of its conjugate base are equal, the pH equals the pK_a value. Thus an acid or a base will only act as a buffer at pH values close to its pK_a. HCl and CH_3COOH will buffer at different pH values because they have different pK_a values. Phosphoric acid will buffer at around three different pH values, each corresponding to one of its pK_a values (figure 5.2).

WORKED EXAMPLE

Calculate the change in pH of water, initially at pH 8·3, after the addition of 0·005 M H^+. How does this value change if Tris, $pK_a = 8{\cdot}3$, is present at 0·1 M?

SOLUTION: In water at pH 8·3, the concentrations of H_3O^+ and OH^- are

$$[H_3O^+] = 10^{-8{\cdot}3} = 5 \times 10^{-9}\,M$$
$$[OH^-] = 10^{-(14-8{\cdot}3)} = 2 \times 10^{-6}\,M$$

These values are several orders of magnitude less than the concentration of H^+ we are adding, so the final concentration of H_3O^+ will be 0·005 M. The pH of the solution will then be

$$pH = -\log_{10}(0{\cdot}005) = 2{\cdot}3$$

Thus, the pH change is 6 units.

Tris is tris(hydroxymethyl)aminomethane. In water, the following equilibrium is established between the base (Tris) and its conjugate acid (Tris–H^+).

$$\underset{\text{Tris–H}^+}{(HOCH_2)_3CNH_3^+} + H_2O \rightleftharpoons \underset{\text{Tris}}{(HOCH_2)_3CNH_2} + H_3O^+$$

At pH = 8·3, the pH equals the pK_a and so the concentration of the base equals the concentration of the conjugate acid

$$[\text{Tris}] = [\text{Tris–H}^+] \qquad \text{and} \qquad [\text{Tris–H}^+] + [\text{Tris}] = 0{\cdot}1\,M$$

then

$$[\text{Tris–H}^+] = [\text{Tris}] = 0{\cdot}05\,M$$

If we add 0·005 mol of H^+ ions to a litre of solution, 0·005 mol of Tris will be converted to Tris–H^+. The new concentrations in solution will then be

$$[\text{Tris–H}^+] = 0{\cdot}055\,M \qquad \text{and} \qquad [\text{Tris}] = 0{\cdot}045\,M$$

From the Henderson–Hasselbalch equation, the new pH is given by

$$pH = pK_a + \log_{10}\left(\frac{[\text{Tris}]}{[\text{Tris–H}^+]}\right) = 8{\cdot}3 + \log_{10}\left(\frac{0{\cdot}045}{0{\cdot}055}\right) = 8{\cdot}213$$

Thus the pH change is 0·087 units.

COMMENT: The presence of Tris, which is a frequently used buffer in biochemical research, reduces the change in pH in this case by 98·5%.

Quantifying buffer strengths

We can quantify the strength of a buffer solution by using the *buffering capacity*, β, where

$$\beta = \frac{\text{Amount of acid or base added}}{\text{Change in pH}}$$

The higher the buffering capacity, the better the solution is as a buffer. It is important to note that the buffering capacity is a property of the solution. The buffering capacity of a solution of any specific acid or base depends strongly on the solution concentration of the acid or base and the pH.

WORKED EXAMPLE

What is the buffering capacity of a 0·1 M solution of Tris at pH 8·3?

SOLUTION: From the previous worked example, adding 0·005 mol H^+ to a litre of 0·1 M Tris at pH 8·3 reduced the pH by 0·087 units. It follows that the buffering capacity is

$$\beta = \frac{\text{Amount of acid added}}{\text{Change in pH}} = \frac{0{\cdot}005}{0{\cdot}087} = 0{\cdot}057\ \text{M H}^+(\text{pH unit})^{-1}$$

COMMENT: There is nothing special about considering the effect of 0·005 M acid. We would get the same result using any arbitrary concentration of acid or base as long as it is much less than the solution concentration of buffer.

The flattest parts of the titration curves occur at the half-equivalence points, when $pH = pK_a$, which corresponds to a 1:1 mixture of the acid and its conjugate base. The buffering capacity is a maximum when the pH equals the pK_a. As the pH moves away from the pK_a, the buffering capacity will decrease. If the concentration of the buffer is increased, then it will require more acid or base to cause neutralisation and so the buffering capacity will increase.

WORKED EXAMPLE

What is the buffering capacity of a 0·1 M solution of Tris at pH 6·3?

SOLUTION: From the Henderson–Hasselbalch equation, the ratio of $[\text{Tris–H}^+]/[\text{Tris}] = 100$ at pH 6·3. The total concentration of Tris is 0·1 M.

$$[\text{Tris–H}^+] + [\text{Tris}] = 0{\cdot}1\,\text{M}$$

therefore

$$[\text{Tris–H}^+] = 0{\cdot}09901\,\text{M} \quad \text{and} \quad [\text{Tris}] = 0{\cdot}00099\,\text{M}$$

If we add 0·0001 mol of H^+ ions to 1 litre of the solution, 0·0001 mol of Tris will be converted to Tris$-H^+$. The new concentrations in solution will then be

$$[\text{Tris–H}^+] = 0{\cdot}09911\,\text{M} \quad \text{and} \quad [\text{Tris}] = 0{\cdot}00089\,\text{M}$$

From the Henderson–Hasselbalch equation, the new pH is given by

$$\text{pH} = \text{pK}_a + \log_{10}\left(\frac{[\text{Tris}]}{[\text{Tris–H}^+]}\right) = 8.3 + \log_{10}\left(\frac{0{\cdot}00089}{0{\cdot}09911}\right) = 6{\cdot}253$$

The change in pH on adding 0·0001 M of acid is 0·047 units. Thus, the buffering capacity is

$$\beta = \frac{\text{Amount of acid added}}{\text{Change in pH}} = \frac{0{\cdot}0001}{0{\cdot}047} = 0{\cdot}0021\,\text{M H}^+(\text{pH unit})^{-1}$$

COMMENT: This value compares with $\beta = 0{\cdot}057\,\text{M H}^+\,(\text{pH unit})^{-1}$ for a 0·1 M solution of Tris at pH 8·3, showing that the buffering capacity is reduced thirty times when the solution pH is two units away from the pK_a. In practice the useful pH range for a buffer is within 1 pH unit of the pK_a.

Plots of buffering capacity versus pH are shown in figure 5.4 for a variety of acids and bases. Any acid or base can act as a buffer and will give the same shape of curve for β versus pH at any given concentration, the only thing that changes is the position of the maximum.

Regulation of pH by ion transport

At first sight it would appear that it should be possible to control the pH in a cell simply by pumping H^+ ions into or out of the cell. However, this is too simplistic an approach for two reasons.

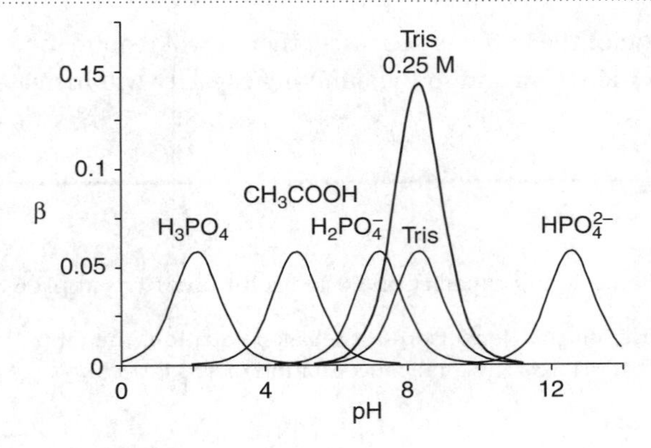

Fig. 5.4 Plots of buffering capacity (β) versus pH for a variety of acids with different pK_a values. All plots are for acids at 0·1 M, except for Tris which is shown at two concentrations. Each acid gives an identical shaped curve, centred on its pK_a.

First, the transfer of H^+ from one compartment to another leads to a separation of charge that prevents the movement of more than an insignificant number of protons. The region that the H^+ ions are pumped from becomes negatively charged and the region that they are pumped to becomes positively charged. This makes it increasingly difficult to pump more H^+ ions. To avoid this problem other ions have to move across the same membrane to balance the charge and to ensure that electroneutrality is maintained on both sides of the membrane. The existence of membrane potentials shows that there is usually a net excess of one charge or another inside a membrane-bound compartment (see Chapter 7) but this excess is very small indeed and it is therefore appropriate to assume that electroneutrality is maintained during pH regulation by ion transport.

Second, the effect of the proton movement on the pH of the two compartments depends on their composition. While H^+ ions can undoubtedly be transferred from one compartment to another in parallel with charge-compensating movements of other ions, H^+ ions are in equilibrium with a large number of weak acids in biological solutions and the positions of these equilibria will alter on addition of H^+. Thus the proton concentration in a solution is a dependent variable, i.e. its value is determined by the values of other variables. These other variables are called independent variables, because their values are independent of each other, and it is changes in the independent variables that give rise to changes in a dependent variable such as $[H^+]$.

To illustrate this point, consider the case of a solution that has been prepared by mixing solutions of HCl and NaOH. The proton concentration has to satisfy two conditions

$$[H^+][OH^-] = K_w$$

$$[Na^+] + [H^+] = [Cl^-] + [OH^-]$$

The Na^+ and Cl^- ions remain fully dissociated irrespective of the composition of the solution whereas the H^+ and OH^- ions can participate in equilibria. On this basis the Na^+ and Cl^- ions are called strong ions and the H^+ and OH^- ions are called weak ions. We can now define the so-called strong ion difference, [SID], as the net concentration of the charges carried by the strong ions and use this to rewrite the electroneutrality equation, giving

$$[H^+] - [OH^-] + [SID] = 0$$

For the case above $[SID] = [Na^+] - [Cl^-]$. Combining this equation with the equation for the ionic product of water leads to the quadratic

$$[H^+]^2 + [SID][H^+] - K_w = 0$$

which can be solved to give

$$[H^+] = \frac{-[SID] \pm \sqrt{[SID]^2 + 4K_w}}{2}$$

Similarly

$$[OH^-] = \frac{[SID] \pm \sqrt{[SID]^2 + 4K_w}}{2}$$

Thus the dependent variables, $[H^+]$ and $[OH^-]$, are determined by K_w and the independent variable [SID], and if a further aliquot of HCl or NaOH is added to the solution then the change in pH is determined by the change in [SID] rather than by the addition of the H^+ or OH^- ions.

When this approach is extended to the solutions that occur inside cells, the full set of independent variables consists of [SID], the total weak acid concentration $[A_{tot}]$, and the partial pressure of CO_2. In this situation the equation determining $[H^+]$ takes the form of a fourth-order polynomial

$$[H^+]^4 + a[H^+]^3 + b[H^+]^2 + c[H^+] + d = 0$$

where a, b, c and d are determined by the dissociation constants for the various equilibria involving water, the weak acids and CO_2, as well as by [SID], $[A_{tot}]$ and pCO_2. This equation can be solved numerically and it can provide useful insights into the relationship between the pH of a biological solution, such as blood plasma or the solution found in the xylem vessels of plants, and its composition. Such calculations emphasise the fact that the movement of H^+ between compartments can only contribute to pH regulation under conditions in which it leads to a change in [SID].

Measurement of pH

The pH of a solution can be determined either by the direct determination of $[H_3O^+]$ or by measuring some property that depends on pH.

Direct measurements of [H_3O^+] can be made with a pH electrode and details of this method are given in Chapter 6. This is usually the method of choice for measuring the pH of solutions in the laboratory, as it is the most accurate. Microelectrodes can be inserted directly into relatively large cells, for example plant cells where they can be used to measure cytoplasmic and vacuolar pH values.

Indirect measurements of pH depend on measuring the pH-dependent property of a component of the solution. The detected component can act as a pH indicator if it dissociates with a pK_a that is close to the pH of the solution and if the measured property changes when the indicator dissociates. Spectroscopic techniques are particularly convenient for detecting the indicator and pH-dependent properties that can be used for pH measurements include: (i) colour; (ii) fluorescence; and (iii) nuclear magnetic resonance (NMR) frequencies.

The pH-dependent colour of a weak acid or base is exploited in pH paper. Here the colour of the indicator molecule depends on its ionisation state

$$\underset{\textit{colour 1}}{\text{indicator} - \text{H}} + H_2O \rightleftharpoons \underset{\textit{colour 2}}{\text{indicator}^-} + H_3O^+$$

The colour change occurs when pH $\approx$ pKa and it takes place gradually over a small range of pH on either side of the pK_a (Table 5.2).

An indicator of this kind can be used to establish whether the pH is above or below the pH range for the colour change, but by using several indicators it is possible to narrow down the pH of the solution. For example, a solution that turns methyl violet blue must be above pH 1·6, but if it also turns litmus red then the pH must be between 1·6 and 5·0. pH paper consists of a mixture of such indicators absorbed on to paper so that the overall colour changes continuously with pH over a defined pH range.

Fluorescence and NMR measurements are also widely used for determining pH and they are particularly useful for measuring intracellular pH values. For example the second pK_a at 6·8 causes the frequency of the ^{31}P NMR signal from the inorganic phosphate ion to be pH-dependent between pH 5·6 and 8·0. Inorganic phosphate is ubiquitous in living systems and the NMR signal is readily detected *in vivo*. This allows intracellular pH values to be measured non-invasively in organisms and tissues in defined physiological states.

TABLE 5.2

Indicator	Colour of acid	pH range for colour change	Colour of base
Methyl violet	yellow	0·0–1·6	blue
Litmus	red	5·0–8·0	blue
Phenolphthalein	colourless	8·2–10·0	red

FURTHER READING

R. de Levie, 1999, 'Aqueous acid-base equilibria and titrations', Oxford University Press—*a more advanced and rigorous treatment of acid/base solutions.*

P.A. Stewart, 1983, 'Modern quantitative acid–base chemistry', in Canadian Journal of Physiology and Pharmacology, vol. 61, pp. 1444–61—*a rigorous introduction to the concept of the strong ion difference.*

PROBLEMS

1. ΔH° for the reaction $H^+(aq) + OH^-(aq) \rightarrow H_2O$ is $-56.8\,kJ\,mol^{-1}$. Given that the ionic product of water, K_w, is 0.61×10^{-14} at 291K, calculate K_w and hence determine neutral pH at 310 K.

2. Calculate the pH of the following solutions

(a) 0·05 M HCl
(b) 0·1 M acetic acid ($K_a = 1.75 \times 10^{-5}$ (M))
(c) 0·1 M aniline ($K_b = 3.82 \times 10^{-10}$ (M))
(d) 0·1 M acetic acid plus 0·001 M HCl
(e) 0·16 M acetic acid plus 0·044 M sodium acetate

3. Using the data given in the chapter, which of the following amino acid side chains

Arg, Asp, Cys, Glu, His, Lys, Tyr

will be

(a) acidic or basic at pH 5·0?
(b) acidic or basic at pH 10·0?

4. The C-terminal domain of the protein colicin E1 has two histidine residues. The data below give the ^1H NMR resonance frequencies (δ) of one of the ring protons for each His residue as a function of pH.

pH	4·75	5·25	5·75	6·25	6·75	7·25	7·75	8·25	8·75
δ(His A)	8·596	8·589	8·566	8·499	8·338	8·071	7·820	7·682	7·627
δ(His B)			8·285	8·009	7·824	7·743	7·714	7·704	7·701

Determine the pK_a values for these two residues and comment on the results.

5. What do you understand by the term isoelectric point, or pI? Prove that the pI value of a zwitterion, such as glycine, is given by the average of the pKa values of the titratable groups. Using the pK_a values in the text, calculate the pI value for glutamic acid.
[Hint: when pH = pI, the concentration of positively charged molecules must equal the concentration of negatively charged molecules.]

6. Determine the volumes of 0·1 M NaH_2PO_4 and 0·1 M Na_2HPO_4 that need to be mixed together to produce 100 cm^3 of a buffer at pH 7·0. The second pK_a of H_3PO_4 is 7·2.

7. Demonstrate, by calculating the buffering capacity of a 0·5 M solution of Tris at pH 8·3, that a fivefold increase in the concentration of a buffer causes a fivefold increase in the buffering capacity of the solution. The pK_a of Tris is 8·3.

8. The hydrolysis of ATP at pH 8·0 is given by the following reaction

$$ATP^{4-} + H_2O \rightarrow ADP^{3-} + HPO_4^{2-} + H^+$$

Calculate the change in pH for the enzymatic hydrolysis of a 1 mM solution of ATP in the presence of (a) 0·1 M Tris buffer, pH 8·0; (b) 0·01 M Tris buffer, pH 8·0; and (c) no buffer, pH 8·0. The pK_a of Tris is 8·3.

9. ΔH° for the protonation of Tris is $-46{\cdot}0\ kJ\ mol^{-1}$ and for the phosphate dianion it is $-4{\cdot}2\ kJ\ mol^{-1}$. What is the ratio of K_a at 273 K to that at 298 K for these two buffers?

10. The [SID] values for three ionic solutions were 41 mM, 1 mM and −39 mM. Calculate the pH values for these solutions before and after the further addition of HCl to a concentration of 2 mM. Comment on the results that you have calculated. Assume that $K_w = 4{\cdot}4 \times 10^{-14}$ (M).

Oxidation–reduction reactions and electrochemistry

6

KEY POINTS:

- Oxidation–reduction (*redox*) reactions involve the transfer of electrons between reactants. Loss of electrons corresponds to oxidation, gain of electrons to reduction. Redox reactions play a key role in energy metabolism.
- Redox reactions can be made to do electrical work by setting up electrochemical cells. These cells are composed of two half-cells corresponding to the separate oxidation and reduction processes, each of which has an *electrode potential*.
- The *Nernst equation* describes the relationship between the electrode potential and the concentrations of the redox components.
- Electrochemical measurements can be made under reversible conditions, allowing the thermodynamic parameters (ΔG, ΔS and ΔH) of a redox reaction to be measured precisely.

Many chemical reactions involve electron transfer between the reactants, and in such processes one reactant loses electrons and is therefore oxidised, while another component gains electrons and is therefore reduced. These reactions are known as oxidation–reduction ('redox') reactions, and they are of particular importance here for two reasons. First, in a thermodynamic context, redox reactions can be used to do electrical work by directing the electrons through an electrical circuit. The electrochemical measurements that can be made with these circuits are very precise and they can also be done under reversible conditions, making it possible to generate a considerable body of useful thermodynamic information. Second, in a biochemical context, electron transfer reactions play a pivotal role in the energy transduction processes that occur during photosynthesis and oxidative phosphorylation, as well as in the many redox reactions that occur in metabolism. Accordingly, this chapter examines the thermodynamics of such processes and provides an introduction to the electrochemical nomenclature that is used to describe them.

Oxidation–reduction reactions

The electron transfer that occurs between zinc and copper in aqueous solution provides a simple example of an oxidation–reduction reaction.

$$Zn(s) + Cu^{2+}(aq) \rightleftharpoons Zn^{2+}(aq) + Cu(s)$$

Here the oxidation state of the zinc increases from zero to two, corresponding to the loss of two electrons, while the oxidation state of the copper decreases from two to zero, corresponding to the gain of two electrons. Similarly an example of a redox reaction of great biochemical importance is

$$2NADH + O_2 + 2H^+ \rightleftharpoons 2NAD^+ + 2H_2O$$

In this case the underlying electron transfer processes can be highlighted by splitting the reaction into two half-reactions, one corresponding to the reduction of the oxygen molecule

$$O_2 + 4H^+ + 4e^- \rightarrow 2H_2O$$

and the other corresponding to the oxidation of the NADH

$$NADH \rightarrow NAD^+ + H^+ + 2e^-$$

The thermodynamics of these reactions can be analysed in exactly the same way as any other chemical process, but it turns out to be convenient to disguise the Gibbs free energy term (ΔG) in the form of a redox potential (E). The reason for this is that electron transfer reactions can be studied in electrochemical cells and in this situation the driving force of the chemical reaction gives rise to a measurable voltage, the electromotive force of the cell. Since electron transfer reactions are very important in biochemistry, and since electrochemical cells provide the opportunity to study the energetics of such reactions under thermodynamically reversible conditions, it is necessary to have some understanding of the principles of electrochemistry.

Electrochemical cells

Electron transfer processes can occur randomly, by simple contact between the reactants, or non-randomly, by providing a specific pathway for the electrons between the separated reactants. Consider the electron transfer between zinc and copper. If zinc is added to a solution of copper sulphate (figure 6.1a), the zinc will dissolve and copper will be precipitated. This process is energetically favourable ($\Delta G^\circ = -213 \cdot 2\ kJ\ mol^{-1}$ at 298 K) and the useful energy will be lost as heat. Alternatively, by placing the reactants in separate containers (figure 6.1b), the electrons can be made to pass from the zinc to the copper ions

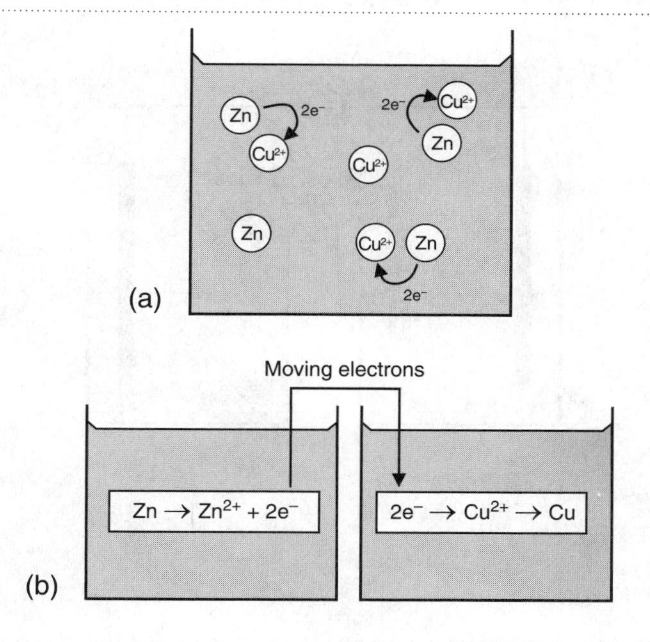

Fig. 6.1 (a) Electron transfer reactions (redox reactions) can take place spontaneously in solution through random collision of reactants with any available energy being lost as heat. (b) Alternatively, if the oxidation and reduction reactions are carried out in physically separate locations electrons can flow between the two locations and the available energy harnessed as an electrical current.

along a connecting wire and the resulting electric current, which is again a manifestation of the favourable free energy change, can be used to do work. This is the principle of the electrochemical cell.

The electrochemical cell splits the redox reaction into two components

$$\mathrm{Zn(s)} \rightarrow \mathrm{Zn^{2+}(aq)} + 2e^-$$
$$\mathrm{Cu^{2+}(aq)} + 2e^- \rightarrow \mathrm{Cu(s)}$$

These two half-reactions cannot occur to any significant extent in isolation because of the difficulty of stabilising the free electrons. For example, consider what happens when a piece of zinc is added to a solution containing Zn^{2+} ions. If some of the zinc dissolves, then the electrons will be left on the metal, and the resulting potential difference between the negatively charged zinc and the positively charged solution will limit the extent to which the zinc can dissolve. Alternatively, if some of the zinc ions precipitate from the solution on to the metal, then the reverse charge separation will occur and the process will be limited by the potential difference between the positively charged zinc and the negatively charged solution. It is important to realise that both of these processes take place when zinc is added to the solution of its ions, but that the extent to which they occur is not necessarily the same. Thus, depending on the concentration of the Zn^{2+} and the energetics of the reaction, the zinc will acquire a net charge and thus an electric (Nernst) potential.

A similar process occurs when a piece of copper is added to a solution of Cu^{2+} with the copper acquiring a net charge and an electric potential. Since it is unlikely that the potentials on the zinc and the copper will be the same,

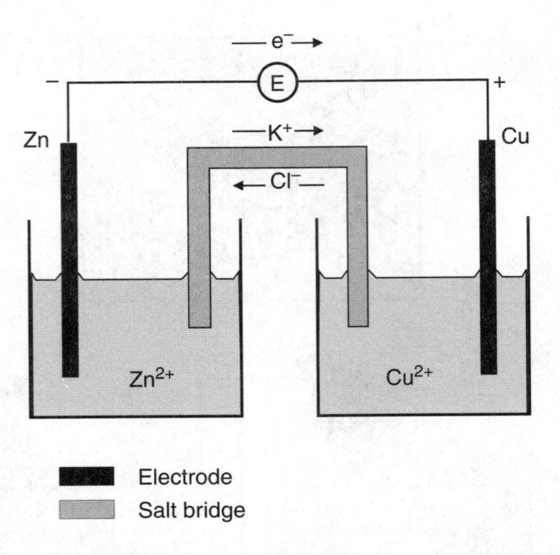

Fig. 6.2 An electrochemical cell consists of two separate redox reactions, in this case the oxidation of Zn and the reduction of Cu^{2+}, connected by a wire to allow electrons (e^-) to flow between the two. A salt bridge of concentrated KCl is used to complete the electrical circuit. KCl is used because K^+ and Cl^- have similar current-carrying properties (mobilities), so that the overall current can be carried by ions moving in either direction with equal facility.

other than by chance, there will be an electric potential difference (E) between the zinc and copper and so a current will flow when the zinc and copper are connected in an electrochemical cell (figure 6.2). The electrons flow from one compartment to the other to give a net cell reaction

$$Zn(s) + Cu^{2+}(aq) \rightleftharpoons Zn^{2+}(aq) + Cu(s)$$

Thus on the left-hand side of the cell zinc is oxidised to Zn^{2+}, while on the right-hand side of the cell Cu^{2+} is reduced to copper. The electrons travel via the zinc and copper electrodes, and the electric circuit is completed by adding a salt bridge containing a concentrated solution of KCl. When electrons move from left to right, there is a compensating movement of charge through the salt bridge with Cl^- moving into the left-hand compartment and K^+ moving to the right.

If the Zn and Cu electrodes are connected, then electrons flow from the Zn to the Cu as Zn^{2+} ions are formed and Cu^{2+} ions are reduced. As the reaction proceeds towards equilibrium the driving force (ΔG) falls and so the amount of electrical work obtained from the cell decreases. However, suppose that we apply a suitable potential difference across the electrodes so as to oppose the effect of the electrochemical cell, as shown in figure 6.3. As the applied potential difference is increased, the current flowing in the circuit will decrease. At a certain point, the null point, no current flows becase the applied potential difference is exactly equal to that of the electrochemical cell. Under these conditions the applied potential difference is equal to the electromotive force (e.m.f.) of the cell. If the applied potential difference is increased beyond this value, the current will reverse its direction, i.e. the cell reaction will proceed in the opposite direction $Cu + Zn^{2+} \rightarrow Cu^{2+} + Zn$.

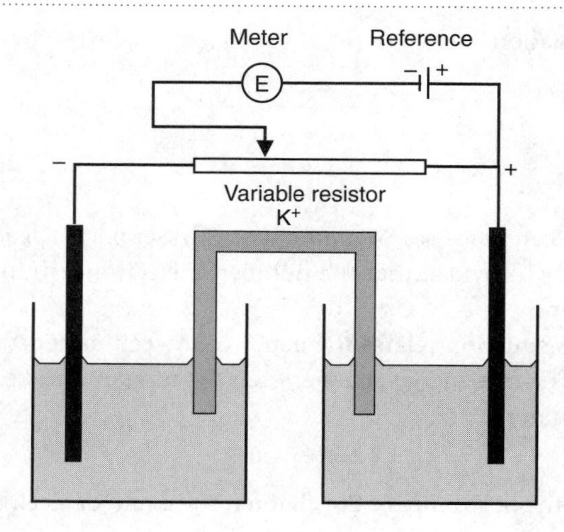

Fig. 6.3 An electrochemical cell for measuring cell e.m.f.s. Applying an opposing potential to an electrochemical cell decreases the current through the wire. By adjusting the potential, in this case using a variable resistor, the current can be halted. Under these conditions the applied potential equals the electromotive force of the cell.

The thermodynamics of reversible cells

The first law of thermodynamics states that

$$\Delta U = \Delta q + \Delta w$$

For an electrochemical cell, Δw, the work done on the system, consists of work done by the surroundings ($-P\Delta V$) and the electrical work done ($-\Delta w_{elec}$). So

$$\Delta U = \Delta q - P\Delta V - \Delta w_{elec}$$

Now at constant pressure

$$\Delta H = \Delta U + P\Delta V$$

and so

$$\Delta H = \Delta q - \Delta w_{elec}$$

Since the electrochemical cell process is reversible at the null point, $\Delta q = T\Delta S$ from the second law of thermodynamics and so

$$\Delta H = T\Delta S - \Delta w_{elec}$$

Since $\Delta G = \Delta H - T\Delta S$ it follows that $\Delta G = -\Delta w_{elec}$ for electrochemical cell processes at constant pressure. Thus ΔG measures the amount of useful work obtainable from the electrochemical cell.

Suppose the reaction involves the transfer of n electrons (e.g. $n = 2$ in the case of $Zn + Cu^{2+} \rightarrow Zn^{2+} + Cu$). Then the electrical work done in transporting n electrons through a potential difference E is given by $\Delta w_{elec} = nFE$ where F, the Faraday, is a conversion constant equal to $96{\cdot}5\ kJ\ V^{-1}\ mol^{-1}$. Thus we arrive at the equation

$$\Delta G = -nFE$$

for all reversible cell processes. Measurement of the e.m.f. (E) of a redox process gives ΔG directly, provided that the number of electrons (n) involved in the process is known.

An analogous equation relates the e.m.f. of the cell under standard conditions (E°) with the free-energy change when the reactants and products are in their standard states (ΔG°)

$$\Delta G^{o} = -nFE^{o}$$

As indicated at the beginning of the chapter, the e.m.f. of a cell (E) is merely a disguise for the free-energy change of the cell reaction (ΔG). However, the disguise is slightly more subtle than it first appears becase ΔG is an extensive state function, whereas E is intensive. It follows that values of E cannot be added and subtracted in the same way as values of ΔG. There are some circumstances in which E values have to be subtracted (see below), but for thermodynamic calculations in which the aim is to work out the energetics of a redox reaction it is essential to convert E values to ΔG before adding and subtracting the components of the overall reaction.

WORKED EXAMPLE

Calculate E° for the electrochemical cell with the reaction

$$\mathbf{Zn(s) + Cu^{2+}(aq) \rightleftharpoons Zn^{2+}(aq) + Cu(s) \quad \Delta G^{\circ} = -213{\cdot}2\,kJ\,mol^{-1}}$$

SOLUTION: As discussed above, the reaction involves the transfer of two electrons and so

$$E^{\circ} = -\frac{\Delta G^{\circ}}{nF} = \frac{213{\cdot}2 \times 10^{3}}{2 \times 96\,500} = 1{\cdot}1\,V$$

COMMENT: Note that a favourable (negative) ΔG° corresponds to a positive value of E°.

Cells and half-cells

In principle, any reaction that involves the transfer of electrons can be used as the basis of an electrochemical cell. Any reaction of this kind can be broken down into an oxidation process and a reduction process, and the cell is arranged so that these processes occur in separate compartments. The reaction that occurs in each compartment is known as a half-cell reaction, and by convention half-cell reactions are always written as reduction processes

$$\mathrm{Ox} + n\mathrm{e}^- \rightarrow \mathrm{Red}$$

where Ox and Red refer to the oxidised and reduced species respectively.

Thus for the zinc and copper electrochemical cell discussed above, the two half-cell reactions are

$$Zn^{2+}(aq) + 2e^- \rightarrow Zn(s)$$

$$Cu^{2+}(aq) + 2e^- \rightarrow Cu(s)$$

The cell is made by combining the two half-cells, with one on the left-hand side (L) and one on the right-hand side (R). By convention the cell reaction is then given by subtracting the left-hand half-cell reaction from the right-hand half-cell reaction (R – L). Thus for the cell illustrated in figure 6.2, with the zinc half-cell on the left, the overall cell reaction is given by

$$Cu^{2+}(aq) - Zn^{2+}(aq) \rightleftharpoons Cu(s) - Zn(s)$$

which can be rearranged to give

$$Zn(s) + Cu^{2+}(aq) \rightleftharpoons Zn^{2+}(aq) + Cu(s)$$

Note that the balanced cell reaction contains no free electrons.

WORKED EXAMPLE

What happens to the e.m.f. of the electrochemical cell in figure 6.2 if the half-cells are rearranged with the copper on the left and the zinc on the right?

SOLUTION: The chemical nature of the half-cells is unchanged, but their relative position is different and applying the convention R – L gives a new cell reaction

$$Zn^{2+}(aq) + Cu(s) \rightleftharpoons Zn(s) + Cu^{2+}(aq)$$

This reaction is the reverse of the reaction considered in the previous worked example and so the sign of ΔG° must change, i.e. $\Delta G^\circ = +213{\cdot}2\ \mathrm{kJ\ mol^{-1}}$. The linear relation between ΔG and E also implies that the sign of E° must change and so the standard e.m.f. of the rearranged cell is $-1{\cdot}1\ \mathrm{V}$.

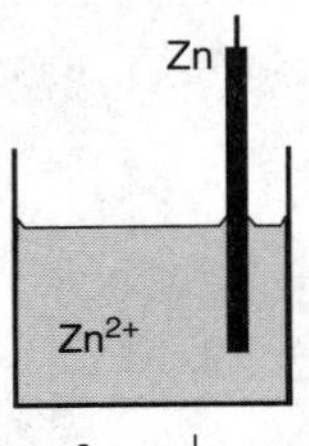

$Zn^{2+}(aq) \mid Zn(s)$

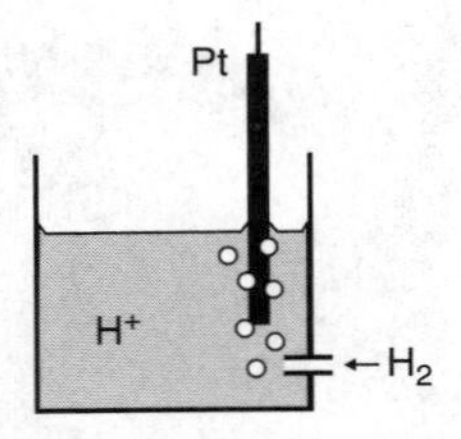

$H^{+}(aq) \mid Pt(s), H_2(g)$

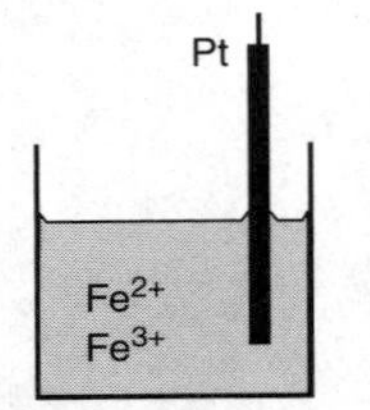

$Fe^{3+}(aq), Fe^{2+}(aq) \mid Pt(s)$

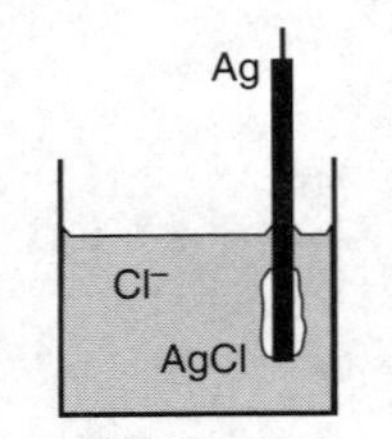

$Cl^{-}(aq) \mid AgCl(s), Ag(s)$

Fig. 6.4 Schematic diagrams of some typical half-cells, together with their shorthand notation.

Cell and half-cell nomenclature

A standard, shorthand notation can be used to represent electrochemical cells and this takes the form

Left-hand electrode | Left-hand solution || Right-hand solution | Right-hand electrode

where | represents the boundary between two phases (e.g. solid and liquid) and || represents the boundary between two half-cells (e.g. a salt bridge). Thus the cell in figure 6.2 can be written as

$$Zn(s) \mid Zn^{2+}(aq) \parallel Cu^{2+}(aq) \mid Cu(s)$$

As explained above, it is important to write down the cell in the correct orientation. Although the cells $Zn(s) \mid Zn^{2+}(aq) \parallel Cu^{2+}(aq) \mid Cu(s)$ and $Cu(s) \mid Cu^{2+}(aq) \parallel Zn^{2+}(aq) \mid Zn(s)$ have the same physical construction, the equation for the cell reaction is reversed and so the thermodynamic quantities (ΔG and E) have opposite signs.

Types of half-cells

Here we summarise some of the more important types of half-cell (figure 6.4), in each case assuming that the half-cell is the right-hand component of a full cell.

(1) *A metal electrode in a solution of its ions*

Many examples of this are known, for example

$$Zn^{2+} + 2e^{-} \rightarrow Zn \qquad Zn^{2+}(aq) \mid Zn(s)$$

$$Ag^{+} + e^{-} \rightarrow Ag \qquad Ag^{+}(aq) \mid Ag(s)$$

(2) *Gas in contact with a solution of its ions*

The best known example of this is the hydrogen electrode

$$2H^{+} + 2e^{-} \rightarrow H_2 \qquad H^{+}(aq) \mid H_2(g), Pt$$

Hydrogen gas is bubbled over a Pt-black electrode. This causes the hydrogen molecules to dissociate and leads to the formation of a monatomic layer of hydrogen in contact with the solution of H^{+} ions. Chlorine and oxygen electrodes can be made in the same way.

(3) *Two different oxidation states of the same species*

Here an inert electrode, such as Pt, allows electrons to enter or leave the solution. For example

$Fe^{3+} + e^- \rightarrow Fe^{2+}$	$Fe^{3+}(aq), Fe^{2+}(aq) \mid Pt$
$Sn^{4+} + 2e^- \rightarrow Sn^{2+}$	$Sn^{4+}(aq), Sn^{2+}(aq) \mid Pt$
$Fe(CN)_6^{3-} + e^- \rightarrow Fe(CN)_6^{4-}$	$Fe(CN)_6^{4-}(aq), Fe(CN)_6^{3-}(aq) \mid Pt$

(4) *Metal in contact with its insoluble salt*

Here a metal is coated with a thin layer of one of its insoluble salts and this is in contact with a solution containing the anion of the insoluble salt. The two most common examples of this type are

$AgCl(s) + e^- \rightarrow Ag(s) + Cl^-$	$Cl^-(aq) \mid AgCl(s), Ag(s)$
$Hg_2Cl_2(s) + 2e^- \rightarrow 2Hg(l) + 2Cl^-$	$Cl^-(aq) \mid Hg_2Cl_2(s), Hg(l)$

The latter is known as the *calomel electrode*, and it is often used as a reference electrode (see below).

WORKED EXAMPLE

Write down the half-cell reactions and the overall reaction for the following electrochemical cell

$$\mathbf{Pt \mid Fe^{3+}(aq), Fe^{2+}(aq) \parallel Sn^{4+}(aq), Sn^{2+}(aq) \mid Pt}$$

SOLUTION: Each half-cell is written as a reduction reaction.

Left-hand half-cell (L):	$Fe^{3+}(aq) + e^- \rightarrow Fe^{2+}(aq)$
Right-hand half-cell (R):	$Sn^{4+}(aq) + 2e^- \rightarrow Sn^{2+}(aq)$

The balanced equation for the cell reaction is obtained by applying the R − L convention and eliminating the free electrons

$$Sn^{4+}(aq) + 2Fe^{2+}(aq) \rightleftharpoons Sn^{2+}(aq) + 2Fe^{3+}(aq)$$

WORKED EXAMPLE

Devise a cell for studying the reduction of Co^{3+} to Co^{2+} by hydrogen.

SOLUTION: The two half-reactions are $Co^{3+}(aq) + e^- \rightarrow Co^{2+}(aq)$ and $2H^+(aq) + 2e^- \rightarrow H_2$. Neither of these reactions involves a conducting solid that could be used as an electrode and so an inert electrode must be used in both cases. Thus the required cell is

$$Pt \,|\, Co^{3+}(aq),\ Co^{2+}(aq) \,\|\, H^+(aq) \,|\, H_2(g),\ Pt$$

Electrode potentials

The e.m.f. of a cell (E) is equal to the potential difference between the electrodes in the two half-cells, and from the convention for the cell reaction it follows that

$$E = E_R - E_L$$

where E_R and E_L are the electrode potentials on the right and left electrodes respectively. Unfortunately, there is no way of measuring the potential difference between an electrode and the surrounding electrolyte directly, and so it is not possible to measure the absolute values of E_R and E_L. However, it is possible to measure the *difference* between the potentials of two half-cells when they are linked to form an electrochemical cell, and so a series of relative values can be obtained if each half-cell is combined with a standard half-cell. The standard hydrogen electrode has been chosen as the standard half-cell and its standard electrode potential (E°) has been arbitrarily set to 0 V. The half-cell reaction for this electrode is

$$2H^+(aq) + 2e^- \rightarrow H_2(g)$$

and when each component is in its standard state (H_2, 1 atm; H^+, a 1 M ideal solution) the electrode potential is defined as 0 V.

The standard electrode potential of any other electrode is defined as the e.m.f. of a cell in which the half-cell of interest, with each component in the standard state, is combined with a standard hydrogen electrode on the left-hand side. A representative selection of E° values is given in Table 6.1.

Note that the sign of E° refers to the reaction as written in Table 6.1. Thus a negative E° value, which is equivalent to a positive value of $\Delta G°$, implies that

TABLE 6.1

Electrode	Reaction	E^{o}(V)
$Na^{+}(aq) \mid Na(s)$	$Na^{+}(aq) + e^{-} \rightarrow Na(s)$	−2·71
$Zn^{2+}(aq) \mid Zn(s)$	$Zn^{2+}(aq) + e^{-} \rightarrow Zn(s)$	−0·76
$H^{+}(aq) \mid H_2(g)$, Pt	$2H^{+}(aq) + 2e^{-} \rightarrow H_2(g)$	0·00
$Cu^{2+}(aq) \mid Cu(s)$	$Cu^{2+}(aq) + 2e^{-} \rightarrow Cu(s)$	0·34
$Fe^{3+}(aq)$, $Fe^{2+}(aq) \mid Pt$	$Fe^{3+}(aq) + e^{-} \rightarrow Fe^{2+}(aq)$	0·77
$(aq) \mid Cl_2(g)$, Pt	$Cl_2(g) + 2e^{-} \rightarrow 2Cl^{-}(aq)$	1·36

the oxidized form is favoured relative to the reduced form under standard conditions. For instance, hydrogen will not reduce Zn^{2+} to zinc. Correspondingly, a positive E°, which is equivalent to a negative value of ΔG°, means that the reduced state is favoured relative to the oxidised state under standard conditions. Thus hydrogen would reduce Cu^{2+} to copper. Electrode potentials are also referred to as redox potentials and, in general, the stronger the oxidising agent, the more positive the redox potential.

WORKED EXAMPLE

The standard electrode potentials, $E^{o\prime}$, for the NO_3^-/NO_2^- and NO_2^-/NH_4^+ half-cells are 0·42 V and 0·48 V respectively. Calculate $E^{o\prime}$ for the NO_3^-/NH_4^+ redox couple.

SOLUTION: As mentioned above, $E^{o\prime}$ is an intensive state function and so values of $E^{o\prime}$ cannot always be simply added or subtracted. The key is to use $\Delta G^{o\prime} = -nFE^{o\prime}$ to convert the electrochemical data into free energies.

$$NO_3^- + 2H^+ + 2e^- \rightleftharpoons NO_2^- + H_2O \quad \Delta G^{o\prime}_{NO_3^-/NO_2^-} = -2FE^{o\prime}_{NO_3^-/NO_2^-}$$

$$NO_2^- + 8H^+ + 6e^- \rightleftharpoons NH_4^+ + 2H_2O \quad \Delta G^{o\prime}_{NO_2^-/NH_4^+} = -6FE^{o\prime}_{NO_2^-/NH_4^+}$$

For the overall reaction

$$NO_3^- + 10H^+ + 8e^- \rightleftharpoons NH_4^+ + 3H_2O \quad \Delta G^{o\prime}_{NO_3^-/NH_4^+} = -8FE^{o\prime}_{NO_3^-/NH_4^+}$$

$\Delta G^{o\prime}$ is given by the sum of the $\Delta G^{o\prime}$ values for the two separate reactions

$$\Delta G^{o\prime}_{NO_3^-/NH_4^+} = \Delta G^{o\prime}_{NO_3^-/NO_2^-} + \Delta G^{o\prime}_{NO_2^-/NH_4^+}$$

$$-8FE^{o\prime}_{NO_3^-/NH_4^+} = -2FE^{o\prime}_{NO_3^-/NO_2^+} - 6FE^{o\prime}_{NO_2^-/NH_4^+}$$

Hence

$$E^{o\prime}_{NO_3^-/NH_4^+} = \frac{-2FE^{o\prime}_{NO_3^-/NO_2^-} - 6FE^{o\prime}_{NO_2^-/NH_4^+}}{-8F} = 0{\cdot}465\,V$$

COMMENT: Electrode potentials cannot be added and subtracted in thermodynamic calculations without reference to the number of electrons transferred. This problem is avoided by converting the electrode potentials into the corresponding extensive property, ΔG.

The Nernst equation

In the same way as ΔG for a reaction is a function of the concentrations of the reactants and products, so E for an electron transfer process is also a function of the concentrations of the participants. Consider a generalised half-reaction

$$Ox + ne^- \rightarrow Red$$

and assume that the participants in this electron transfer process are allowed to equilibrate with an inert electrode (M).

At equilibrium (see Chapter 3)

$$\bar{\mu}_{ox}(aq) + n\bar{\mu}_{e^-} = \bar{\mu}_{red}(aq)$$

Note that we have to use the electrochemical potential function ($\bar{\mu}$) here (see Chapter 7), because any net movement of charge to or from the electrode will generate a potential difference between the electrode and the solution. Expanding this equation using the definition of $\bar{\mu}$ gives

$$\mu^{\circ}_{ox} + RT\log_e[ox] + z_{ox}F\Phi_{soln} + n\mu_{e^-}(M) - nF\Phi_M = \mu^{\circ}_{red} + RT\log_e[red] + z_{red}F\Phi_{soln}$$

where Φ_{soln} and Φ_M are the electric potentials of the solution and the inert electrode, and where z_{ox} and z_{red} are the charges carried by the oxidised and reduced species.

Rearranging, by grouping the electric potential and concentration terms,

$$z_{red}F\Phi_{soln} - z_{ox}F\Phi_{soln} + nF\Phi_M = \mu^{\circ}_{ox} - \mu^{\circ}_{red} + n\mu_{e^-}(M) + RT\log_e\frac{[ox]}{[red]}$$

and noting that $z_{ox} - z_{red} = n$, leads to

$$\Phi_M - \Phi_{soln} = \frac{\mu^{\circ}_{ox} - \mu^{\circ}_{red} + n\mu_{e^-}(M)}{nF} + \frac{RT}{nF}\log_e\frac{[ox]}{[red]}$$

However, by definition, $\Phi_M - \Phi_{soln} = E$, the Nernst potential, and since $E = E^\circ$ when the oxidised and reduced species are in their standard states

$$E = E^\circ + \frac{RT}{nF}\log_e\frac{[ox]}{[red]}$$

This is the Nernst equation for the concentration dependence of the redox potential corresponding to the generalised half-reaction. A similar equation can be derived for the concentration dependence of the e.m.f. of an electrochemical cell. One approach is to note that the e.m.f. is the net result of two half-reactions, each giving rise to a redox potential, E_L or E_R, with a concentration dependence given by the Nernst equation. So if the right-hand half-cell corresponds to

$$A + ne^- \rightarrow C$$

and the left-hand half-cell corresponds to

$$D + ne^- \rightarrow B$$

Then recalling that $E = E_R - E_L$ leads to

$$E = E^\circ + \frac{RT}{nF}\log_e\frac{[A][B]}{[C][D]}$$

Alternatively, we can take the expression for the concentration dependence of the free-energy change for the reaction $A + B \rightleftharpoons C + D$

$$\Delta G = \Delta G^\circ + RT\log_e\frac{[C][D]}{[A][B]}$$

and substitute $\Delta G = -nFE$ to obtain

$$E = E^\circ + \frac{RT}{nF}\log_e\frac{[A][B]}{[C][D]}$$

This alternative approach can also provide a short cut to the Nernst equation for the redox potential of a half-cell, since the latter is identical to the redox potential of a cell in which the half-cell is connected to a standard hydrogen electrode.

WORKED EXAMPLE

The standard electrode potential for the Fe^{3+}(aq), Fe^{2+}(aq) | Pt half-cell is 0·77 V. What is the value of E for the half-cell when $[Fe^{3+}]=0{\cdot}2$ M and $[Fe^{2+}]=0{\cdot}05$ M (T = 298K)?

SOLUTION: The half-cell reaction is

$$Fe^{3+}(aq) + e^- \rightarrow Fe^{2+}(aq)$$

and the Nernst equation, assuming ideality, takes the form

$$E = E^\circ + \frac{RT}{F}\log_e\frac{[Fe^{3+}]}{[Fe^{2+}]}$$

Hence

$$E = 0{\cdot}77 + \frac{RT}{F}\log_e\frac{0{\cdot}2}{0{\cdot}05}$$

$$= 0{\cdot}77 + \frac{8{\cdot}31 \times 298}{96\,500}\log_e 4$$

$$= 0{\cdot}77 + 0{\cdot}036 = 0{\cdot}806\text{ V}$$

WORKED EXAMPLE

The standard e.m.f. of the electrochemical cell Pt, H_2 | H^+, Cl^- | AgCl, Ag is 0·2225 V. What is the e.m.f. when the HCl concentration is 10^{-3} M and the partial pressure of the hydrogen gas is 1 atm (T = 298 K)?

SOLUTION: The left-hand half-cell reaction is

$$2H^+(aq) + 2e^- \rightarrow H_2(g)$$

and the right-hand cell reaction is

$$AgCl(s) + e^- \rightarrow Ag(s) + Cl^-(aq)$$

giving an overall cell reaction of

$$2AgCl(s) + H_2(g) \rightleftharpoons 2Ag(s) + 2H^+(aq) + 2Cl^-(aq)$$

with n = 2. The Nernst equation for this cell reaction is

$$E = E^\circ + \frac{RT}{2F}\log_e\frac{P_{H_2}}{[H^+]^2[Cl^-]^2}$$

Note that Ag and AgCl do not appear in this expression because they are both solids and hence in their standard state.
Substituting $[H^+] = [Cl^-] = 10^{-3}$ M and $P_{H_2} = 1$ atm gives

$$E = 0{\cdot}2225 + \frac{8{\cdot}31 \times 298}{2 \times 96\,500}\log_e 10^{12} = 0{\cdot}577\text{ V}$$

WORKED EXAMPLE

What is the value of E for the half-cell NAD^+, H^+, NADH | Pt at pH 7, given that E° is −0·11 V (T = 298 K)? Assume that NAD^+ and NADH are both present at a concentration of 1 M.

SOLUTION: The half-cell reaction is

$$NAD^+(aq) + H^+(aq) + 2e^- \rightarrow NADH(aq)$$

and the Nernst equation takes the form

$$E = E^\circ + \frac{RT}{2F}\log_e\frac{[NAD^+][H^+]}{[NADH]}$$

Since $[H^+] = 10^{-7}$ M at pH 7

$$E = -0{\cdot}11 + \frac{8{\cdot}31 \times 298}{2 \times 96\,500}\log_e 10^{-7} = -0{\cdot}32\,V$$

COMMENT: The conditions in this half-cell correspond to the biochemical standard state (see Chapter 2), i.e. all the components of the reaction are present in their standard states, except for H^+, which is present at 10^{-7} M, corresponding to pH 7. In the same way that the standard free-energy change under these conditions is denoted by $\Delta G^{\circ\prime}$, so the symbol $E^{\circ\prime}$ is used to refer to standard electrode potentials at pH 7. Thus the standard redox potential for the NAD^+/NADH couple at pH 7 is −0·32 V.

Potentiometric titrations

Redox potentials provide a quantitative measure of the strength of oxidising and reducing agents. Strong oxidising agents, such as the lanthanide ion Ce^{4+} and oxygen, have strongly positive redox potentials—the E° values for Ce^{4+}/Ce^{3+} and O_2/H_2O are 1·61 V and 1·23 V respectively—and they can be used to oxidise the reduced forms of ions with less positive potentials. For example, the E° value for the Fe^{3+}/Fe^{2+} couple is only 0·77 V and so Ce^{4+} will tend to oxidise Fe^{2+} to Fe^{3+} in an aqueous solution

$$Ce^{4+}(aq) + Fe^{2+}(aq) \rightleftharpoons Ce^{3+}(aq) + Fe^{3+}(aq)$$

In the same way that a pH titration can be used to follow the reaction of an acid with a base, it is possible to use an oxidising agent to titrate a reducing agent and to follow the process by measuring the electric potential sensed by an electrode in the reaction mixture. Figure 6.5 shows the result of titrating a

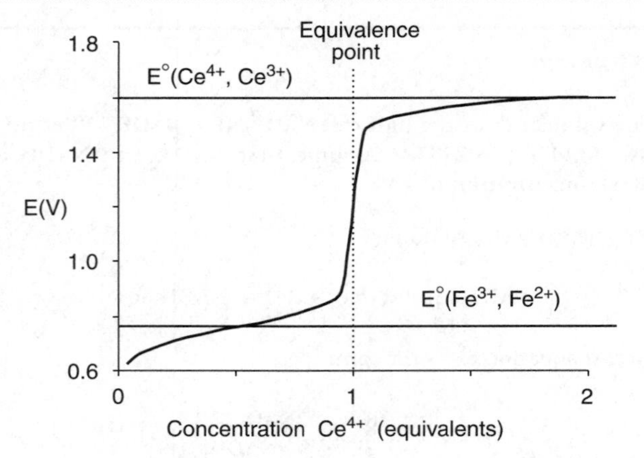

Fig. 6.5 A plot of E versus amount of Ce^{4+} added for the potentiometric titration of a solution of Fe^{2+} by Ce^{4+}. There is a rapid change in E when the concentration of Ce^{4+} added equals the initial concentration of Fe^{2+}, i.e. at the equivalence point.

solution of Fe^{2+} with Ce^{4+}. The potential detected by the electrode depends on the concentrations of the ions in the solution, but at every point on the curve this potential is given by

$$E = E^{\circ}_{Ce^{4+}/Ce^{3+}} + \frac{RT}{F}\log_e\frac{[Ce^{4+}]}{[Ce^{3+}]} = E^{\circ}_{Fe^{3+}/Fe^{2+}} + \frac{RT}{F}\log_e\frac{[Fe^{3+}]}{[Fe^{2+}]}$$

The reason for this is that while the electrode can act as either a source or a sink for every redox couple in the solution, it can only be at one potential under a particular set of conditions. This potential is the Nernst potential for each couple provided all the ions in the solution have equilibrated.

The E° values for the redox couples involved in the titration in figure 6.5 can be obtained form the titration curve by noting that $E = E^{\circ}_{Fe^{3+}/Fe^{2+}}$ when $[Fe^{3+}] = [Fe^{2+}]$ and that $E = E^{\circ}_{Ce^{4+}/Ce^{3+}}$ when $[Ce^{4+}] = [Ce^{3+}]$. If the volume of the Ce^{4+} corresponding to the equivalence point of the titration is x ml, then the first condition is satisfied when the volume equals x/2 and the second when the volume equals 2x. These points are indicated in figure 6.5.

Concentration cells

As we have seen, an electrochemical cell can be made out of any two half-cells. If two identical half-cells are connected, for example to give Zn(s) | Zn^{2+}(aq) || Zn^{2+}(aq) | Zn(s), then E° for the cell would be zero, because the standard half-cell potentials are identical. However, it can be seen from the Nernst equation that the e.m.f. of the cell will be non-zero, if the concentration term is not zero. This occurs when the ratio of the reactant and product concentrations is not the same in both half-cells, and the resulting cell is called a concentration cell (figure 6.6).

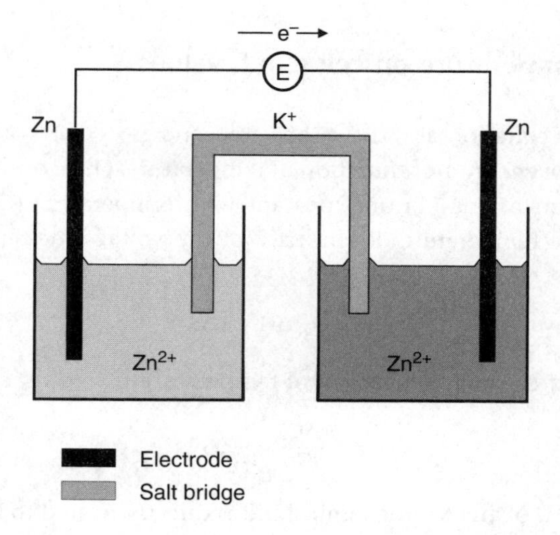

Fig. 6.6 A Zn^{2+} concentration cell consists of two $Zn(s)\,|\,Zn^{2+}(aq)\|$ half-cells. When the concentration of $Zn^{2+}(aq)$ is different in the two half-cells, then electrons flow from one electrode to the other.

WORKED EXAMPLE

Calculate the e.m.f. of a Ag, AgCl concentration cell in which $[Cl^-] = 0{\cdot}1$ M in the left-hand half-cell and 0·01 M in the right-hand half-cell (T = 298 K).

SOLUTION: The half-cell reactions are

$$AgCl(s) + e^- \rightarrow Ag(s) + Cl^-(aq)$$

and the cell reaction is

$$AgCl(s)(R) + Ag(s)(L) + Cl^-(aq)(L) \rightleftharpoons AgCl(s)(L) + Ag(s)(R) + Cl^-(aq)(R)$$

where L and R indicate the left- and right-hand compartments of the cell. The Nernst equation for the cell reaction is

$$E = E^\circ + \frac{RT}{F}\log_e \frac{[AgCl(s)]_R[Ag(s)]_L[Cl^-(aq)]_L}{[AgCl(s)]_L[Ag(s)]_R[Cl^-(aq)]_R}$$

All the solid species are in the standard state and so the Nernst equation reduces to

$$E = E^\circ + \frac{RT}{F}\log_e \frac{[Cl^-(aq)]_L}{[Cl^-(aq)]_R}$$

and since $E^\circ = 0$

$$E = \frac{8{\cdot}31 \times 298}{96\,500}\log_e \frac{0{\cdot}1}{0{\cdot}01} = 0{\cdot}059\ \text{V}$$

Effect of temperature on cell e.m.f. values

The defining equation for the Gibbs free energy, $G = H - TS$, shows that changes in free energy are a function of temperature. One consequence of this is the variation of equilibrium constant with temperature (Chapter 3) and another is the temperature dependence of the voltage measured in electrochemical cells.

$$\Delta G = \Delta H - T\Delta S$$

and if ΔH and ΔS are independent of temperature

$$\frac{\delta(\Delta G)}{\delta T} = -\Delta S$$

Since the e.m.f. of an electrochemical cell is directly related to the free-energy change of the cell reaction

$$\Delta S = -\frac{\delta(\Delta G)}{\delta T} = nF\frac{\delta E}{\delta T}$$

This relationship shows that the entropy change for a reaction can be calculated from the temperature dependence of the e.m.f. of the corresponding electrochemical cell.

Calculation of thermodynamic quantities from electrochemical data

Electrochemical cells are very useful devices for investigating the thermodynamics of cell reactions because they can be operated reversibly. The following relations exist between measurements of E or E° and ΔG, ΔS, ΔH and the equilibrium constant K

$$\Delta G = -nFE$$

$$\Delta S = nF\frac{\delta E}{\delta T}$$

$$\Delta H = \Delta G + T\Delta S = -nFE + nTF\frac{\delta E}{\delta T}$$

$$K = \exp\frac{-\Delta G^{o}}{RT} = \exp\frac{nFE^{o}}{RT}$$

The direct link between the electrochemical measurements and the thermodynamic parameters for a reaction can provide a convenient route to the calculation of such quantities as equilibrium constants. Moreover, as the next example shows, it is not necessary to construct hypothetical cells when seeking to handle such data.

WORKED EXAMPLE

Use the Nernst equation to calculate the equilibrium constant for the oxidation of malate^{2-} by NAD$^+$ given that the E° values for NAD$^+$/NADH and oxaloacetate^{2-}/malate^{2-} are −0·113 and 0·239 V respectively.

SOLUTION: The equilibrium reaction is

$$\text{malate}^{2-}(aq) + NAD^+(aq) \rightleftharpoons \text{oxaloacetate}^{2-}(aq) + NADH(aq) + H^+(aq)$$

In principle we could imagine setting up a hypothetical cell in which the NAD$^+$/NADH reaction occurs on the right-hand side and the oxaloacetate^{2-}/malate^{2-} reaction occurs on the left-hand side, and then work out the equilibrium constant from the E° values. However, an equivalent but simpler approach is to suppose that an inert electrode has been inserted into an equilibrium mixture of the reactants and products. This electrode would act as a sink or source of electrons for each half-reaction in the mixture, and since the mixture is at equilibrium the resulting potential on the electrode would be the same for every half-reaction involved. Accordingly, if E is this potential

$$E = E^{\circ}_{oxal/mal} + \frac{RT}{2F}\log_e\frac{[\text{oxaloacetate}^{2-}][H^+]^2}{[\text{malate}^{2-}]} = E^{\circ}_{NAD^+/NADH} + \frac{RT}{2F}\log_e\frac{[NAD^+][H^+]}{[NADH]}$$

and rearranging gives

$$E^{\circ}_{NAD^+/NADH} - E^{\circ}_{oxal/mal} = \frac{RT}{2F}\log_e\frac{[\text{oxaloacetate}^{2-}][NADH][H^+]}{[\text{malate}^{2-}][NAD^+]} = \frac{RT}{2F}\log_e K$$

where K is the equilibrium constant for the oxidation of malate. Substituting the E° values

$$-0{\cdot}113 - 0{\cdot}239 = \frac{8{\cdot}31 \times 298}{2 \times 96\,500}\log_e K$$

leading to

$$\log_e K = -27{\cdot}43$$
$$K = 1{\cdot}22 \times 10^{-12}\ (M)$$

The effect of non-ideality

So far we have assumed that all components in an electrochemical cell behave ideally. However, as will be discussed in Chapter 8 the assumption of ideality can lead to errors in the thermodynamic analysis of solutions and this is particularly true for electrolyte solutions. Consider the cell

$$Pt,\ H_2(1\,atm)\,|\,H^+,\ Cl^-(\text{conc. } c)\,|\,Hg_2Cl_2,\ Hg$$

The cell reaction is

$$H_2(g) + Hg_2Cl_2(s) \rightleftharpoons 2Hg(l) + 2H^+(aq) + 2Cl^-(aq)$$

and the Nernst equation for the cell e.m.f. takes the form

$$E = E_R - E_L = E^o - \frac{RT}{F} \log_e [H^+][Cl^-]$$

This equation can be used to calculate the apparent E° as a function of the HCl concentration(c) from the measured values of E. The following results are obtained

c (M)	E (V)	E° (V)
10^{-1}	0.4046	0.2866
10^{-2}	0.5099	0.2739
10^{-3}	0.6239	0.2699
10^{-4}	0.7406	0.2686

The calculated value of E° varies becase of the non-ideality of the solution. It is possible to account for this departure from ideality by substituting activities for concentrations in the Nernst equation. The activity coefficients can be calculated from the Debye–Hückel theory for dilute electrolyte solutions (Chapter 8) and when this is done E° has a constant value (0·2680 V). Alternatively the same value can be obtained by extrapolating the E° values in the table to zero ionic strength (i.e. infinite dilution).

Coupled oxidation–reduction processes

It should be clear from this description of the thermodynamics of redox process that e.m.f. values can be thought of as entirely analogous to free-energy changes. So in the same way that ΔG° values for individual reactions can be used to predict the equilibrium position for a coupled reaction, it is also possible to use E° values to calculate the equilibrium constants for reactions of the kind

$$(Ox)_1 + (Red)_2 \rightleftharpoons (Red)_1 + (Ox)_2$$

One of the most important examples of such processes is the mitochondrial electron transport chain, where the presence of several redox components allows electrons to be transferred from NADH to molecular oxygen via a series of intermediate molecules. The electron carriers are arranged randomly in the plane of the membrane, but each has a specific orientation with respect to the inner and outer surfaces to allow proton translocation to accompany electron flow (figure 6.7). At each step in the chain the electrons move to a carrier with a higher redox potential. This corresponds to a spontaneous change with a release of free energy, since $\Delta G = -nFE$, and much of this energy is stored in a proton electrochemical gradient $\Delta\bar{\mu}_{H+}$ across the membrane. This gradient is then used to synthesise ATP in the process of oxidative phosphorylation (see Chapter 7 for more details).

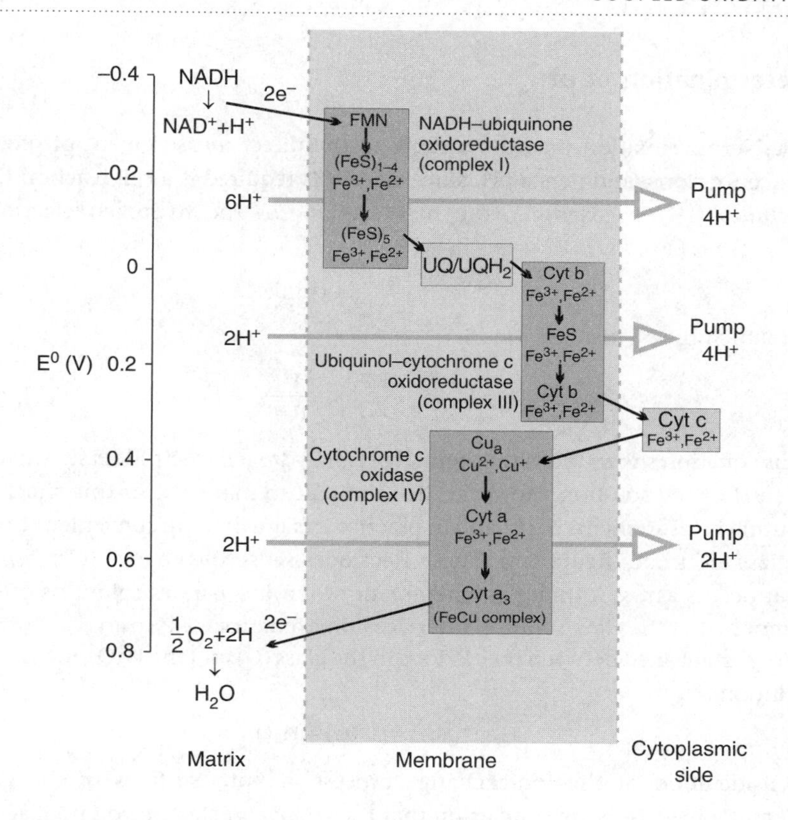

Fig. 6.7 A schematic representation of the mitochondrial electron transport chain. Electrons are transferred from NADH to O_2 by a series of membrane-bound carriers. Most of these carriers are based on the Fe^{3+}, Fe^{2+} redox reaction. The E° of each carrier is such that each successive carrier has a higher E° than the previous one. As electrons pass down this chain, some of the energy released is used to pump protons across the membrane. This stores the released energy as an electrochemical proton gradient (dealt with in more detail in Chapter 7).

WORKED EXAMPLE

Comment on the fact that while E° for Fe^{3+}/Fe^{2+} is 0·77 V, E° for the same redox couple in cytochrome c is 0·22 V.

SOLUTION: E° values are directly related to ΔG° values and hence to the equilibrium position of the redox reaction. Here the redox couple of interest is

$$Fe^{3+}(aq) + e^- \rightarrow Fe^{2+}(aq)$$

and the altered value for E° in the protein indicates that the polypeptide environment alters the relative stability of the +2 and +3 oxidation states of iron. In particular, the lower value for E° in cytochrome c indicates that Fe^{3+} is more stable relative to Fe^{2+} than in a simple aqueous environment, i.e. Fe^{3+} is not such a strong oxidising agent in the microenvironment created by the polypeptide.

COMMENT: This effect is crucial to the operation of the mitochondrial electron transport chain. Several of the redox centres in this pathway include iron, but the differences in the microenvironment of the iron result in several different values for the redox potential (figure 6.7). This allows the energy of the electrons derived from NADH to be released in a number of stages, maximising the energy that can be stored in the proton electrochemical gradient.

Determination of pH

The Nernst equation provides a route to the direct measurement of proton concentrations, and hence pH, since all that is required is a cell reaction that includes H^+. For example, the half-reaction for the hydrogen electrode, $H^+(aq) \mid H_2(g)$, Pt, is

$$2H^+(aq) + 2e^- \rightarrow H_2(g)$$

for which the Nernst equation is

$$E = E^\circ_{H^+/H_2} + \frac{RT}{2F}\log_e \frac{[H^+]^2}{p_{H_2}}$$

This equation shows that the potential of the hydrogen electrode is sensitive to the pH of the solution. However, it is difficult to make use of this effect for routine measurements of pH and in practice it is much more convenient to use a glass electrode (figure 6.8). These electrodes sense the change in potential that occurs across a thin glass membrane when it separates solutions of different pH. At the glass/solution interface, ion exchange leads to an equilibrium being established between the H^+ ions in the glass (G) and the H_3O^+ ions in the solution

$$H_3O^+(aq) \rightleftharpoons H^+(G) + H_2O$$

Consideration of the ion-exchange process at both surfaces of the glass membrane leads to the conclusion that the change in electric potential across the membrane is directly proportional to the difference in the pH of the two solutions.

The change in potential across the glass membrane is measured by combining the glass electrode with a reference electrode

Ag(s), AgCl(s) | HCl(aq) | glass | solution of unknown pH | reference electrode

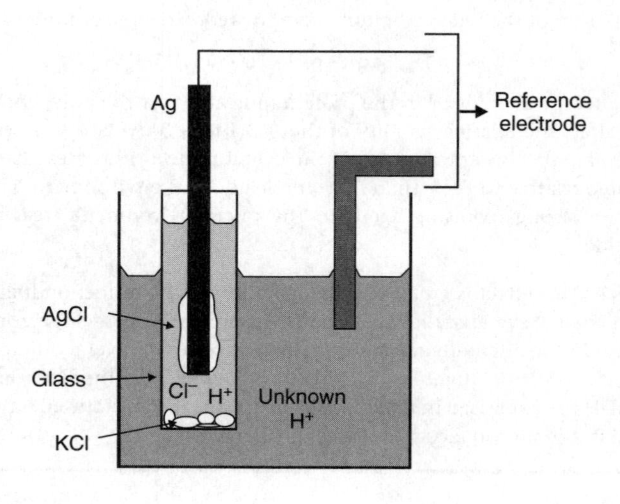

Fig. 6.8 Schematic diagram of a pH electrode. A AgCl(s), Ag(s) | Cl^-(aq) electrode in glass is put into the solution of interest and connected to a reference electrode. The H^+(aq) in solution equilibrates with the glass, leading to a change in the e.m.f. of the cell that is proportional to $[H^+(aq)]$. The solid KCl is present to keep the concentration of Cl^- ions constant within the AgCl(s), Ag(s) | Cl^-(aq) electrode.

The reference electrode is usually a calomel electrode, and the two electrodes of the cell are often combined in a single unit that is known as a combination electrode. The e.m.f. of the cell depends on the pH values of the HCl solution and the unknown solution, and a change in the pH of the unknown solution (ΔpH) leads to a change in e.m.f. (ΔE) given by

$$\Delta E = \frac{2{\cdot}303RT\Delta pH}{F}$$

Thus, the change in e.m.f. is directly proportional to ΔpH and it is a straightforward matter to calibrate the voltmeter so that it reads out the pH of the solution.

FURTHER READING

P.W. Atkins, 1998, 'Physical chemistry', Oxford University Press—*excellent coverage of electrochemistry from a chemical perspective.*

R.G. Compton and G.H.W. Sanders, 1996, 'Electrode potentials', Oxford University Press—*a useful introduction to electrode potentials.*

D.G. Nicholls and S.J. Ferguson, 1992, 'Bioenergetics 2', Academic Press—*the standard textbook on bioenergetics, putting redox potentials into their biochemical context.*

PROBLEMS

1. Write down the redox reactions for the following half-cells.

(a) Zn^{2+} | Zn
(b) H^+ | H_2, Pt
(c) Pt | Co^{3+}, Co^{2+}
(d) AgBr, Ag | Br^-
(e) Hg_2Cl_2, Hg | Cl^- (the calomel electrode)
(f) Pt | fumarate, H^+, succinate^{2-}
(g) Pt | cytochrome c (Fe^{3+}), cytochrome c (Fe^{2+})
(h) Pt | CO_2, H^+, formate$^-$
(i) Pt | NAD^+, H^+, NADH

2. Write down the half-cell reactions and the overall cell reactions for each of the following electrochemical cells.

(a) Cu | Cu^{2+} || Zn^{2+} | Zn
(b) Pt, H_2 | H^+ || Ag^+ | Ag
(c) Pt, H_2 | HCl | AgCl, Ag
(d) Pt, H_2 | H^+ || Fe^{3+}, Fe^{2+} | Pt
(e) Pt | NAD^+, H^+, NADH || oxaloacetate^{2-}, H^+, malate^{2-} | Pt

Calculate E° for cells (a) –(d) and E°′ for cell (e) from the following redox potentials.

$E^{\circ}_{Cu^{2+}/Cu} = 0{\cdot}34\,V$ $E^{\circ}_{Zn^{2+}/Zn} = -0{\cdot}76\,V$ $E^{\circ}_{Ag^{+}/Ag} = 0{\cdot}80\,V$
$E^{\circ}_{Cl^{-}/AgCl,\,Ag} = -0{\cdot}22\,V$ $E^{\circ}_{Fe^{3+},\,Fe^{2+}/Pt} = 0{\cdot}77\,V$ $E^{\circ\prime}_{oxal,\,mal/Pt} = -0{\cdot}17\,V$
$E^{\circ\prime}_{NAD^{+},\,NADH/Pt} = -0{\cdot}32\,V$

3. Write down the electrochemical cells corresponding to the following reactions.

(a) $Sn^{2+} + Pb \rightleftharpoons Pb^{2+} + Sn$
(b) $lactate^{-} + NAD^{+} \rightleftharpoons pyruvate^{-} + NADH + H^{+}$

4. Calculate the standard e.m.f. of the electrochemical cell corresponding to the reaction

$$Fe + Cu^{2+} \rightleftharpoons Fe^{2+} + Cu$$

given that $\Delta H^{\circ} = -148{\cdot}8\,kJ\,mol^{-1}$ and $\Delta S^{\circ} = 8{\cdot}8\,J\,K^{-1}\,mol^{-1}$ for the reaction at 298 K.

5. Write down the cell reaction for the following cell

$$Zn\,|\,Zn^{2+}\,||\,Fe^{3+},\,Fe^{2+}\,|\,Pt$$

Calculate values of ΔG°, ΔS° and ΔH° given that the standard e.m.f. of the cell is 1·53 V at 298 K and 1·55 V at 323 K. State the assumptions involved in your calculations.

6. The standard electrode potentials (E°′) for the acetaldehyde/ethanol and $pyruvate^{-}/lactate^{-}$ half-cells are −0·163 V and −0·190 V respectively.

A solution at 30°C initially contains the following reactants

ethanol	100 mM
pyruvate	100 mM
lactate	10 mM
acetaldehyde	1 mM

(a) Predict the direction of the redox reaction that occurs in this solution.
(b) Calculate the concentrations of the four components once equilibrium has been established at pH = 7·0.

7. The standard redox potential for the cytochrome c Fe^{3+}/Fe^{2+} couple is 0·21 V. Use the Nernst equation to calculate the equilibrium ratio of the oxidised and reduced forms of cytochrome c at the following potentials: 0·1, 0·15, 0·2, 0·25 and 0·3 V (T = 298 K).

8. Calculate the E°′ value for the $oxaloacetate^{2-}/malate^{2-}$ redox couple at a temperature of 300 K given that the E° value is 0·239 V.

9. Use the Nernst equation to calculate the equilibrium constant for the oxidation of NADH by molecular oxygen given the following E° values (T = 303 K).

$$O_2 + 2H^+ + 2e^- \rightarrow H_2O_2 \quad E^\circ = 0{\cdot}682\ V$$
$$H_2O_2 + 2H^+ + 2e^- \rightarrow 2H_2O \quad E^\circ = 1{\cdot}776\ V$$
$$NAD^+ + H^+ + 2e^- \rightarrow NADH \quad E^\circ = -0{\cdot}11\ V$$

10. The following oxidation–reduction couples are involved in the photosynthetic chain

$$\text{Cyt-b}(Fe^{3+}),\ \text{Cyt-b}(Fe^{2+}) \quad E^{\circ\prime} = 0{\cdot}06\ V$$
$$\text{Cyt-f}(Fe^{3+}),\ \text{Cyt-f}(Fe^{2+}) \quad E^{\circ\prime} = 0{\cdot}36\ V$$

(a) What is $\Delta G^{o\prime}$ for the coupled reaction between Cyt-b(Fe^{2+}) and Cyt-f(Fe^{3+})?
(b) Does this reaction in the standard state provide sufficient free energy to drive the formation of ATP from ADP, given that $\Delta G^{o\prime}$ for the hydrolysis of ATP is $-30{\cdot}5\ kJ\ mol^{-1}$?
(c) How would your conclusion be altered if we considered two electrons flowing from Cyt-b to Cyt-f?
(d) Is the above analysis likely to be meaningful in the context of ATP production within a real cell?

[We will return to the analysis of this system in the problems at the end of Chapter 7.]

7 Chemical potentials and the properties of solutions

KEY POINTS

- The chemical potential of the solvent in a solution can be used to evaluate the *colligative* properties of the solution (depression of freezing point, elevation of boiling point and osmotic pressure).
- The *water potential* of a solution, defined in terms of the chemical potentials of pure water and the solution, describes the movement of water in plants.
- The chemical potential of solute species can be used to explain the distribution of ions across a membrane and how this is influenced by differences in pH and by the presence of charged macromolecules such as proteins.
- Electrochemical gradients of ions across membranes can store energy. For example, H^+ gradients provide the key link between respiration and ATP synthesis in the process of *oxidative phosphorylation.*

The thermodynamic properties of solutions can be analysed by using chemical potentials to assess the stability of each component in the solution. This stability is a function of the concentration of the component, and so the starting point for the analysis is the equation that describes the concentration dependence of the chemical potential (μ_x) of a component x in an ideal solution.

$$\mu_x = \mu_x^o + RT \log_e[x]$$

This equation was introduced in Chapter 3 and a rigorous derivation will be given in Chapter 8. μ_x^o is the chemical potential of the component x under standard conditions and [x] is the molar concentration in the solution of interest. Equations of this kind can be written for each component in the solution, and we shall use the symbol i to represent the solvent and j, k, etc., to represent the solutes.

The concentration of i in a solution can also be expressed in terms of its mole fraction (N_i)

$$N_i = \frac{n_i}{n_i + n_j + n_k + \cdots}$$

where n_i, n_j, n_k, etc., are the number of moles of each component in the solution. Using N_i as a measure of concentration allows the concentration dependence of the chemical potential of i to be expressed in the form

$$\mu_i = \mu_i^o + RT \log_e N_i$$

This equation is an alternative to the one based on the molar concentration and is particularly useful for considering the thermodynamic properties of the *solvent* in a solution.

It is important to recognise that the equations for the concentration dependence of μ_i provide a quantitative description of the concentration dependence of the stability of i in the solution. The higher the value of μ_i, the lower the stability of i, and spontaneous changes in i will be associated with a decrease in μ_i. Thus inspection of the equation above shows that when a solute (j) is added to a solvent (i), N_i becomes less than 1 and μ_i becomes less than μ_i^o. Thus dissolving a solute in a solvent to form an ideal solution stabilises the solvent. The reason for this is that the entropy of the solvent increases as it is mixed with the solute to form the ideal solution.

Colligative properties

The stabilisation of the solvent in an ideal solution leads to a number of measurable changes in the properties of the solvent. These changes do not depend on the identity of the solute because the reduction in N_i only depends on the number of solute molecules that are added to the solution. Properties of this kind are known as colligative properties and the two most familiar examples are the changes that occur in the freezing and boiling points of the solvent.

At the freezing point of a solvent i

$$\mu_i(s) = \mu_i(l) = \mu_i^o + RT \log_e x_i$$

The presence of the solute in the solution has no effect on the chemical potential of i in the solid phase, but it does reduce the chemical potential of i in the liquid phase, and consideration of figure 7.1 shows that the effect of this is to depress the freezing point of the solution. Similarly, the presence of the solute has no effect on the chemical potential of i in the gas phase and so the effect of the solute is to raise the boiling point of the solution. The depression of the freezing point has some biological significance, since it contributes to the avoidance of freezing in organisms that have a temperature that varies with the temperature of the surroundings, but a colligative property of much wider biological significance is the property known as osmosis.

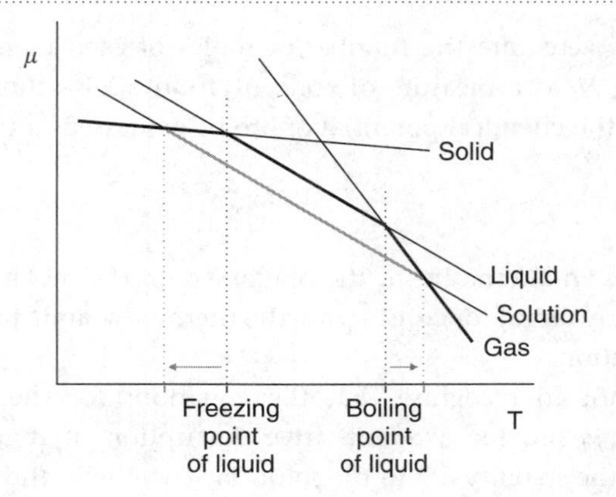

Fig. 7.1 A plot of solvent chemical potential, μ, versus temperature, T, for a solid, pure liquid, ideal solution and gas. The presence of the solute reduces the mole fraction of the solvent and hence reduces its chemical potential. The arrows indicate the effect of the solute on the freezing and boiling points of the solution.

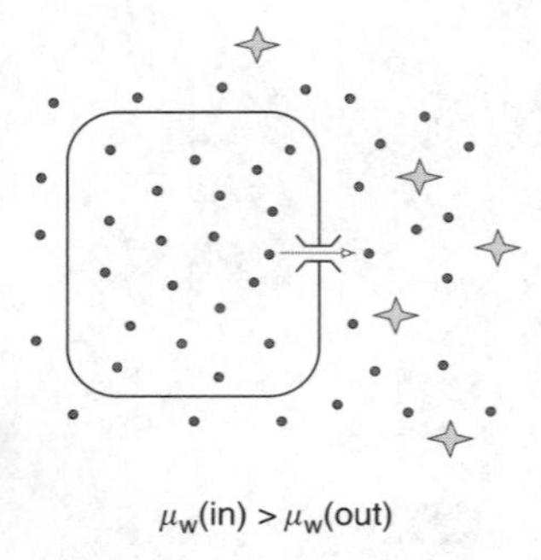

Fig. 7.2 A schematic representation of a two-compartment system, with pure water (•) on the inside and water plus a solute (✧) on the outside. The arrow indicates the direction of the net flux of the water.

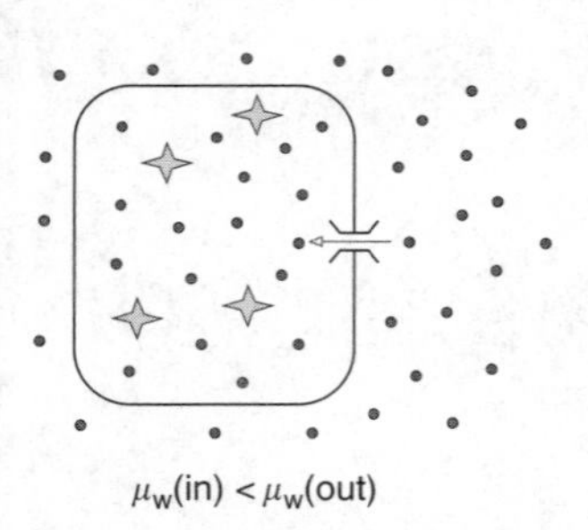

Fig. 7.3 A schematic representation of a two-compartment system, with water (•) plus a solute (✧) on the inside and pure water on the outside. The arrow indicates the direction of the net flux of the water.

Osmosis

Water moves spontaneously from regions of high chemical potential to low chemical potential, and one manifestation of this is osmosis, the net flux of water that can occur across membranes. Consider a two-compartment system in which the contents of the inner compartment are separated from the outer compartment by a barrier that is only permeable to water, and suppose that the inner compartment contains pure water while the outer compartment contains an aqueous solution of some unspecified solute (figure 7.2).

In the inner compartment, the chemical potential of the water (μ_w) is given by

$$\mu_w(\text{in}) = \mu_w^o(\text{l})$$

while in the outer compartment

$$\mu_w(\text{out}) = \mu_w^o(\text{l}) + \text{RT} \log_e N_w$$

Since N_w is less than 1,

$$\mu_w(\text{in}) > \mu_w(\text{out})$$

allowing a net flow of water from the inner compartment to the outer compartment. This spontaneous process will only stop when

$$\mu_w(\text{in}) = \mu_w(\text{out})$$

but since this condition can never be satisfied, no matter how dilute the solution in the outer compartment becomes, the water efflux will continue until the inner compartment is empty.

Now suppose that the inner compartment contains an aqueous solution of an unspecified solute while the outer compartment contains pure water (figure 7.3). This is exactly the reverse of the first situation, and consideration of the chemical potential of the water in the two compartments shows that

$$\mu_w(\text{out}) > \mu_w(\text{in})$$

In this situation the net flow of water occurs from the outer to the inner compartment, and it will continue until all the water has moved into the inner compartment or until the membrane bursts.

The chemical potential of water has an important bearing on the physiology of living systems. In the kidney, for example, the reabsorption of water from the urine depends on establishing a chemical potential gradient that favours the spontaneous movement of water from the lumen of the renal tubule. More generally, the permeability of cell membranes to water means that organisms have had to develop a variety of strategies for controlling the water content of their cells, since the uncontrolled efflux of water from a cell will lead to plasmolysis and the uncontrolled influx of water will lead to the cell bursting. In most cases the aim is to balance water influx and water efflux by eliminating the chemical potential difference for water across the membrane, and this can be achieved either by controlling the composition of the internal and external volumes or by enclosing the cell in a rigid cell wall. The first approach is an important strategy for mammalian cells, while the second approach is very important for plants and bacteria.

Osmotic pressure

So far we have assumed that the pressure in the two compartments is the same. However, if the membrane is inelastic, then the movement of water into the inner compartment in figure 7.3 will increase the pressure inside the membrane and this will oppose the continued influx of the solvent.

Suppose that the system equilibrates when the pressure reaches $(\Pi + 1)$ atm (figure 7.4). At equilibrium

$$\mu_w(\text{out}) = \mu_w(\text{in})$$

and since

$$\mu_w(\text{out}) = \mu_w^\circ(\text{l}, 1\,\text{atm})$$

and

$$\mu_w(\text{in}) = \mu_w^\circ(\text{l}, \Pi + 1\,\text{atm}) + \text{RT}\log_e N_w$$

it follows that

$$\mu_w^\circ(\text{l}, 1\,\text{atm}) = \mu_w^\circ(\text{l}, \Pi + 1\,\text{atm}) + \text{RT}\log_e N_w$$

To proceed further it is necessary to consider the pressure dependence of the chemical potential. As we shall see in Chapter 8, the defining equation for the

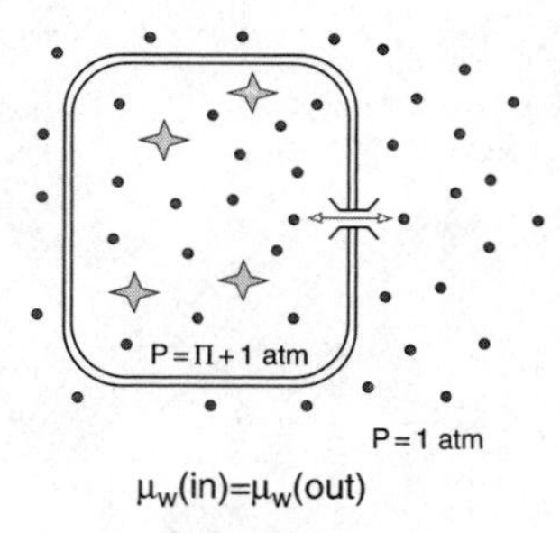

Fig. 7.4 A schematic representation of a two-compartment system, with water (•) plus a solute (✧) on the inside and pure water on the outside. The inside is at a higher pressure than the outside, sufficient to prevent further influx of water down its concentration gradient. Under these conditions, equilibrium is established with no net flux of water.

Gibbs free energy, G = H − TS, leads to the following expression for the pressure dependence of G at constant temperature

$$\delta G = V\delta P \quad \text{or} \quad \left(\frac{\delta G}{\delta P}\right)_T = V$$

The analogous equation for the standard chemical potential of water is

$$\left(\frac{\delta \mu_w^o}{\delta P}\right)_T = \bar{V}_w$$

where $\bar{V}_w$ is the partial molar volume of water in the system of interest. $\bar{V}_w$ is equal to the volume of 1 mole of water for a system comprising pure water, and it is often very close to this value for a solution.

This equation can be integrated

$$\int_1^{\Pi+1} d\mu_w^o = \bar{V}_w \int_1^{\Pi+1} dP$$

to give

$$\mu_w^o(1, \Pi + 1\,\text{atm}) = \mu_w^o(1, 1\,\text{atm}) + \Pi\bar{V}_w$$

This equation can now be substituted into the equilibrium condition to give

$$\Pi\bar{V}_w = -RT\log_e N_w$$

This equation relates the excess pressure in the inner compartment, Π, the so-called osmotic pressure, to the concentration of the solution in the same compartment. However, it is not very convenient to express the concentration in terms of the mole fraction of the solvent, and so it is usual to rearrange it in the following way.

First, from the definition of the mole fraction

$$N_w = 1 - N_j$$

where N_j is the mole fraction of the solute, and if the solution is dilute then

$$\Pi\bar{V}_w = -RT\log_e N_w \equiv -RT\log_e(1 - N_j) \approx RTN_j$$

Moreover, for a dilute solution

$$N_j = \frac{n_j}{n_w + n_j} \approx \frac{n_j}{n_w}$$

and

$$n_w\bar{V}_w = V$$

where V is the volume of the solution. Substituting in $\Pi\bar{V}_w = RTN_j$ gives

$$\Pi V = n_j RT$$

which can be rewritten as

$$\Pi = c_j RT$$

where c_j is the molar concentration of the solute in the solution.

This equation is known as the van't Hoff equation, and it provides a direct relation between the concentration of the solution in the inner compartment and the excess pressure that is required to prevent the movement of water into the solution from the outer compartment. This pressure is readily measured in an osmometer and when combined with the van't Hoff equation, such measurements can be used to determine the molecular weight of the solute.

WORKED EXAMPLE

Calculate the osmotic pressures of 0·1 M solutions of mannitol, sodium chloride and potassium sulphate at 20°C.

SOLUTION: According to the van't Hoff equation, $\Pi = c_j RT$ where c_j is the sum of the molar concentrations of the solutes. For the mannitol solution, $c_j = 0{\cdot}1$ M and so

$$\Pi = 0{\cdot}1 \times 8{\cdot}314 \times 293 = 243{\cdot}6\,\text{kPa}$$

For the sodium chloride solution, $c_j = 0{\cdot}2$ M, since each ion contributes separately to the total solute concentration, and so

$$\Pi = 0{\cdot}2 \times 8{\cdot}314 \times 293 = 487{\cdot}2\,\text{kPa}$$

For the potassium sulphate (K_2SO_4) solution, $c_j = 0{\cdot}3$ M and so

$$\Pi = 0{\cdot}3 \times 8{\cdot}314 \times 293 = 730{\cdot}8\,\text{kPa}$$

COMMENT: It is important to remember that osmotic pressure is a colligative property and so depends on the total number of solute species in solution. It is therefore necessary to allow for the dissociation of salts in these solutions. The answer comes out in kPa when c_j is expressed in mol l^{-1} because $1\,\text{Pa} \equiv 1\,\text{N m}^{-2} \equiv 1\,\text{J m}^{-3}$. Pascals are relatively unfamiliar and it may be helpful to note that 1 atm $\sim 0{\cdot}1$ MPa.

WORKED EXAMPLE

The following osmotic pressure data were obtained at 278 K for a protein dissolved at its isoelectric point.

Protein concentration/g dm^{-3}	7·3	18·4	27·6	42·1	57·4
Osmotic pressure/kPa	0·211	0·533	0·804	1·236	1·701

Calculate the molecular weight of the protein.

SOLUTION: If the protein concentration in g dm^{-3} is denoted x and the molecular weight M, then the van't Hoff equation for the osmotic pressure can be written as

$$\Pi = \frac{x}{M}RT$$

Plotting Π against x gives a straight line with a gradient of $2{\cdot}92 \times 10^{-2}$ kPa dm^3 g^{-1}. This gradient is given by RT/M and so

$$M = \frac{8{\cdot}314 \times 278}{2{\cdot}92 \times 10^{-2}} = 79\,000\ \text{Da}$$

COMMENT: The protein carries a net charge of zero at the isoelectric point (see Chapter 5) and so it is not necessary to consider the contribution of any counter ions to the measured values of the osmotic pressure.

Water potentials

Water movement through plants can also be analysed in terms of the chemical potential, but plant physiologists prefer to use a quantity known as the water potential, Ψ. This quantity has the units of pressure (megapascals; MPa) and is defined by

$$\Psi = \frac{\mu_W - \mu_W^{\circ}}{\bar{V}_W}$$

where μ_W is the chemical potential of the water fraction of interest, μ_W° is the chemical potential of pure water, and $\bar{V}_W$ is the partial molar volume of water in the solution. Water flows through the vascular tissues from regions of high Ψ to low Ψ, and the thermodynamic driving force $\Delta\Psi$ is determined by the factors that influence μ_W. Thus the concentration of solutes, the hydrostatic pressure (P; conventionally expressed as the difference between the actual hydrostatic pressure and atmospheric pressure), and the effect of the gravitational field all have to be taken into account in the thermodynamic analysis of the movement of water from the roots to the leaves of a plant. The hydrostatic pressure, otherwise known as the turgor pressure, is the internal pressure that the cell exerts on the surrounding cell wall.

Consideration of these factors leads to the equations

$$\mu_W = \mu_W^{\circ} - \bar{V}_W\Pi + \bar{V}_W P + m_W gh$$

and

$$\Psi = P - \Pi + \rho_W gh$$

where m_W is the molar mass of water, g is the acceleration due to gravity, h is the height relative to an arbitrary reference point, and ρ_W is the density of water. The gravitational term is particularly significant for the movement of water in trees, the gravitational contribution to the water potential increases by ~0·1 MPa when water moves upwards by 10 m.

WORKED EXAMPLE

Typical measurements indicate that the water potential in xylem vessels of a root near the surface of the ground is −0·6 MPa, whereas the water potential in the xylem vessels of a leaf at a height of 10 m above the ground is −0·8 MPa. Comment on these values.

SOLUTION: The first point is that the water potential in the leaves is lower than in the roots and this is consistent with the upward movement of water in the transpiration system. The second point is that this occurs despite the unfavourable gravitational contribution to Ψ. The gravitational contribution is given by $\rho_w gh$, and substituting the values $\rho_w = 1000\,\mathrm{kg\,m^{-3}}$, $g = 9{\cdot}81\,\mathrm{m\,s^{-2}}$ and $h = 10\,\mathrm{m}$ gives

$$\rho_w gh = 1000 \times 9{\cdot}81 \times 10 \approx 10^5\,\mathrm{kg\,m^{-1}s^{-2}} \equiv 0{\cdot}1\,\mathrm{MPa}$$

Thus the observed difference in the water potential values indicates that there must be a decrease of 0·3 MPa in the term $(p - \Pi)$.

COMMENT: There is unlikely to be any difference in the composition of the xylem between the base of the plant and the leaf, and therefore no difference in Π. Thus the driving force for the upward movement of the water in the stem must come from a decrease in the hydrostatic pressure. This negative gradient in P is established through the evaporation of water from the leaf surface.

Chemical potential of the solute

So far the discussion has been restricted to the properties of the solvent in a solution, but it is a straightforward matter to extend the analysis to the solutes because the concentration dependence of the chemical potential of a solute (μ_j) in a solution takes exactly the same form as the corresponding equation for the solvent. However, it turns out to be more convenient to use [j] as a measure of concentration, rather than N_j, and so the concentration dependence of the chemical potential of a solute is usually expressed as

$$\mu_j = \mu_j^o + RT \log_e[j]$$

where [j] is the molar concentration. Note that as usual in thermodynamic equations, [j] has to be expressed relative to the standard state, which in this case corresponds to the solute in an ideal solution at a concentration of 1 M. It follows that the concentration of j should always be converted to molar units before it is inserted into the equation.

The chemical potential of the solute can be used to assess the stability of the solute in a compartmented system in exactly the same way as was done for the solvent. Thus if

$$\mu_j(\mathrm{in}) = \mu_j(\mathrm{out})$$

in a two-compartment system in which the inner compartment is bounded by a membrane that is permeable to j, then the system is at equilibrium and there is no tendency for a net flux across the membrane. Alternatively if

$$\mu_j(\text{in}) > \mu_j(\text{out})$$

then the thermodynamic driving force causes a net efflux of j from the inner compartment, and if

$$\mu_j(\text{out}) > \mu_j(\text{in})$$

then the net flux is reversed and j moves into the inner compartment.

Determination of pH using permeable weak acids and bases

One method for determining intracellular pH involves measuring the equilibrium distribution of a radioactively labelled weak acid (HA) across the plasma membrane. This method is particularly convenient for measuring the cytoplasmic pH in simple unicellular organisms and it is based on the assumption that the membrane is permeable to HA, but impermeable to H^+ or A^-. Thus at equilibrium

$$\mu_{HA}(\text{in}) = \mu_{HA}(\text{out})$$

and since $\mu_{HA} = \mu^o_{HA} + RT\log_e[HA]$ it follows that

$$[HA(\text{in})] = [HA(\text{out})]$$

Now from the definition of the acid dissociation constant

$$[HA(\text{in})] = \frac{[H^+(\text{in})][A^-(\text{in})]}{K_a}$$

and since

$$[A^-] = [\text{Total acid}] - [HA]$$

$$[HA(\text{in})] = \frac{[H^+(\text{in})][\text{Total acid(in)}]}{K_a + [H^+(\text{in})]}$$

An identical equation can be written for [HA(out)] and this allows the equilibrium condition to be rewritten as

$$\frac{[H^+(\text{in})][\text{Total acid(in)}]}{K_a + [H^+(\text{in})]} = \frac{[H^+(\text{out})][\text{Total acid(out)}]}{K_a + [H^+(\text{out})]}$$

which can be rearranged to give an expression for the intracellular proton concentration

$$[H^+(\text{in})] = \frac{[H^+(\text{out})][\text{Total acid(out)}]K_a}{(K_a + [H^+(\text{out})])[\text{Total acid(in)}] - [H^+(\text{out})][\text{Total acid(out)}]}$$

The important point about this ungainly expression is that everything on the right-hand side of the equation is either known or can be measured. Thus K_a is known, $[H^+(out)]$ can be measured with a pH electrode, and the intracellular and extracellular values of [Total acid] can be determined by spinning down the cells and measuring the radioactivity of the pellet and suspending medium. Thus the equilibration of the weak acid allows the determination of the intracellular pH.

The expression for the intracellular proton concentration can be simplified if a weak acid is chosen with a pK_a value that is much lower than the intracellular and extracellular pH values. In this situation

$$K_a \gg [H^+(in)], [H^+(out)]$$

and the equation for $[H^+(in)]$ reduces to

$$[H^+(in)] = \frac{[\text{Total acid(out)}][H^+(out)]}{[\text{Total acid(in)}]}$$

This method is one of several methods for measuring intracellular pH (see Chapter 5) and it works well provided that the underlying assumptions are satisfied. These assumptions are: (i) that HA is the only permeable species; (ii) that there is no active transport of HA; (iii) that HA does not bind to any component in the intracellular or extracellular solutions; (iv) that HA is not metabolised; (v) that HA does not have any toxic effect on the cells; and (vi) that the effective acid dissociation constant is the same in the intracellular and extracellular solutions. The method is also much less useful if the cell is subdivided into compartments with different pH values.

WORKED EXAMPLE

A 1 ml suspension of rat liver mitochondria in 0·25 M sucrose at pH 6·0 was incubated with succinate and [^{3}H]acetate for 2 minutes, and then the mitochondria were rapidly separated from the incubation medium. Subsequent analysis showed that one-sixth of the radiolabel was present in the mitochondria. Calculate the pH gradient across the mitochondrial membrane, given that the mitochondria contained 5 mg protein and the mitochondrial volume was 0·4 ml g^{-1} protein.

SOLUTION: The internal volume of the mitochondria is $0{\cdot}4 \times 0{\cdot}005 = 0{\cdot}002$ ml and this volume contains one-sixth of the acetate. The remaining five-sixths of the acetate are in the suspending medium, which has a volume of $1 - 0{\cdot}002 = 0{\cdot}998$ ml. Substituting this information into the simplified equation for $[H^+(in)]$ gives

$$[H^+(in)] = \frac{5 \times 0{\cdot}002 \times 6}{6 \times 0{\cdot}998}[H^+(out)] = 0{\cdot}01[H^+(out)]$$

Thus the internal proton concentration is only one-hundredth of the external concentration, indicating a pH gradient of 2 pH units and a mitochondrial pH of 8·0.

COMMENT: Measurements of this kind provide important information on the contribution of the pH gradient across the inner mitochondrial membrane to the overall proton motive force that drives ATP synthesis in oxidative phosphorylation (see below).

Equilibration of mobile solutes in the presence of charged macromolecules: the Donnan effect

The equilibration of ions across a membrane is influenced by the charges carried by impermeable macromolecules. This effect is known as the Donnan or Gibbs–Donnan effect and it can be important in studies of macromolecules that involve dialysis or osmotic pressure measurements.

Consider the equilibration of a system comprising two compartments, one, the outer compartment, containing a solution of sodium chloride, and the other, the inner compartment, containing the sodium salt of a macromolecule (figure 7.5). If the membrane surrounding the inner compartment is impermeable to the macromolecule, then the only ions that can equilibrate across the membrane are Na^+ and Cl^- and so at equilibrium

$$\mu_{Na^+}(\text{in}) + \mu_{Cl^-}(\text{in}) = \mu_{Na^+}(\text{out}) + \mu_{Cl^-}(\text{out})$$

Substituting the expression for the concentration dependence of the chemical potential leads to

$$[Na^+(\text{in})]_{eq}[Cl^-(\text{in})]_{eq} = [Na^+(\text{out})]_{eq}[Cl^-(\text{out})]_{eq}$$

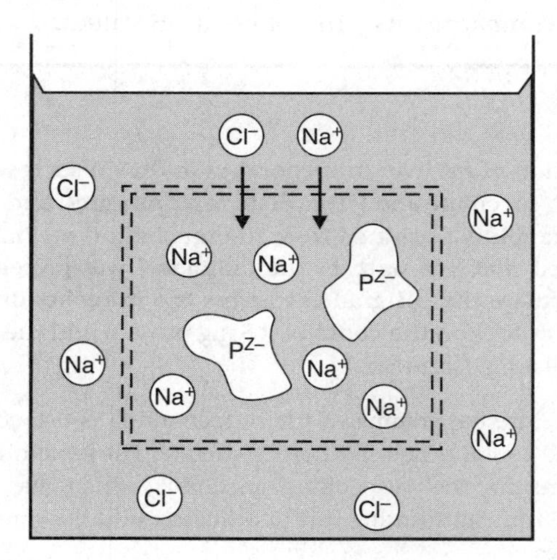

Fig. 7.5 Initially, the inner compartment contains the Na salt of a protein and the outer compartment contains NaCl. These compartments are separated by a membrane permeable to Na^+ and Cl^- but not to the protein. Cl^- ions will move from outside to in, down their concentration gradient. To preserve electroneutrality, one Na^+ ion will also have to move from outside to in with each Cl^- ion (shown by the arrows). At equilibrium, there will be a Cl^- ion concentration gradient with $[Cl^-(\text{out})] > [Cl^-(\text{in})]$ and an equal and opposite Na^+ ion concentration gradient with $[Na^+(\text{out})] < [Na^+(\text{in})]$.

This equation describes the relationship between the concentrations of the sodium and chloride ions in the two compartments at equilibrium, and it can be used to calculate the redistribution of the ions as the system equilibrates.

Suppose that the initial concentration of the sodium chloride in the outer compartment is x_2 M and that the concentration of the macromolecule, with its negative charge of z_m, is x_1 M. Equilibration occurs by the movement of equal numbers of Na^+ and Cl^- ions in the same direction, preserving electroneutrality (figure 7.5), and suppose that this reduces the sodium chloride concentration in the outer compartment by y M. This leads to the following equilibrium concentrations

$$[Na^+(in)]_{eq} = z_m x_1 + y \qquad [Na^+(out)]_{eq} = x_2 - y$$
$$[Cl^-(in)]_{eq} = y \qquad [Cl^-(out)]_{eq} = x_2 - y$$

Substituting these values in the equation linking the equilibrium concentrations leads to

$$y = \frac{x_2^2}{z_m x_1 + 2x_2}$$

This expression allows the equilibrium concentrations to be determined in terms of the initial composition of the two compartments. It can also be used to show

$$\frac{[Cl^-(out)]_{eq}}{[Cl^-(in)]_{eq}} = 1 + \frac{z_m x_1}{x_2} \equiv 1 + \frac{[Na^+(in)]_{initial}}{[Na^+(out)]_{initial}}$$

In the absence of the charged macromolecule, $[Cl^-(in)]_{eq}$ would equal $[Cl^-(out)]_{eq}$ and the ratio would be 1; but in the presence of the charged macromolecule, the Donnan effect occurs and the ratio differs from 1, i.e. the partitioning of the mobile ions in the system is influenced by the charge on the impermeable macromolecules.

WORKED EXAMPLE

Calculate the equilibrium concentrations for the system illustrated in figure 7.5 if the macromolecule carries a net negative charge of 3 and if the values of x_1 and x_2 are 3 mM and 50 mM respectively.

SOLUTION: As above, assume that equilibration occurs by the movement of y mM of sodium chloride into the compartment containing the macromolecule. Substituting the given information into

$$y = \frac{x_2^2}{z_m x_1 + 2x_2}$$

gives y = 22·94 mM. Hence at equilibrium: $[Na^+(in)]_{eq}$ = 31·94 mM; $[Na^+(out)]_{eq}$ = 27·06 mM; $[Cl^-(in)]_{eq}$ = 22·94 mM; and $[Cl^-(out)]_{eq}$ = 27·06 mM.

The Donnan effect is a potential complication in any experimental procedure involving the equilibration of ions across semipermeable membranes in the presence of charged macromolecules. For example, the binding of small charged ligands to proteins can be investigated using equilibrium dialysis. A solution containing the protein is placed on one side of a membrane, and a solution of the ligand is added to the other side. After equilibration, the total amount of ligand in the two compartments is measured. Assuming that the concentration of unbound ligand is the same on both sides of the membrane, the difference between the two measurements gives the amount of ligand bound to the protein. However, it is important to realise that the Donnan effect, as well as ligand binding, can influence the distribution of the ligand across the dialysis membrane. Fortunately, the size of the Donnan effect can be reduced by using a low protein concentration and a high ligand concentration, and it can be eliminated completely by conducting the experiment at the pH at which the protein carries no net charge, i.e. at the isoelectric point, assuming that the protein still binds to the ligand at this pH.

Similar considerations apply to osmotic pressure measurements on solutions of macromolecules, since differences in the ionic concentrations across the membrane will themselves contribute to Π. Accordingly, molecular weight determinations of proteins by this method should be carried out at the isoelectric point.

Charged solutes and electric fields

The chemical potential of a component in a solution is a measure of the stability of the component relative to its standard state. Several factors can influence the stability and so far we have considered the effects of concentration, hydrostatic pressure and the gravitational field. However, there is a further factor that plays a major role in determining the thermodynamic properties of ions in biological systems, and this is the effect of an electric field.

Cations are destabilised in regions of positive electric potential and stabilised in regions of negative electric potential, while anions are stabilised in regions of positive electric potential and destabilised in regions of negative electric potential (figure 7.6). This effect can be taken into account by adding an electrical work term to the expression for the chemical potential, to give

$$\bar{\mu}_j = \mu_j + z_j F \Phi$$

where $\bar{\mu}_j$, the electrochemical potential, is a new molar free energy term, z_j is the charge on the solute j, F is a constant, the Faraday, equal to 96 490 C mol^{-1}, and Φ is the electric potential. This equation may be rewritten in an expanded form

$$\boxed{\bar{\mu}_j = \mu_j^o + RT \log_e[j] + z_j F \Phi}$$

to emphasise that the free energy of the ion, and hence its stability, depends on both the electric potential and the concentration of the ion. Note that the stability decreases as expected when the electrical term $z_jF\Phi$ is positive.

The electrochemical potential of a component in a compartmented system can be interpreted in exactly the same way as the chemical potential. Thus if

$$\bar{\mu}_j(\text{in}) = \bar{\mu}_j(\text{out})$$

in a two-compartment system in which the inner compartment is bounded by a membrane that is permeable to j, then the system is at equilibrium and there is no tendency for a net flux across the membrane. Alternatively, if

$$\bar{\mu}_j(\text{in}) > \bar{\mu}_j(\text{out})$$

then the thermodynamic driving force causes a net efflux of j from the inner compartment, and if

$$\bar{\mu}_j(\text{out}) > \bar{\mu}_j(\text{in})$$

then the net flux is reversed and j moves into the inner compartment.

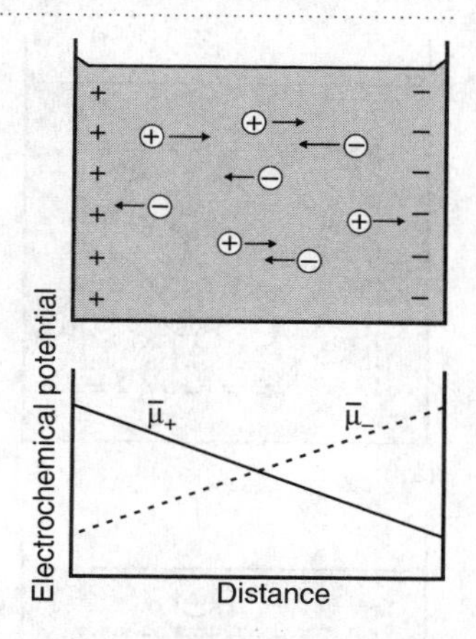

Fig. 7.6 The electrochemical potential of an ion depends on the electric field that it experiences. An ion with a positive charge has a lower electrochemical potential, and hence is more stable, in a region with a more negative electric field. Conversely, an ion with a negative charge has a higher electrochemical potential in a region with a more negative electric field. Ions placed in an electric field gradient will move to minimise their electrochemical potential (shown by the arrows).

Membrane potentials

The electrochemical potential is relevant to the energetics of cellular systems because of the naturally occurring differences in electric potential that occur across many cell membranes. Thus the electric potential in the cytosol of a cell often differs from the electric potential outside the cell, and this difference in potential gives rise to a membrane potential across the plasma membrane. Membrane potentials can be measured directly, if the cell is large enough to allow a microelectrode with a tip diameter of 1–2 μm to be inserted without disrupting the cell, and a typical potential difference across the plasma membrane of a root cell, for example, would be – 100 mV.

A plasma membrane potential of – 100 mV indicates that the cytosol is negatively charged relative to the external medium. The link between the membrane potential and the unbalanced charge within the cell depends on the capacitance of the plasma membrane, i.e. the extent to which the membrane can store charge, and the size and shape of the cell. For a spherical cell with a radius of 30 μm and a membrane capacitance of 10 mF m^{-2}, it can be shown that an unbalanced anion concentration of 1 μM is sufficient to generate a membrane potential of – 100 mV. Given that the total concentration of negative charge in the cell could easily be 100 mM, it can be seen that it requires only a very small imbalance in the charges inside a cell to create a significant electric potential.

The existence of such potential differences is closely related to the ion transport mechanisms that allow ions to move from one compartment to another. The important point here is that while there are likely to be many different passive, i.e. free-energy releasing, and active, i.e. free-energy requiring, processes for transporting ions across membranes, there is no requirement

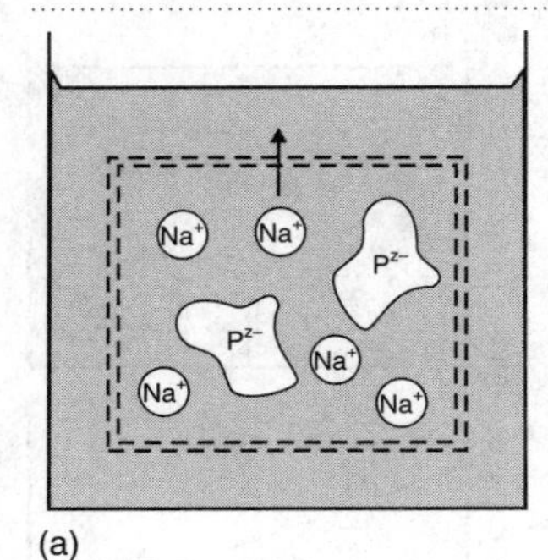

(a)

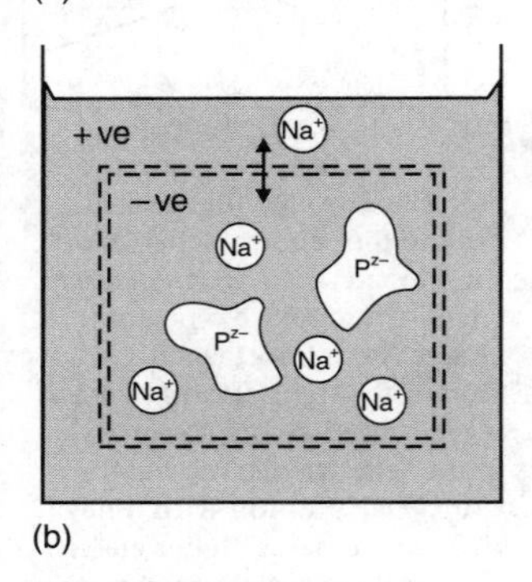

(b)

Fig. 7.7 (a) If the Na salt of a protein is placed in an inner compartment, separated from the outer compartment by a membrane permeable to Na^+ but not to the protein, then Na^+ ions will move from in to out down their concentration gradient. (b) As no counter ions can cross the membrane with the Na^+ ions, the inner compartment will acquire a net negative charge and the outer compartment a net positive charge. This potential difference will tend to work against further efflux of Na^+ ions. At equilibrium, the Na^+ ion concentration gradient is exactly balanced by the electric potential difference.

that the net effect of these processes, in combination with the fixed charges in the system, such as those associated with membranes and non-transportable macromolecules, should lead to a net charge of zero when the cell is in a steady state.

The logic of this argument can be seen by considering ion transport in a hypothetical two-compartment system. Suppose that one compartment, the inner compartment, contains the sodium salt of a macromolecule, while the other compartment, the outer compartment, contains only water (figure 7.7a). In this state the net charge in each compartment is zero and the energetics of the various components can be analysed in terms of their chemical potential.

Now if it is assumed that the membrane is permeable to Na^+, then sodium ions will be able to move spontaneously from the inside to the outside, since $\mu_{Na+}(\text{in}) > \mu_{Na+}(\text{out})$. However, if it is also assumed that the membrane is impermeable to every other species in the system, then there will be no compensating movement of anions and the net result of the sodium efflux will be to create a positive electric potential outside and a negative potential inside (figure 7.7b). This potential difference will tend to work against the efflux of the sodium ions and an equilibrium will be established when

$$\bar{\mu}_{Na^+}(\text{in}) = \bar{\mu}_{Na^+}(\text{out})$$

Substituting the expression for the electrochemical potential leads to

$$\mu^{o}_{Na^+} + RT\log_e[Na^+(\text{in})] + z_{Na^+}F\Phi(\text{in}) = \mu^{o}_{Na^+} + RT\log_e[Na^+(\text{out})] + z_{Na^+}F\Phi(\text{out})$$

and after noting that $z_{Na+} = 1$, this can be rearranged to give

$$\Delta\Phi \equiv \Phi(\text{in}) - \Phi(\text{out}) = \frac{RT}{F}\log_e\left(\frac{[Na^+(\text{out})]}{[Na^+(\text{in})]}\right)$$

The potential difference, $\Delta\Phi$, is the membrane potential arising from the passive diffusion of the sodium ions across the membrane and it is therefore described as a diffusion potential. Real membrane potentials have contributions from all the ion transport processes that occur across the membrane, but the point of the derivation is that it emphasises that there is no reason to suppose that the net effect of these processes will lead to an exact balance of positive and negative charge in a given compartment. Thus membrane potentials are an inevitable result of the differential permeability of membranes to ions and they have a major influence on the energetics of ion transport between membrane-bound compartments.

The equation for the diffusion potential is often referred to as the Nernst equation for the equilibrium distribution of the ion. It can be used to predict the equilibrium distribution of an ion across a membrane in the presence of a membrane potential.

WORKED EXAMPLE

Calculate the intracellular concentration of K^+ that would be in equilibrium with an external concentration of 1 mM for a unicellular organism with a plasma membrane potential of − 116 mV at 20°C.

SOLUTION: The answer is obtained by substituting the given values into the Nernst equation, taking care to express the membrane potential in volts and the concentrations in molar units, giving

$$-0{\cdot}116 = \frac{8{\cdot}31 \times 293}{96\,490} \log_e \frac{0{\cdot}001}{[K^+(in)]}$$

This can be rearranged to give

$$[K^+(in)] = 0{\cdot}001 \times e^{4{\cdot}6} = 0{\cdot}099\,M$$

Thus the predicted concentration at equilibrium is 99 mM.

COMMENT: Note that while some ions do equilibrate across membranes, reflecting the existence of an efficient pathway for shuttling the ion back and forth across the membrane, many others do not. If an ion does not equilibrate, then either it is unable to cross the membrane at a significant rate or the cell is doing work to maintain a non-equilibrium distribution. Ions that fall into the latter category turn out to be critical for many important cellular functions.

Electrochemical gradients for ions

If an ion does not equilibrate between the inside and outside of a two-compartment system then

$$\bar{\mu}_j(in) \neq \bar{\mu}_j(out)$$

and

$$\Delta\bar{\mu}_j \equiv \bar{\mu}_j(in) - \bar{\mu}_j(out) \neq 0$$

$\Delta\bar{\mu}_j$ is the thermodynamic driving force that would allow the ion to equilibrate if a pathway were available and an explicit expression for $\Delta\bar{\mu}_j$ can be obtained by substituting for $\bar{\mu}_j(in)$ and $\bar{\mu}_j(out)$

$$\Delta\bar{\mu}_j = \mu_j^o + RT\log_e[j(in)] + z_jF\Phi(in) - \mu_j^o - RT\log_e[j(out)] - z_jF\Phi(out)$$

This can be simplified to give

$$\Delta\bar{\mu}_j = RT\log_e\frac{[j(in)]}{[j(out)]} + z_jF\Delta\Phi$$

An important point about this equation is that it contains both a concentration term and an electrical term, i.e. the direction of spontaneous movement for an ion is not just determined by the concentration of the ion in the two compartments. In the case of a neutral molecule, the thermodynamic driving force is only determined by the concentrations of the molecule and the direction of spontaneous movement is from high to low concentration. In contrast, for an ion, the direction of spontaneous movement can be in either direction depending on the electrical contribution to $\Delta\bar{\mu}_j$. In other words ions can move spontaneously from low to high concentrations provided there is a favourable electrical term. Thus in the worked example above, the higher potassium concentration inside the cell reflects the favourable electrical term corresponding to the negative membrane potential. If the actual intracellular concentration is only 70 mM, then if the potassium were to equilibrate, the concentration would have to increase to 99 mM and this would require the spontaneous movement of potassium from low to high concentration. Thus the electrical terms play an essential role in determining the energetics of ion transport.

WORKED EXAMPLE

The pH gradient across the plasma membrane of a cell was 1·2 pH units with the outside acidic and the membrane potential was 80 mV with the inside negative. Calculate the electrochemical potential difference across the membrane for the proton at 25°C and identify the direction in which H^+ transport is favoured.

SOLUTION: The charge on the proton is +1 and so from the analysis of the thermodynamic driving force

$$\Delta\bar{\mu}_{H^+} \equiv \bar{\mu}_{H^+}(\text{in}) - \bar{\mu}_{H^+}(\text{out}) = RT\log_e\frac{[H^+(\text{in})]}{[H^+(\text{out})]} + F\Delta\Phi$$

Recalling that $pH = -\log_{10}[H^+]$, the equation for the proton electrochemical gradient can be rewritten as

$$\Delta\bar{\mu}_{H^+} = F\Delta\Phi - 2{\cdot}303RT\Delta pH$$

where $\Delta pH = pH(\text{in}) - pH(\text{out})$. Substituting $\Delta pH = 1{\cdot}2$ and $\Delta\Phi = -0{\cdot}08$ V into this equation leads to

$$\Delta\bar{\mu}_{H^+} = -96\,490 \times 0.08 - 2.303 \times 8.31 \times 298 \times 1.2 = -14.56\ \text{kJ mol}^{-1}$$

COMMENT: The negative value for the electrochemical potential difference indicates that the proton is more stable inside the cell and so the thermodynamic driving force favours the spontaneous influx of H^+. In fact the concentration and electrical terms both favour transport in the same direction in this case and so the overall effect is intuitively obvious.

Electrochemical gradients as energy stores

The electrochemical potential difference for an ion across a membrane, $\Delta\bar{\mu}_j$, is a measure of the energy that is stored in the ionic gradient. Thus, in the previous worked example, if a mole of protons were to move into the cell (without changing ΔpH or ΔΦ) then 14·56 kJ of free energy would be released. This energy is available to do work and it can be tapped by coupling the thermodynamically favourable equilibration of the ion to some other energetically unfavourable process. For example, the release of energy as an ion moves across a membrane can be used to transport another ion against its electrochemical gradient. Examples of such coupled transport processes include: (i) the proton co-transport of phosphate and nitrate into plant roots, where the energy available from the movement of protons into the root cells drives the accumulation of phosphate and nitrate; and (ii) the sodium co-transport of glucose into the epithelial cells of the intestine, where the energy available from the movement of sodium into the cell drives the uptake of glucose from the intestinal lumen.

WORKED EXAMPLE

Assuming a cytoplasmic pH of 7·6 and a vacuolar pH of 5·5, calculate the maximum concentration ratio of vacuolar to cytoplasmic sodium that could be maintained by a Na^+/H^+ antiport in the vacuolar membrane.

SOLUTION: The ion transport process can be represented by the equation

$$Na^+(cyt) + H^+(vac) \rightleftharpoons Na^+(vac) + H^+(cyt)$$

and at equilibrium

$$\bar{\mu}_{Na^+}(cyt) + \bar{\mu}_{H^+}(vac) = \bar{\mu}_{Na^+}(vac) + \bar{\mu}_{H^+}(cyt)$$

This can be rearranged to give

$$\bar{\mu}_{Na^+}(cyt) - \bar{\mu}_{Na^+}(vac) = \bar{\mu}_{H^+}(cyt) - \bar{\mu}_{H^+}(vac)$$

Substituting for $\bar{\mu}$ leads to

$$RT\log_e\frac{[Na^+(cyt)]}{[Na^+(vac)]} + F\Delta\Phi_{cv} = RT\log_e\frac{[H^+(cyt)]}{[H^+(vac)]} + F\Delta\Phi_{cv}$$

where $\Delta\Phi_{cv}$ is the membrane potential across the vacuolar membrane, the tonoplast.

Recalling that $pH = -\log_{10}[H^+]$, this equation can be simplified to give

$$\log_{10}\frac{[Na^+(vac)]}{[Na^+(cyt)]} = pH(cyt) - pH(vac)$$

and substituting the pH values leads to

$$\frac{[Na^+(vac)]}{[Na^+(cyt)]} = 125{\cdot}9$$

Thus the energy required to move sodium into the vacuole is exactly equal to the energy available when protons move out of the vacuole when the vacuolar concentration of sodium is 125·9 times the cytoplasmic concentration.

COMMENT: This conclusion is independent of the value of the tonoplast membrane potential because the ion transport process is not electrogenic, i.e. there is no net movement of charge during the operation of the antiport.

Oxidative phosphorylation and photophosphorylation

The free energy released when an ion moves down an electrochemical potential gradient can also be coupled to energy-requiring chemical reactions. The most important examples of this occur in the processes of oxidative phosphorylation and photophosphorylation, where energy that has been stored in the form of a proton gradient ($\Delta\bar{\mu}_{H+}$) is used to drive the synthesis of ATP. Thus in the case of oxidative phosphorylation, energy derived from the oxidation of NADH and $FADH_2$ is stored in the form of a proton gradient across the inner mitochondrial membrane; while in the case of photophosphorylation, energy derived from sunlight is stored in a proton gradient across the thylakoid membrane of the chloroplasts. In the usual way the energy stored in these gradients can be expressed in terms of ΔpH and ΔΦ

$$\Delta\bar{\mu}_{H^+} = F\Delta\Phi - 2{\cdot}303RT\Delta pH$$

Experiments on isolated liver mitochondria typically yield a $\Delta\Phi \equiv \Phi(\text{matrix}) - \Phi(\text{suspending medium})$ of $-170\,\text{mV}$ and a $\Delta pH \equiv pH(\text{matrix}) - pH(\text{suspending medium})$ of 0·5, and substituting these values into the equation leads to

$$\Delta\bar{\mu}_{H^+} = -19{\cdot}25\,\text{kJ mol}^{-1}$$

Thus the movement of protons back into the matrix is energetically favourable and this provides the driving force for ATP synthesis. Note that many bioenergeticists prefer to express this driving force as a proton motive force (Δp) defined by

$$\Delta p = -\frac{\Delta\bar{\mu}_{H^+}}{F} = \frac{2{\cdot}303RT\Delta pH}{F} - \Delta\Phi$$

This has the effect of expressing the driving force for ATP synthesis as a positive number in volts.

So far we have used the inside–outside convention for differences in thermodynamic quantities across membranes. However, the equation for the proton motive force is frequently encountered in the form

$$\Delta p = \Delta\Phi - \frac{2{\cdot}303RT\Delta pH}{F}$$

where $\Delta\Phi$ is defined as Φ(positive phase) – Φ(negative phase) and ΔpH is defined as pH(positive phase) – pH(negative phase). The positive (P) phase is the compartment receiving the protons during proton pumping and on this convention the data for the liver mitochondria should be expressed as $\Delta\Phi$ = 170 mV and ΔpH = −0·5. However, it is not unknown for the same data to be reported with different conventions for Φ and pH, and in this situation it is important to be aware of the conventions that are being used before substituting the values into an equation for Δp.

The data above for the liver mitochondria show that the proton gradient is dominated by the electrical term, with the difference in pH only contributing around 15% of the Δp value of 200 mV. In contrast, in chloroplasts proton pumping leads to a negligible membrane potential across the thylakoid membrane and Δp is determined entirely by the pH gradient. The reason for this is that the thylakoid membrane is readily permeable to Mg^{2+} and Cl^- ions, and so any attempt to establish a charge separation across the membrane by proton pumping is cancelled out by compensating movements of other charges. This contrast between chloroplasts and mitochondria emphasises the inherent flexibility of using ion gradients as cellular energy stores.

Stoichiometry of proton pumping and ATP synthesis

Establishing the stoichiometries that underpin the utilisation of the proton gradient that drives ATP synthesis in mitochondria proved to be a major challenge in bioenergetics but consensus values are now available. The key questions are: (i) how many protons are pumped out of the mitochondrial matrix for each oxygen atom consumed (the H^+/O ratio) or for each electron pair transferred (the $H^+/2e^-$ ratio); and (ii) how many protons have to flow back into the matrix to produce an ATP molecule from ADP (the H^+/ATP ratio)? These ratios are directly related to the P/O ratio, which is the number of ATP molecules that can be made from ADP and phosphate for each oxygen atom consumed by the mitochondria during the process of oxidative phosphorylation.

The current view is that the H^+/O ratio is 10 and the H^+/ATP ratio is 4 for mitochondria oxidising matrix NADH. This corresponds to a P/O ratio of 2·5, which is less than the value of 3 that was assumed to be correct for many years. A detailed analysis indicates that four of the 10 protons are transferred through complex I, another four through complex III, and the remaining two through complex IV (figure 6.7). It has also been established that three of the four protons required for ATP synthesis are translocated through the F_o component of the ATP synthase and that the fourth proton is translocated as a result of the turnover of the phosphate and adenine nucleotide translocators.

The H^+/O and H^+/ATP ratios can be used to calculate the maximum possible yield of ATP from the complete oxidation of substrates such as glucose and palmitate. For example, the maximum possible yield of ATP from the conversion of glucose to carbon dioxide via glycolysis and the TCA cycle is now considered to be 31 ATP molecules per glucose. This is significantly lower than the value of 38 molecules of ATP that was accepted for many years.

WORKED EXAMPLE

Calculate the cytoplasmic ATP yield from the complete oxidation of palmitate.

SOLUTION: Inspect the stoichiometry of the reaction in a standard textbook and identify all the steps that generate NADH, $FADH_2$ and GTP (by substrate level phosphorylation in the TCA cycle), as well as the steps that consume ATP. Then assume that the ATP/$2e^-$ ratio is 2·5 for the oxidation of NADH and 1·5 for the oxidation of $FADH_2$. Finally recall that the export of an ATP molecule from the matrix requires a proton and that this will reduce the energy yield from the production of GTP in the matrix. Consideration of all these points leads to the following table

Reaction step	ATP production
Conversion of palmitate to palmitoyl CoA	−2
β-oxidation—NADH production	$7 \times 2{\cdot}5 = 17{\cdot}5$
β-oxidation—$FADH_2$ production	$7 \times 1{\cdot}5 = 10{\cdot}5$
TCA cycle—NADH production	$24 \times 2{\cdot}5 = 60$
TCA cycle—$FADH_2$ production	$8 \times 1{\cdot}5 = 12$
TCA cycle—GTP production	$8 \times 0{\cdot}75 = 6$
Complete oxidation of palmitate	104

The yield of 104 molecules of ATP may be contrasted with the value of 129 that was accepted for many years. In fact the actual yield of ATP in a cell under normal physiological conditions would be considerably less than 104 molecules of ATP per palmitate because the proton motive force is not used exclusively for ATP synthesis. Moreover, the proton gradient is also dissipated by the passive leakage of protons across the inner mitochondrial membrane and it has been estimated that this accounts for about 26% of the oxygen consumption rate of resting liver cells. Overall it is thought that only about 52% of the oxygen used by these cells is used for ATP synthesis, giving an experimentally determined P/O ratio of only 1·18. This may be compared with the value of 2·26, i.e. 104/46, where 46 is the number of NADH and $FADH_2$ molecules produced in the complete oxidation of palmitate, that can be obtained from the calculation in the table.

FURTHER READING

P.W. Atkins, 1998, 'Physical chemistry', Oxford University Press—*a more advanced treatment of colligative properties.*

D.A. Harris, 1995, 'Bioenergetics at a glance', Blackwell Science—*an easy introduction to the principles of bioenergetics.*

D.G. Nicholls and S.J. Ferguson, 1992, 'Bioenergetics 2', Academic Press—*a full description of the thermodynamics underlying biochemical energy transduction.*

P.S. Nobel, 1991, 'Physicochemical and environmental plant physiology', Academic Press—*a rigorous description of the thermodynamic properties of water in a plant physiological context.*

PROBLEMS

1. At the freezing point of a solution, the chemical potentials of the solvent in the solid and liquid phases are equal. This equilibrium condition can be expressed in the form

$$\mu_i(s) = \mu_i^o(l) + RT \log_e x_i$$

where x_i is the mole fraction of the solvent in the solution. Since $x_i < 1$, the freezing point of a solution is less than the pure solvent. Consideration of the temperature dependence of the chemical potential leads to the following expression for the depression of the freezing point (ΔT) of a dilute solution

$$\Delta T \approx \frac{RT_m^2}{L_f} x_s$$

where T_m is the melting point of the pure solvent, L_f is the latent heat of fusion, i.e. the molar enthalpy change associated with the transition from the solid to liquid phase, and x_s is the mole fraction of the solute in the solution.

An antifreeze protein, with a molecular weight of 17 000 Da, was found to be present at a concentration of 10 g dm^{-3} in the serum of an Antarctic fish. Calculate the expected depression of the freezing point and compare your result with (i) the observed depression of 0·6 K and (ii) the depression of 0·0014 K observed for a 10 g dm^{-3} solution of lysozyme, molecular weight of 14 500 Da. For pure water, $L_f = 6{\cdot}02$ kJ mol^{-1} and $T_m = 273{\cdot}15$ K.

2. A solution containing 0·5 g ribonuclease in 0·1 dm^3 of 0·2 M NaCl exerted an osmotic pressure of 0·983 kPa at 298 K. Determine the molecular weight of ribonuclease on the assumption that the membrane in the osmometer was permeable to everything except the protein. Why might the result have been different if the measurement has been carried out in the presence of a much smaller amount of sodium chloride?

3. Calculate the difference in osmotic pressure at normal body temperature between the blood plasma of a diabetic patient and that of a healthy individual, on the assumption that the difference is solely due to the higher level of glucose in the diabetic patient. The glucose levels are 1·80 and 0·85 g dm^{-3} respectively.

4. The sodium salt of a protein, molecular weight of 60 000 Da, carrying six negative charges was dissolved in a 0·15 M solution of NaCl at a concentration of 60 g dm^{-3}. The solution was enclosed in a membrane that was permeable to everything except the protein, and an equal volume of water was added to the outside of the membrane. Calculate the equilibrium concentrations of the ions on both sides of the membrane. Comment on the fact that in the kidney, the observed concentrations of Na^+ and Cl^- are 120 mM on the plasma side of the glomerular membrane and 6 mM in the filtrate.

5. A cell with a hydrostatic pressure P_i and an osmotic pressure Π_i is in equilibrium with pure water at atmospheric pressure. How would P_i and Π_i respond to a gradual increase in the external osmotic pressure Π_o? Assume that the cell water remains in equilibrium with the external solution throughout, and ignore any changes in cell volume.

6. A bacterial cell suspension was resuspended in a solution containing a 10 mM concentration of a non-metabolisable permeant weak base with a pK_a of 9·0. After the cells had equilibrated, the pH of the suspending medium was 8·5 and the intracellular pH was 7·6. Calculate the intracellular concentration of weak base.

7. The cytoplasmic inorganic phosphate (P_i) concentration in maize roots grown on a full nutrient medium is approximately 6·5 mM. Predict the expected vacuolar Pi concentration at 20°C on the assumption that (a) $H_2PO_4^-$ and (b) HPO_4^{2-} reaches electrochemical equilibrium across the tonoplast. Calculate the P_i distribution for tonoplast membrane potentials ($\Delta\Phi_{vc}$) of 20, 30 and 40 mV (vacuolar contents positive) and comment on the results in the light of the measured vacuolar P_i concentration of 6·5 mM. Assume that the cytoplasmic and vacuolar pH values are 7·6 and 5·5 respectively, and that the second pK_a of P_i is 6·8.

8. Measurements on the squid axon gave the following intracellular concentrations

$$[K^+] = 400\,mM \qquad [Na^+] = 50\,mM \qquad [Cl^-] = 100\,mM$$

under conditions where the extracellular concentrations were

$$[K^+] = 20\,mM \qquad [Na^+] = 440\,mM \qquad [Cl^-] = 540\,mM$$

Predict the direction of the thermodynamic driving force acting on each ion at 18°C given that the measured membrane potential was -75 mV.

9. A suspension of mitochondria (5 g protein dm^{-3}) was incubated with [^3H]-acetate, $^{86}Rb^+$, valinomycin (an ionophore that makes the membrane permeable to Rb^+), ADP, P_i and succinate, for a sufficient period to permit the ATP level to reach a constant value. At the end of this incubation the mitochondria were filtered rapidly from the suspension and counted for radioactivity from [^3H]-acetate and ^{86}Rb. It was found that the amount of [^3H]-acetate in the mitochondria was one-eleventh of the total acetate added, while one-sixth of the total ^{86}Rb was also found in the mitochondria. The supernatant was analysed for ATP, ADP and P_i. These concentrations were 1 mM, 0·02 mM and 1 mM respectively. Are these data compatible with the view that two protons must pass across the mitochondrial membrane for each ATP molecule synthesised? Assume that the internal volume of mitochondria is 0·4 $cm^3\,g^{-1}$ protein. ($T = 300$ K. $\Delta G^{o\prime}$ for the reaction ADP + $P_i \rightleftharpoons$ ATP is 30·5 kJ mol^{-1}.)

10. In the photosynthetic chain the following redox reaction occurs.

$$\text{Cyt-b}(Fe^{2+}) + \text{Cyt-f}(Fe^{3+}) \rightleftharpoons \text{Cyt-b}(Fe^{3+}) + \text{Cyt-f}(Fe^{2+})$$

As we saw in problem 10, Chapter 6, $\Delta G^{o\prime}$ for this reaction is -29 kJ mol^{-1} and the (invalid) comparison of this value with $\Delta G^{o\prime}$ for ATP hydrolysis suggested that one electron passed along the chain would not be sufficient to drive ATP synthesis.

Given that ΔG for the hydrolysis of ATP within a cell is -50 kJ mol^{-1} and that ATP synthesis requires the movement of four protons across the thylakoid membrane, determine whether the passage of one electron through this redox reaction can contribute to the generation of a proton electrochemical gradient sufficient to drive ATP synthesis.

8 Ideal and non-ideal solutions

KEY POINTS

- *Ideal gases* and *ideal solutions* behave as though all the molecules present interact identically with each other, regardless of type, and are therefore completely random.
- Ideal gases and solutions can be defined (i) by their physical properties, i.e. the ideal gas law or Raoult's law; or (ii) by the linear relationship between the chemical potential and $\log_e$[concentration] for each component.
- A *non-ideal solution* deviates from the above model. This is because the different types of molecules interact with each other in different ways.
- The solute in a non-ideal solution can be treated as having an *effective concentration*, which is the value of the concentration that would need to be used in the ideal model to reproduce the observed behaviour.
- The *activity coefficient* of a solute is the ratio of the effective to actual concentrations. The value of the activity coefficient can be used to measure the deviation of the solution from ideality.
- Dilute ionic solutions deviate from ideality because of the electrostatic interactions between the solute ions. This leads to a non-random arrangement of the ions in solution.
- The Debye–Hückel theory can be used to quantify the non-ideal behaviour of dilute ionic solutions and to calculate activity coefficients in such solutions.
- At higher concentrations of ionic solutes, the Debye–Hückel theory breaks down because additional factors, as well as the direct electrostatic interactions between solute ions, need to be considered.

In the absence of an electric field, the equation relating the chemical potential of a species to its concentration in an ideal solution is

$$\mu_X = \mu_X^o + RT\ \log_e[X]$$

This equation forms the basis of all the thermodynamics dealt with in the previous chapters, including the derivations of the equations for ΔG and K given in Chapter 3. The purpose of this chapter is to give a derivation of the equation for the concentration dependence of μ, to discuss what is meant by an ideal solution and to look at what happens when the equation breaks down. The starting point for this discussion, as outlined in Chapter 2, is the thermodynamic description of an ideal gas.

Ideal gases

Definition. An ideal gas is defined as a gas whose behaviour follows the equation

$$PV = nRT$$

where P is the pressure, V is the volume, n is the number of moles present, R is the gas constant and T is the absolute temperature. This equation is called the *ideal gas law*. For a mixture of ideal gases, the behaviour of each separate component, x, is given by the equation

$$\boxed{P_x V = n_x RT}$$

where P_x is the partial pressure of the component, the total pressure of the mixture being the sum of the partial pressures of all the components, and n_x is the number of moles of the component x.

Physical interpretation. All the molecules present behave as independent point masses. The only interactions that occur are collisions between the molecules and collisions with the walls of the vessel that contains them (figure 8.1). There are no forces of attraction or repulsion between the molecules, and so changing the neighbouring molecules around any given molecule has no effect on its environment. It follows that the behaviour of a given molecule is independent of the composition of the gas. It also follows that the arrangement of molecules in the gas is completely random, and that there is no enthalpy change on mixing the components of an ideal gas, i.e. $\Delta H_{mix} = 0$.

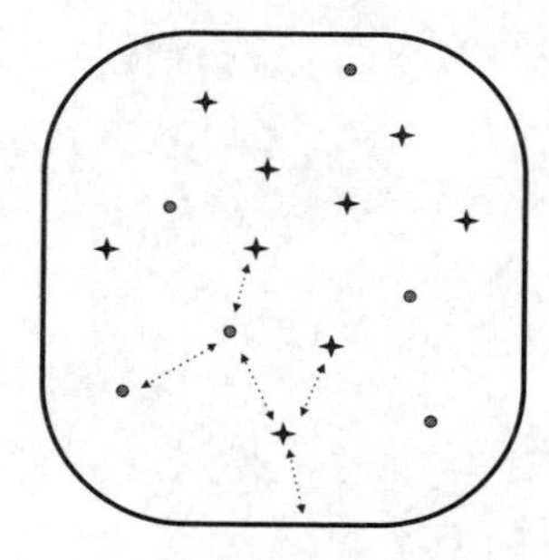

Fig. 8.1 Physical model of a mixture of two components, ✦ and ●, that form an ideal gas. The molecules do not attract or repel each other. The only interactions that occur are collisions (represented by dotted arrows). The properties of a given molecule are independent of composition because it does not interact specifically with any neighbouring molecules.

Thermodynamics of ideal gases

As discussed in Chapter 2, the thermodynamic behaviour of ideal gases can be analysed by combining the ideal gas law with the basic equations of thermodynamics. From the first and second laws of thermodynamics

$$H = U + PV \quad \text{and} \quad G = H - TS$$

Thus

$$G = U + PV - TS$$

Differentiating (taking small increments) we get

$$\delta G = \delta U + V\delta P + P\delta V - S\delta T - T\delta S$$

For a system at equilibrium, and assuming that the only form of work done by the system is the work of expansion, then from the first and second laws

$$\delta U = \delta q - P\delta V \qquad \text{and} \qquad \delta q = T\delta S$$

Substituting for δU in the equation for δG, we get

$$\boxed{\delta G = V\delta P - S\delta T}$$

This equation gives the variation of G with both pressure and temperature. At constant temperature, $\delta T = 0$ and so

$$\delta G = V\delta P$$

From the ideal gas law

$$V = \frac{nRT}{P}$$

and so

$$\delta G = \frac{nRT}{P}\delta P$$

Integrating this between the standard state for the gas, our reference point where $G = G^{\circ}$ and $P = P^{\circ} = 1$ atm, and any arbitrary value of P

$$\int_{G^{\circ}}^{G} dG = \int_{P^{\circ}}^{P} \frac{nRT}{P} dP$$

gives

$$G = G^{\circ} + RT\log_e\left(\frac{P}{P^{\circ}}\right)^n$$

Recalling the definition of the chemical potential (see Chapter 3)

$$\mu = \frac{G}{n}$$

allows the presence dependence of the free energy of the ideal gas to be rewritten as

$$\mu = \mu^{\circ} + RT\log_e\left(\frac{P}{P^{\circ}}\right)$$

Similarly, for a component x in a mixture of ideal gases we need to use μ_x, the partial free energy or chemical potential of x, defined by

$$\mu_x = \frac{G_x}{n_x}$$

to give

$$\boxed{\mu_x = \mu_x^\circ + RT\log_e\left(\frac{P_x}{P^\circ}\right)}$$

These equations can be used as a thermodynamic definition of an ideal gas or an ideal mixture of gases.

Ideal solutions

Definition. Just as ideal gases are defined by an equation relating the partial pressures of the individual components to the composition, so ideal solutions are defined by an equation relating the partial vapour pressures of the individual components to the composition. The partial vapour pressure, P_x, of a component in a solution is the pressure of that component in the gas phase that is in equilibrium with the solution. P_x is related to the composition of an ideal solution by Raoult's law

$$P_x = N_x \times P_x^*$$

where N_x is the mole fraction of x in the solution (see Chapter 7) and P_x^* is the vapour pressure of the pure liquid x.

Physical interpretation. In contrast to an ideal gas, there must be attractive forces between the molecules in an ideal solution to keep them in the liquid phase. However, all the molecules in an ideal solution interact with each other identically (figure 8.2) with the result that the chemical environment of each molecule is independent of the overall composition. Thus an individual molecule has the same properties in the pure liquid as in a solution, and the macroscopic properties of the solution only depend on how many molecules are present. This leads to the colligative properties discussed in the previous chapter. It also follows that in an ideal solution, as in an ideal gas, the arrangement of the molecules is completely random, and that there is no enthalpy change on mixing the components to make the solution, i.e. $\Delta H_{mix} = 0$.

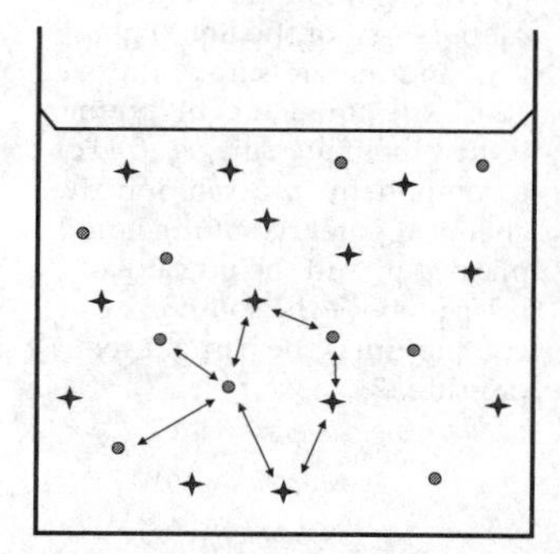

Fig. 8.2 Physical model of an ideal solution of two components. The molecules attract each other (represented by solid arrows). The interactions that occur between all pairs of molecules are the same, having identical magnitudes and distance dependencies. The properties of a specific molecule are independent of composition because changing one of the neighbouring molecules does not affect the intermolecular interactions.

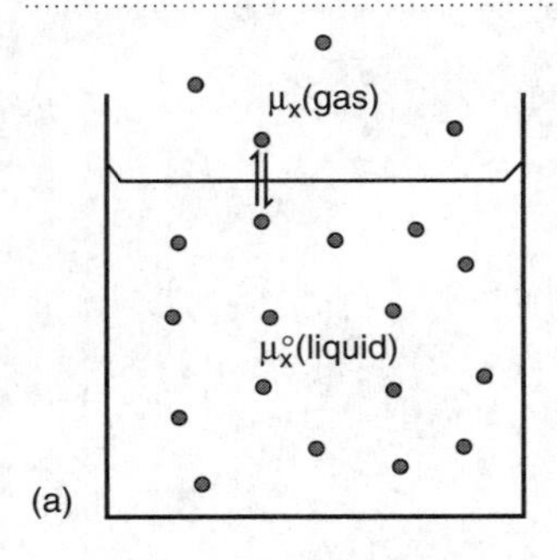

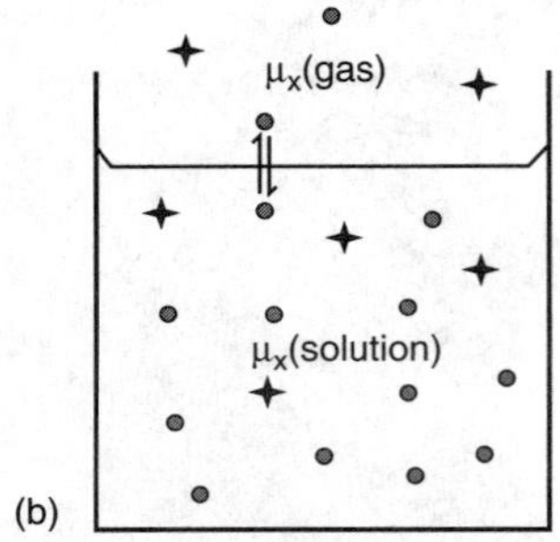

Fig. 8.3 At equilibrium between the liquid and gas phases, the chemical potentials in the two phases must be equal. (a) For a pure liquid the chemical potential of the liquid phase is μ_x^o and the pressure of the gas above the liquid at equilibrium is the vapour pressure, P_x^*. (b) For a component in a solution the chemical potential of the liquid phase is μ_x and the pressure of the gas above the solution at equilibrium is the partial vapour pressure, P_x.

Thermodynamics of ideal solutions

The starting point for analysing the thermodynamic properties of ideal solutions is to consider the equilibrium between the liquid phase and the gas phase (figure 8.3). At equilibrium, a molecule that is present in both phases will have the same chemical potential in each phase

$$\mu_x(\text{solution}) = \mu_x(\text{gas})$$

For a pure liquid, this equilibrium condition can be rewritten as

$$\mu_x^o(\text{liquid}) = \mu_x^o(\text{gas}) + RT\log_e\left(\frac{P_x^*}{P^o}\right)$$

where μ_x^o is the standard chemical potential of the pure liquid and P_x^* is the vapour pressure of x.

Similarly, if x is a component in an ideal solution

$$\mu_x(\text{solution}) = \mu_x^o(\text{gas}) + RT\log_e\left(\frac{P_x}{P^o}\right)$$

where P_x is the partial vapour pressure.

Subtracting the two equations gives

$$\mu_x(\text{solution}) - \mu_x^o(\text{liquid}) = RT\log_e\left(\frac{P_x}{P^o}\right) - RT\log_e\left(\frac{P_x^*}{P^o}\right) = RT\log_e\left(\frac{P_x}{P_x^*}\right)$$

and this expression can be rearranged using Raoult's law to give

$$\mu_x(\text{solution}) = \mu_x^o(\text{liquid}) + RT\log_e N_x$$

This equation describes the concentration dependence of the chemical potential of a component in an ideal solution. An analogous equation can be written in terms of molar concentrations

$$\mu_x = \mu_x^o + RT\ \log_e\left(\frac{[x]}{[x]^o}\right)$$

where $[x]^o$ is the concentration of x in its standard state. This equation is more usually written as

$$\mu_x = \mu_x^o + RT\log_e[x]$$

but it is important to remember that [x] means the concentration of x relative to its concentration in the standard state. The normal standard state is the pure liquid for the solvent and a concentration of 1 M for a solute, but in the case of the biochemical standard state (indicated by a prime ′) the reference concentration of H^+ is set at 10^{-7} M. These equations provide a thermodynamic definition of an ideal solution, and as described in Chapter 7 it is convenient to use the form

$$\mu_i = \mu_i^o + RT\log_e N_i$$

for a solvent i, and

$$\mu_i = \mu_j^o + RT\log_e[j]$$

for a solute j.

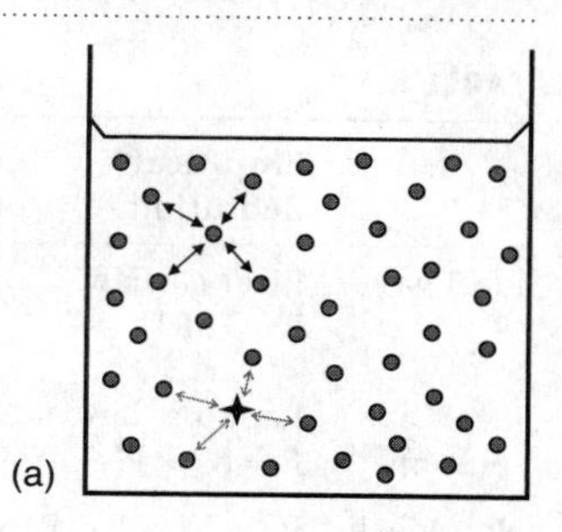

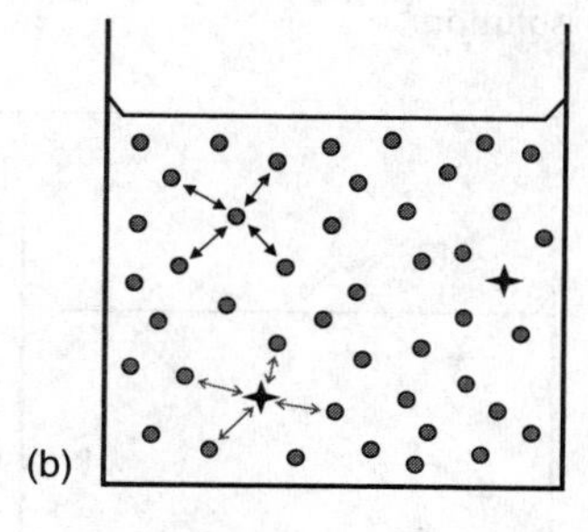

Fig. 8.4 The solvent and solute show ideal behaviour in very dilute solutions, even if the solvent–solvent and solute–solvent interactions are not the same (represented by different types of solid arrows). (a) At very high dilution the solvent molecules ● spend almost all the time interacting with other solvent molecules and the solute molecules ✦ only interact with solvent molecules. (b) Increasing the solute concentration by a small amount does not change the average chemical environment of either ● or ✦.

Dilute solutions

It turns out that very few compounds form ideal solutions over a wide concentration range. However, both the solvent and the solute can be treated as ideal in very dilute solutions (figure 8.4) and this is because the number of solute molecules is very small in such solutions. Thus a solvent molecule in a very dilute solution spends almost the whole time interacting with other solvent molecules. Increasing the concentration of the solute will not alter this until the number of solute molecules is sufficiently high to result in the solvent molecule spending a significant proportion of the time interacting with solute molecules. Until this concentration is reached, altering the solute concentration has no effect on the behaviour of the solvent. Similarly, at very high dilution each solute molecule only interacts with solvent molecules. Increasing the solute concentration will not alter this until the concentration is high enough for a significant number of solute–solute interactions to occur. Thus, at high dilutions the chemical environments of both the solvent and the solute do not change with concentration and this is compatible with the physical model of an ideal solution. It follows that the equations introduced in the previous section are most useful for analysing the thermodynamic properties of dilute solutions.

Non-ideal solutions

Definition. In contrast to ideal solutions, non-ideal solutions are solutions for which

- $P_x \neq N_x \times P_x^*$, i.e. Raoult's law is inapplicable
- $\mu_x \neq \mu_x^o + RT\log_e[x]$
- $\Delta H_{mix} \neq 0$

Physical interpretation. In contrast to an ideal solution, the molecules in a non-ideal solution do not interact with each other identically (figure 8.5). The

TABLE 8.1

	Empirical definition	Enthalpy change on mixing	Chemical potential	Molecular interpretation	
Ideal gas	Ideal gas law $PV = nRT$	$\Delta H_{mix} = 0$	$\mu_X = \mu_X^o + RT \log_e \left(\frac{P_X}{P^o}\right)$	Molecules do not interact except by colliding	Distribution of molecules is random
Ideal solution	Raoult's law $P = N_X \times P_X^*$	$\Delta H_{mix} = 0$	$\mu_X = \mu_X^o + RT \log_e[x]$	All molecules interact identically	Distribution of molecules is random
Non-ideal solution	$P_X \neq N_X \times P_X^*$	$\Delta H_{mix} \neq 0$	$\mu_X \neq \mu_X^o + RT \log_e[x]$	Molecules do not interact identically	Distribution of molecules is not random

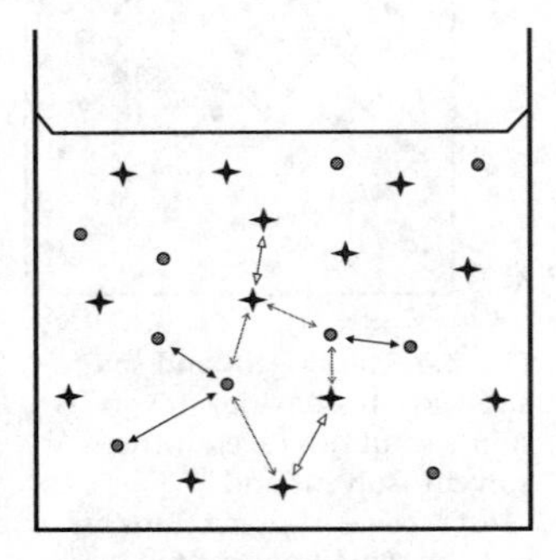

Fig. 8.5 Physical model of a non-ideal solution. The different types of molecules interact with each other in different ways (represented by different types of solid arrows). The properties of a specific molecule are dependent on composition because changing one of the neighbouring molecules alters the intermolecular interactions.

interactions between like and unlike molecules are different, and so the behaviour of a given molecule depends on the composition of the solution. If the neighbouring molecules are changed by altering the relative concentrations of the species present, then the properties of the molecule are altered. It follows that the arrangement of the molecules in a non-ideal solution is non-random. Two molecules in the solution that attract each other strongly will be closer together than predicted by a random distribution, while two molecules that repel each other will be further apart. Thus the solution self-organises to maximise the favourable interactions and to minimise the unfavourable ones, with the result that the distribution of the molecules is no longer random.

Table 8.1 summarises the definitions of ideality and non-ideality. Note that the underlying cause of non-ideality is the existence of preferential interactions between the components in a solution and this is likely to be encountered in many solutions. For example, it is immediately obvious that the components of an ionic solution are going to interact with each other in different ways, with like charges repelling each other and unlike charges attracting each other. Thus non-ideality is frequently observed, and although it disappears at high dilution, it is necessary to extend the thermodynamic analysis to include non-ideal solutions.

Thermodynamics of non-ideal solutions: effective concentrations and activity coefficients

A plot of μ_X versus $\log_e[x]$ does not give a straight line (figure 8.6) for a non-ideal solution because

$$\mu_X \neq \mu_X^o + RT \log_e[x]$$

However, we can use the equation for the concentration dependence of the chemical potential in an ideal solution to calculate the concentration that would be needed to give the same chemical potential as the real solution (figure 8.7). This calculated concentration is called the effective concentration or activity, a_x, of x and it is related to the chemical potential by

$$\mu_x = \mu_x^\circ + RT\log_e a_x$$

It is more useful to use an activity coefficient, γ, defined by

$$\gamma_x = \frac{\text{effective concentration}}{\text{actual concentration}} = \frac{a_x}{[x]}$$

rather than the activity itself and so the equation for the chemical potential is usually written in the form

$$\mu_x = \mu_x^\circ + RT\log_e \gamma_x[x]$$

This equation is always correct, for any solution, because it is always possible to convert the actual concentration into an effective concentration by choosing a suitable value for γ_x.

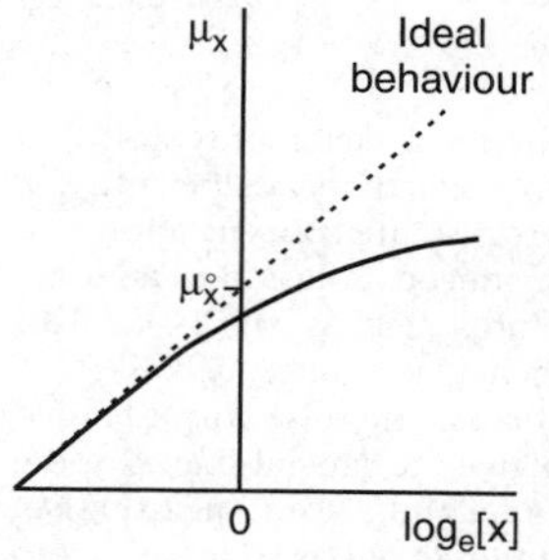

Fig. 8.6 A plot of μ_x versus $\log_e[x]$ for a solute in an ideal solution should be a straight line. Real systems frequently follow this behaviour at very low values of [x], i.e. at very high dilution, but deviate signficantly as [x] becomes larger. In the case shown, μ_x is lower than that predicted for an ideal solution and so x is more stable than expected. The standard state for a solute is defined as a concentration of 1 M assuming that the solution behaves as an ideal solution.

WORKED EXAMPLE

The vapour pressure of pure water is 0·122 atm at 323 K. The vapour pressure above a sucrose solution in which the mole fraction of the water was 0·96 (>2 M sucrose) was found to be 0·1165 atm at the same temperature. Calculate the activity coefficient of the water in the solution.

SOLUTION: We can use Raoult's law to calculate the mole fraction of water (N_w) that would give a vapour pressure of 0·1165 atm for an ideal solution.

$$N_w = \frac{P_x}{P_x^*} = \frac{0{\cdot}1165}{0{\cdot}122} = 0{\cdot}955$$

This value is the *effective* mole fraction of the water since it has been calculated on the basis of an ideal solution. The *actual* mole fraction is 0·96. Thus, the activity coefficient is given by

$$\gamma_w = \frac{\text{effective mole fraction}}{\text{actual mole fraction}} = \frac{0{\cdot}955}{0{\cdot}96} = 0{\cdot}995$$

COMMENT: The activity coefficient is close to one and so even at a high concentration of sucrose we can treat the behaviour of the solvent as approximately ideal.

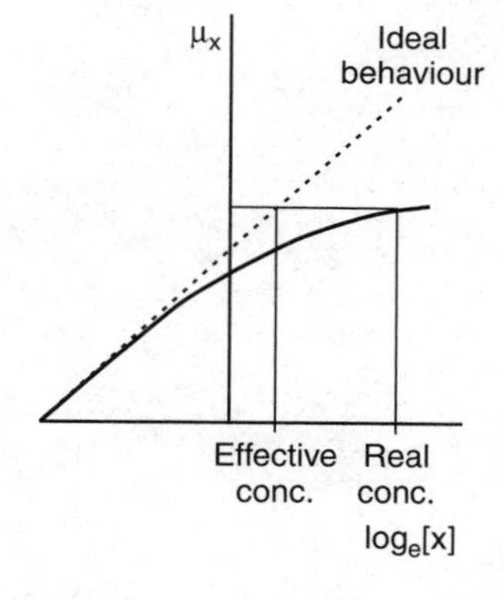

Fig. 8.7 A plot of μ_x versus $\log_e[x]$ for a solute showing the definition of the effective concentration, or activity, of a solute. The effective concentration is the concentration predicted by the ideal model to have the same chemical potential as the actual solution. In this case, the effective concentration is less than the actual concentration, corresponding to an activity coefficient of less than one.

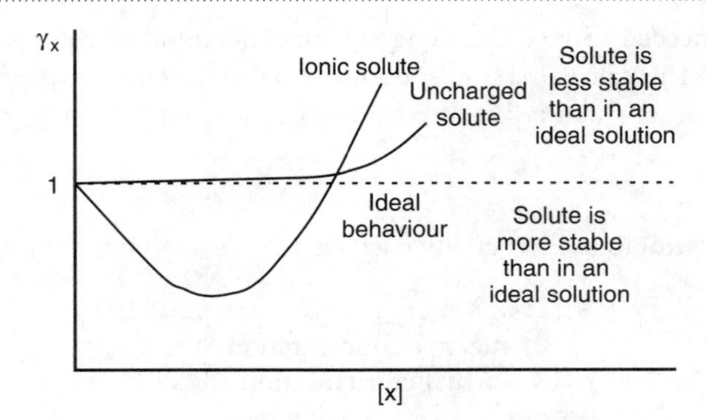

Fig. 8.8 A sketch of typical plots of activity coefficient, γ, versus solute concentration for an aqueous solution of an ionic solute, such as $MgCl_2$, and an uncharged solute, such as glucose. These are only intended to give the general shapes of the curves and are not quantitative. γ tends to one as [x] tends to zero because of the definition of the standard state of the solute.

The activity coefficient converts the real concentration into an effective concentration, and at first sight it appears to be little more than a fudge factor to allow the equation for the chemical potential to be used for non-ideal solutions. However, γ does have a physical significance, and this can be seen by expanding the equation for the chemical potential of a non-ideal solution

$$\mu_x(\text{non-ideal}) = \mu_x^o + \text{RT}\log_e[\text{x}] + \text{RT}\log_e\gamma_x = \mu_x(\text{ideal}) + \text{RT}\log_e\gamma_x$$

Rearranging this equation to give

$$\text{RT}\log_e\gamma_x = \mu_x(\text{non-ideal}) - \mu_x(\text{ideal})$$

shows that γ can be used as a measure of the deviation from ideality. If $\gamma = 1$, the behaviour of x is predicted by the ideal model; if $\gamma < 1$, x is more stable than predicted by the ideal model; and if $\gamma > 1$, x is less stable than predicted.

A further important point is that γ is concentration dependent (figure 8.8), indicating that the deviation from ideality is also concentration dependent, and as expected it tends to 1 in very dilute solutions. The question now arises as to whether the value of γ can be understood in terms of the intermolecular interactions in the solution. The deviation from ideality is particularly important in ionic solutions and so the rest of this chapter explains how γ can be understood in such solutions.

Non-ideality in ionic solutions

Consider an aqueous solution of a fully dissociated ionic compound such as sodium chloride. If the solution is very dilute then as we have seen above, both the solvent and the solute can be treated as ideal, despite the fact that the intermolecular interactions between the various components of the solution are not the same. Water is a strongly hydrogen-bonded liquid (figure 8.9a) and the network of hydrogen bonds will be disrupted by the hydration shells formed as a result of the strong electrostatic interactions between the ions and

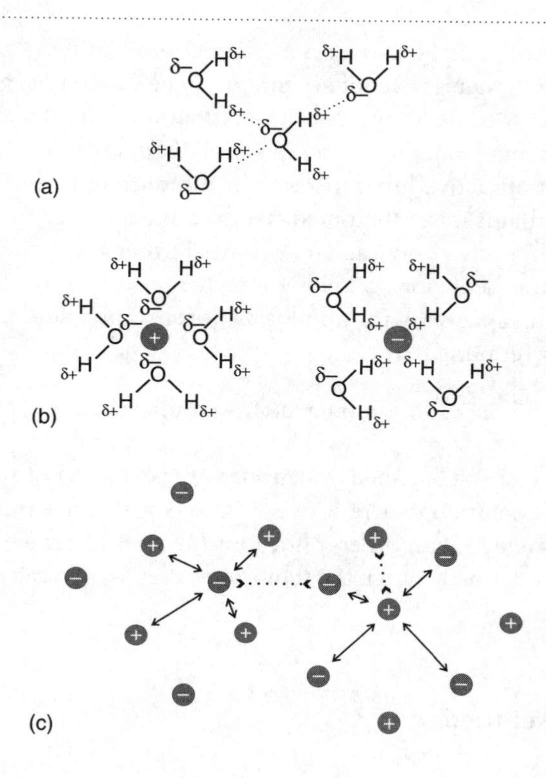

Fig. 8.9 Summary of the electrostatic interactions that occur in dilute aqueous ionic solutions. (a) The water molecules in the pure liquid form a hydrogen-bonded network. These interactions are not significantly perturbed at low solute concentrations and so the solvent behaviour is ideal. (b) Individual ions interact with the water molecules, resulting in the formation of a hydration shell. These are the only interactions of the solute ions at very low concentrations where the solute behaviour is ideal. (c) At higher concentrations, direct electrostatic interactions between solute ions become significant. This results in an ion being surrounded by more ions of opposite charge than of similar charge. The solute distribution is no longer completely random, leading to non-ideal behaviour of the solute.

the water molecules (figure 8.9b). However, when the solute is present at a very low concentration the ions have a negligible effect on the properties of the average water molecule and the solution is ideal.

Although the strong interactions between the ions and the water molecule do not cause non-ideality at low concentrations, they do affect the enthalpy and entropy changes that occur when the sodium chloride is dissolved. The favourable electrostatic interactions between the ions and the surrounding water molecules more than compensate for the disruption of the hydrogen bond network in the vicinity of the ions, and so the enthalpy change on transferring an ion from the gas phase to an aqueous solution is negative (figure 8.10). The water molecules surrounding an ion are also held in a particular orientation, resulting in a decrease in entropy on transferring an ion from the gas phase to an aqueous solution (figure 8.10). In general, smaller or more highly charged ions interact more strongly with the solvent and so lead to larger enthalpy and entropy changes. The standard state of an ionic solute is defined as a concentration of 1 M behaving ideally, i.e. as at infinite dilution, and so all these interactions are present in the standard state and contribute to the value of μ^{o}.

As the concentration of the ionic solute increases, the average distance between the ions decreases and direct electrostatic interactions between the solute ions become more important (figure 8.9c). Like charged ions repel each

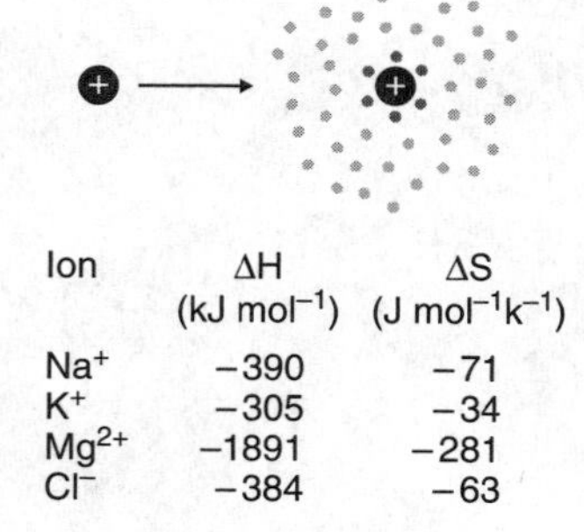

Ion	ΔH ($kJ\ mol^{-1}$)	ΔS ($J\ mol^{-1}k^{-1}$)
Na^{+}	−390	−71
K^{+}	−305	−34
Mg^{2+}	−1891	−281
Cl^{-}	−384	−63

Fig. 8.10 The calculated changes in enthalpy and entropy on transferring one mole of a gaseous ion into water. The electrostatic interactions between the ion and water leads to a favourable ΔH. The ordering of the water molecules around the ion leads to an unfavourable ΔS.

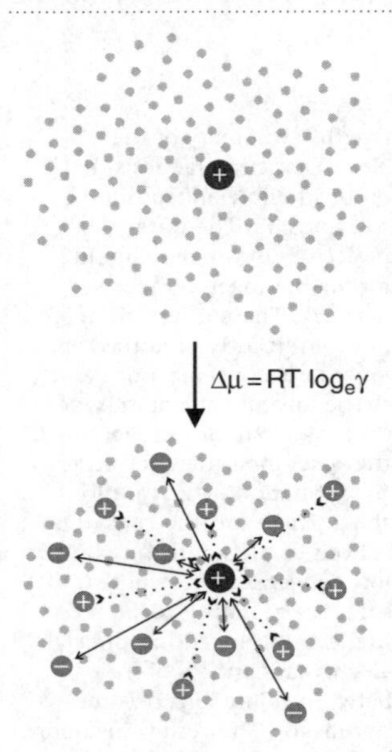

Fig. 8.11 The activity coefficient, γ, of an ion in a non-ideal solution can be calculated from the change in chemical potential on moving the ion from an ideal solution to a non-ideal solution. For a dilute ionic solution, this involves calculating the average solute–solute electrostatic interactions.

other, whereas oppositely charged ions attract each other. This leads to a situation in which there is a tendency for ions to be surrounded by ions of the opposite charge, with the result that the distribution of the solute ions in the solution is no longer random. Each ion is stabilised, relative to the hydrated ion, by the net attractive interactions with the surrounding ions and so its chemical potential is lower than predicted by a model based on ideal behaviour. This corresponds to $\gamma < 1$ and γ decreases further as the concentration is increased and the solute ions get closer together.

This qualitative analysis of the non-ideality of an ionic solution can be made quantitative by recalling that

$$RT \log_e \gamma_X = \mu_X(\text{non-ideal}) - \mu_X(\text{ideal}) \equiv \Delta\mu_X$$

In this case, $\Delta\mu_X$ can be obtained by considering the transfer of an ion from an infinitely dilute solution, where it only interacts with solvent, to a solution where it also interacts with other solute ions (figure 8.11). If we can calculate the average interaction between the solute ions, then we can calculate $\Delta\mu_X$ and hence obtain γ.

Debye–Hückel theory

The Debye–Hückel theory provides a quantitative model for calculating the solute–solute electrostatic interactions in an ionic solution. The initial assumptions of the theory are:

- The solute is present in solution as single solvated ions, so there is no ion pair formation.
- The ions are regarded as point charges.
- The solvent is treated as a continuous medium unaffected by the presence of the ions. The molecular nature of the solvent is ignored.
- Only electrostatic interactions are considered.
- The thermal energy of the solution is much larger than the electrostatic interactions. This assumption is justified because if it were not true the system would be a solid.

Calculating the electrostatic interactions in a static system, such as a crystal, is straightforward because the ions are held in place and so the distances between them are fixed. Solutions are dynamic with ions continuously changing their positions. A completely random distribution of ions in solution results in no net attractive or repulsive interactions because the average separations of like charged and oppositely charged ions are the same. However, as we have seen the distribution of ions in the solution is not random and this has to be taken into account. In order to calculate the average electrostatic interactions in a solution we need to consider all the possible arrangements of

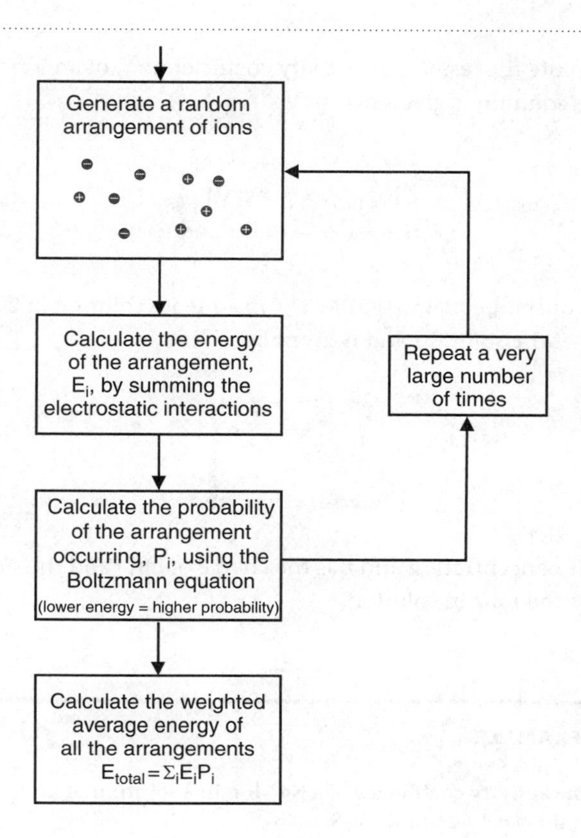

Fig. 8.12 The procedure for calculating the average electrostatic interactions in an ionic solution. This type of procedure, known as a Monte-Carlo calculation, can be performed numerically on a computer. In this case the equations can be solved analytically, leading to the Debye–Hückel equation.

the ions in solution but we also need to recognise that some of these arrangements are more probable than others.

The method for this calculation is shown in figure 8.12. We need two basic equations, one to calculate the energy of a given arrangement of ions, the Poisson equation, and one to calculate the probability of that arrangement occurring, the Boltzmann equation. We can produce a model of a dynamic solution by calculating the energies of a very large number of randomly selected arrangements of ions and then weighting these energies by the probability of each arrangement occurring. This method is an example of a general type of energy calculation called a Monte-Carlo simulation. Instead of trying to perform calculations on a changing system, calculations are performed on a large number of randomly chosen static systems. As long as a large enough number of repetitions is used, the average answer will be the same.

The calculation of the average electrostatic interactions in an ionic solution can be done numerically on a computer; however, it is also possible to solve the combination of the Poisson and Boltzmann equations to give an analytical result. We will not do this here because the mathematics involved is tedious and gives no further insight into the physical model behind the calculation, so we will just quote the result. The activity coefficient, γ_x, of an ion with charge z_x in an ionic solution is given by

$$\log_{10} \gamma_x = -Az_x^2\sqrt{I}$$

where A is a constant equal to $0{\cdot}51\,M^{-1/2}$ in aqueous solution at 298 K. I is the *ionic strength* of the solution and is given by

$$I = \frac{1}{2}\sum_j c_j z_j^2$$

where c_j is the concentration and z_j is the charge of ion j and the summation is taken over all the ions in solution.

WORKED EXAMPLE

Calculate the activity coefficient of a Na^+ ion in a solution of (a) 50 mM NaCl and (b) 50 mM NaCl + 10 mM Na_2SO_4.

SOLUTION: The activity coefficient of a Na^+ ion is given by

$$\log_{10} \gamma_{Na} = -Az_{Na}^2\sqrt{I}$$

where $A = 0{\cdot}51\,M^{-1/2}$ and $z_{Na} = 1$. The ionic strength of the solution is given by a summation over all the ions in the solution.

(a) The solution contains 50 mM Na^+ ions and 50 mM Cl^- ions. The ionic strength is given by

$$I = \tfrac{1}{2}\{0{\cdot}05 \times (1)^2 + 0{\cdot}05 \times (-1)^2\} = 0{\cdot}05\,M$$

and so $\gamma_{Na} = 0{\cdot}77$.

(b) The solution contains 70 mM Na^+ ions, 50 mM Cl^- ions and 10 mM SO_4^{2-} ions. The ionic strength is given by

$$I = \tfrac{1}{2}\{0{\cdot}07 \times (1)^2 + 0{\cdot}05 \times (-1)^2 + 0{\cdot}01 \times (-2)^2\} = 0{\cdot}08\,M$$

and so $\gamma_{Na} = 0{\cdot}72$.

COMMENT: For a 1:1 electrolyte, such as NaCl, the ionic strength equals the concentration. Electrolytes with ions of charge more than one have a larger impact on the value of the ionic strength.

Comparison of the Debye–Hückel theory with experiment

It is not possible to measure γ for an ion in the absence of its counter ion, since we cannot add a positive ion to the solution without a balancing negative ion. However, we can measure the average value of γ for both ions, $\gamma_\pm$, which is given by

$$\log_{10}\gamma_\pm = -A|z_+||z_-|\sqrt{I}$$

where z_+ and z_- are the charges on the cation and anion respectively. This equation predicts that

- $\gamma_\pm$ is always less than one, i.e. ions in solution always have a lower chemical potential than expected for an ideal solution, reflecting the stabilising influence of the non-random distribution of the ions.
- $\gamma_\pm$ decreases as the concentration of ions in solution increases.
- the value of $\gamma_\pm$ only depends on the charges of the ions present, not on their chemical identity, i.e. addition of NaCl to a solution has the same effect on $\gamma_\pm$ as addition of the same concentration of KBr.
- a plot of $\log_{10}\gamma_\pm$ versus $\sqrt{I}$ should be a straight line, the slope depending on the charges of the ions being measured.

All these predictions are confirmed by experiment if $\sqrt{I}$ is less than $\approx 0{\cdot}2\,M^{1/2}$, (figure 8.13), demonstrating the validity of the Debye–Hückel model for dilute ionic solutions.

Concentrated ionic solutions

At higher values of $\sqrt{I}$ the predicted behaviour no longer matches the experimental data (figure 8.13). In fact, the Debye–Hückel theory predicts that $\gamma_\pm$ will continue decreasing at high values of $\sqrt{I}$, whereas the experimental

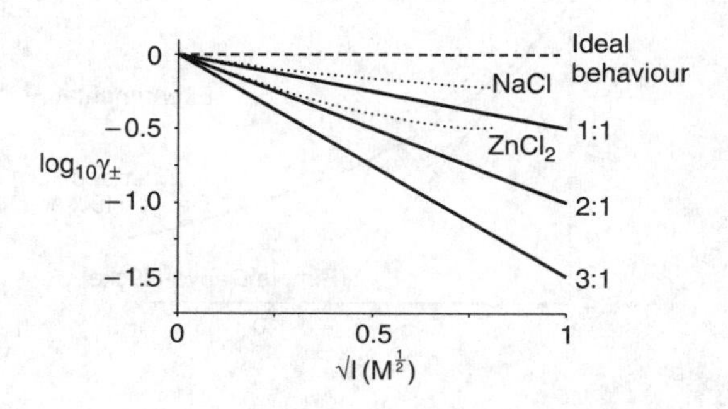

Fig. 8.13 Calculated plots of $\log_{10}\gamma_\pm$ versus $\sqrt{I}$ for a 1:1 electrolyte, such as NaCl, a 2:1 electrolyte, such as $ZnCl_2$ and a 3:1 electrolyte, such as $LaCl_3$ (solid lines). The experimental data for NaCl and $ZnCl_2$ are shown as dotted lines.

data show $\gamma_{\pm}$ reaching a minimum and then increasing to be greater than one (figure 8.8). To explain the behaviour of ionic solutions at higher concentrations we need to look at the assumptions underlying the theory and question their validity.

First, the assumption that the ions can be regarded as point charges is incorrect. Ions have a finite size and there is a minimum distance of approach for two ions that is equal to the sum of their van der Waals radii (figure 8.14a). Fortunately, this assumption is easy to remove from the solution to the Poisson and Boltzmann equations, giving

$$\log_{10}\gamma_x = \frac{-Az_x^2\sqrt{I}}{1 + rB\sqrt{I}}$$

where B is a constant equal to $3{\cdot}29 \times 10^9\,\mathrm{m}^{-1}\,\mathrm{M}^{-1/2}$ in aqueous solution at 298 K, and r is the sum of the van der Waal's radii of the ions. This equation is called the *extended* Debye–Hückel equation. At low values of $\sqrt{I}$, when $rB\sqrt{I} \ll 1$, the extended equation reduces to the simple Debye–Hückel equation, whereas at high values of $\sqrt{I}$, when $rB\sqrt{I} \gg 1$, the extended equation becomes

$$\log_{10}\gamma_x = \frac{-Az_x^2}{rB}$$

This suggests that γ should reach a minimum limiting value with increasing $\sqrt{I}$; but while the extended Debye–Hückel equation gives good agreement with the experimental data up to $\sqrt{I} \approx 0{\cdot}3\,\mathrm{M}^{1/2}$, it still does not predict the increase in γ that is seen at higher ionic strengths (figure 8.14b).

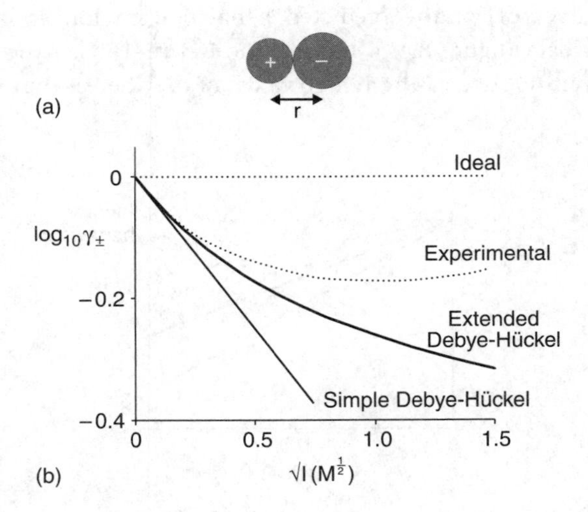

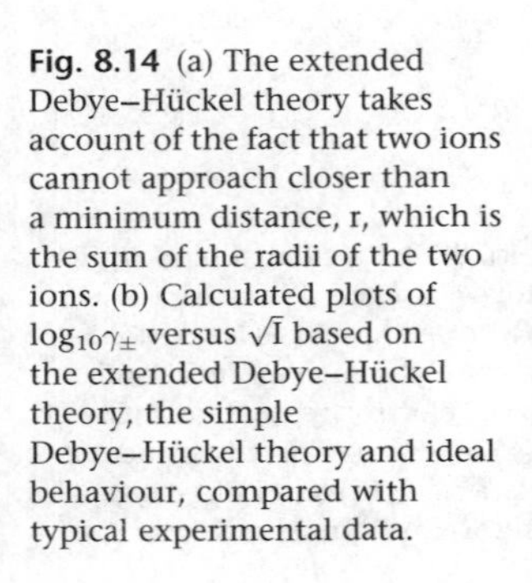

Fig. 8.14 (a) The extended Debye–Hückel theory takes account of the fact that two ions cannot approach closer than a minimum distance, r, which is the sum of the radii of the two ions. (b) Calculated plots of $\log_{10}\gamma_{\pm}$ versus $\sqrt{I}$ based on the extended Debye–Hückel theory, the simple Debye–Hückel theory and ideal behaviour, compared with typical experimental data.

Second, the assumption that the solvent can be treated as a continuous medium that is unaffected by the presence of the ions is also incorrect. Water is composed of molecules that hydrogen bond to each other and form hydration spheres around the ions. Hydrating the large number of ions present at high ionic strength causes extensive disruption of the hydrogen bond network; and at still higher ionic strengths there may be insufficient water molecules to hydrate all the ions completely (figure 8.15). Both these factors lead to a destabilisation of the solution and so to an increase in γ for both the solutes and the solvent. The Debye–Hückel theory only deals with electrostatic solute–solute interactions and so cannot be modified to take account of these factors, but extra terms can be added to the equation for $\log_{10}\gamma$. The result is the Robinson–Stokes equation, which we will not quote here, but which provides an accurate description of the experimental data over a large range of ionic strengths (figure 8.16).

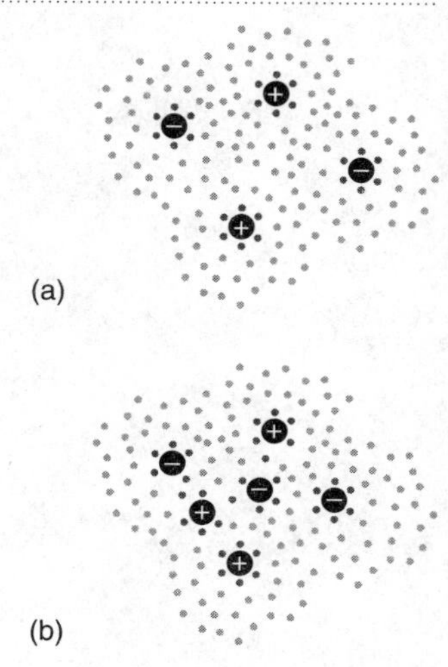

Fig. 8.15 (a) In a dilute ionic solution, all ions are completely solvated and only a small fraction of the solvent molecules are involved in forming hydration spheres. (b) In a more concentrated ionic solution, a large fraction of the solvent molecules are involved in forming hydration spheres and some ions are incompletely solvated.

Solutions of uncharged solutes

Dipolar solutes

The electrostatic interactions between dipolar solutes are much smaller than those between charged solutes and can be ignored to a first approximation. Thus the Debye–Hückel equation does not apply. Dipolar solutes such as glucose form hydrogen bonds to water similar to those that occur between two water molecules. These solutions often behave as ideal, with γ close to one, over significant concentration ranges. At higher concentrations, significant perturbations of the hydrogen bond network of water occur and there can be a lack of water to hydrate the solute. In this concentration range γ becomes

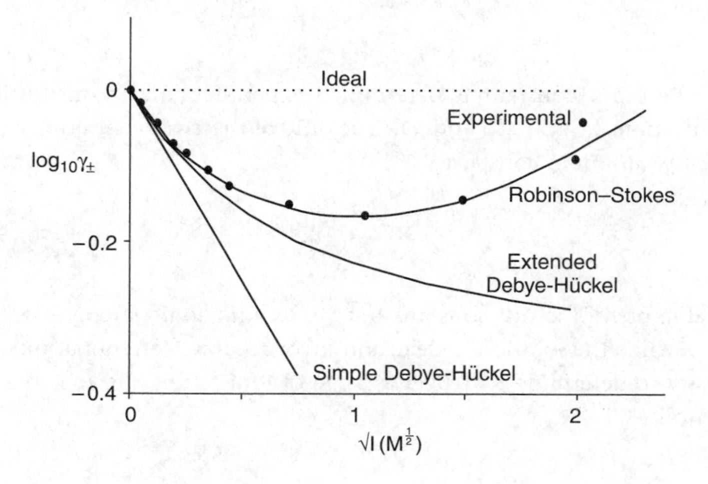

Fig. 8.16 Calculated plot of $\log_{10}\gamma_{\pm}$ versus $\sqrt{I}$ based on the Robinson–Stokes equation, the extended Debye–Hückel equation, the simple Debye–Hückel equation and ideal behaviour, compared with typical experimental data (points). [Based on figure 1.5 from J. Robbins, 1972, *Ions in solution 2*, Clarendon Press, Oxford.]

greater than one, as is observed for ionic solutes. The overall behaviour is shown schematically in figure 8.8.

Non-polar solutes

The situation for non-polar solutes is slightly different. Again electrostatic solute–solute interactions are insignificant but in this case the solute–water hydrogen bonds are much weaker than water–water hydrogen bonds. This gives rise to the *hydrophobic effect,* which is discussed in more detail in Chapter 18, and can lead to the solute becoming insoluble at higher concentrations.

Properties of non-ideal solutions

Activities should always be used instead of concentrations in a rigorous investigation of the thermodynamic and kinetic properties of a non-ideal solution.

Real and apparent equilibrium constants

Equilibrium constants should be expressed in terms of activities rather than concentrations. Thus, for

$$A + B \overset{K}{\rightleftharpoons} C + D$$

the equilibrium constant is given by

$$K = \frac{a_C \times a_D}{a_A \times a_B} = \frac{\gamma_C[C] \times \gamma_D[D]}{\gamma_A[A] \times \gamma_B[B]}$$

The equilibrium constant is determined experimentally by measuring the concentrations [A], [B], [C] and [D]. The ratio of these concentrations gives an *apparent* equilibrium constant, K^{app}.

$$K^{app} = \frac{[C][D]}{[A][B]}$$

The value of K^{app} is not constant but varies with ionic strength. K^{app} only equals K when the solution is ideal, and so for accurate determination of K it is necessary to determine K^{app} over a range of ionic strengths and then extrapolate to $I = 0$.

WORKED EXAMPLE

Predict the effect of increasing ionic strength on the apparent second pK_a of a dilute solution of phosphoric acid.

SOLUTION: The equilibrium for the second pK_a of phosphoric acid is

$$H_2PO_4^- + H_2O \overset{K_a}{\rightleftharpoons} HPO_4^{2-} + H_3O^+$$

The true equilibrium constant, K_a, is given by

$$K_a = \frac{a_{HPO_4^{2-}} \times a_{H_3O^+}}{a_{H_2PO_4^-}} = \frac{\gamma_{HPO_4^{2-}}[HPO_4^{2-}] \times \gamma_{H_3O^+}[H_3O^+]}{\gamma_{H_2PO_4^-}[H_2PO_4^-]}$$

assuming that H_2O is in the standard state. The apparent equilibrium constant, K_a^{app}, is given by

$$K_a^{app} = \frac{[HPO_4^{2-}] \times [H_3O^+]}{[H_2PO_4^-]}$$

Thus

$$K_a = K_a^{app} \frac{\gamma_{HPO_4^{2-}} \times \gamma_{H_3O^+}}{\gamma_{H_2PO_4^-}}$$

Taking $\log_{10}$ of both sides and changing signs gives

$$pK_a = pK_a^{app} - \log_{10}\gamma_{HPO_4^{2-}} - \log_{10}\gamma_{H_3O^+} + \log_{10}\gamma_{H_2PO_4^-}$$

For a dilute solution, the $\log_{10}\gamma$ term for each ion can be calculated using the simple Debye–Hückel equation.

$$\log_{10}\gamma_x = -Az_x^2\sqrt{I}$$

Substituting into the equation for pK_a then gives

$$\begin{aligned} pK_a &= pK_a^{app} + A(-2)^2\sqrt{I} + A(1)^2\sqrt{I} - A(-1)^2\sqrt{I} \\ &= pK_a^{app} + 4A\sqrt{I} \end{aligned}$$

The value of pK_a is constant. Thus, as the ionic strength increases the value of pK_a^{app} decreases, i.e. increasing ionic strength favours dissociation.

This equation relating pK_a to pK_a^{app} only works quantitatively at low values of $\sqrt{I}$ where the Debye–Hückel equation is valid. However, the pK_a^{app} is less than the pK_a as long as the values of γ for the ions are less than one.

COMMENT: Experimentally, the value of the second pK_a for phosphoric acid is 7·2 and the value of pK_a^{app} at $I = 1{\cdot}0$ M is 6·6. The value calculated by the simple Debye–Hückel theory for $I = 1{\cdot}0$ M is approximately 5·2. The difference between the calculated and experimental values is because the assumptions used for the simple Debye–Hückel theory are no longer valid at this ionic strength.

Salting in and salting out

One specific example of the variation in apparent equilibrium constant is the effect of ionic strength on the solubility of an ionic solute. Consider the equilibrium

$$MX(s) \overset{K_s}{\rightleftharpoons} M^+(aq) + X^-(aq)$$

The equilibrium constant for this reaction, called the *solubility product* K_s, is given by

$$K_s = a_{M^+} \times a_{X^-} = \gamma_{M^+}[M^+] \times \gamma_{X^-}[X^-]$$

If γ_{M^+} and γ_{X^-} decrease, $[M^+]$ and $[X^-]$ must increase to keep K_s constant. Conversely, if γ_{M^+} and γ_{X^-} increase, $[M^+]$ and $[X^-]$ must decrease. At low ionic strengths, an increase in I leads to a decrease in γ and hence to an increase in the solubility of MX, termed *salting in*. At high ionic strengths, an increase in I leads to an increase in γ and hence to a decrease in the solubility of MX, termed *salting out*.

WORKED EXAMPLE

AgCl dissolves in pure water to a concentration of $1{\cdot}26 \times 10^{-5}$ M at 298 K. Calculate the solubility of AgCl in an aqueous solution of ionic strength 0·01 M.

SOLUTION: The dissolution of AgCl in water gives the following equilibrium.

$$AgCl \overset{K_s}{\rightleftharpoons} Ag^+ + Cl^-$$

A concentration of Ag^+ ions and Cl^- ions of $1{\cdot}26 \times 10^{-5}$ M is sufficiently dilute to be treated as ideal. Thus

$$K_s = [Ag^+][Cl^-] = (1{\cdot}26 \times 10^{-5})^2 = 1{\cdot}6 \times 10^{-10}$$

For a solution with an ionic strength of 0·01 M, K_s is given by

$$K_s = 1{\cdot}6 \times 10^{-10} = \gamma_{Ag^+}\gamma_{Cl^-}[Ag^+][Cl^-]$$

The activity coefficients of the ions are given by the simple Debye–Hückel equation.

$$\gamma_{Ag^+} = \gamma_{Cl^-} = 10^{-A\sqrt{I}} = 0{\cdot}89$$

Substitution then gives

$$[Ag^+] = [Cl^-] = 1{\cdot}42 \times 10^{-5}\ M$$

COMMENT: The solubility of AgCl increases from $1{\cdot}26 \times 10^{-5}$ M in an ideal solution to $1{\cdot}42 \times 10^{-5}$ M at an ionic strength of 0·01 M. This is an example of salting in. In practice, addition of many salts to an AgCl solution has a greater effect on $[Ag^+]$ than this, either increasing or reducing it, because the ions may either complex with or precipitate with Ag^+.

The variation of solubility with ionic strength is important for protein solutions (figure 8.17). Many proteins are only sparingly soluble in pure water but are much more soluble in 0·1 M NaCl solution. The variation of solubility can also be exploited in methods for protein separation and protein crystallisation.

Primary kinetic salt effect

We shall deal with reaction kinetics in Chapter 9 and so will only mention the effects of ionic strength on reaction rate in passing. The reaction between species with charges z_A and z_B will give a transition state with charge $(z_A + z_B)$.

$$A^{z_A} + B^{z_B} \rightleftharpoons AB^{\ddagger(z_A+z_B)} \rightarrow \text{Products}$$

Increasing the ionic strength will decrease the activities of A, B and $AB^\ddagger$, depending on the magnitude of their charges. If A and B have charges of the same sign, $AB^\ddagger$ will have a greater charge than the reactants and so will be stabilised relative to the reactants, leading to an increase in rate. If A and B have charges of opposite sign, $AB^\ddagger$ will have a lower charge than the reactants

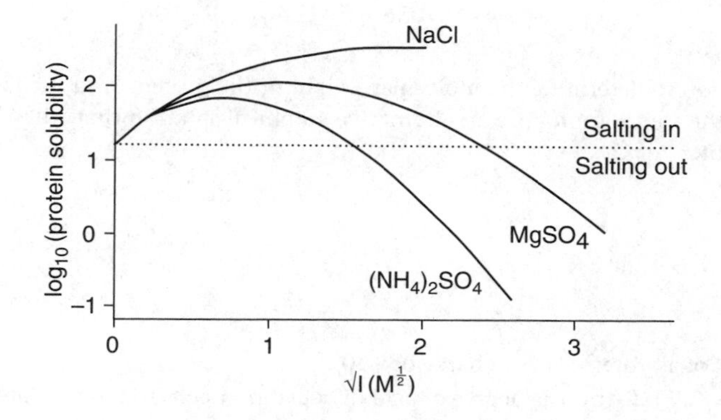

Fig. 8.17 Plots of $\log_{10}$(solubility) versus $\sqrt{I}$ for carbonmonoxyhaemoglobin on addition of NaCl, $MgSO_4$ and $(NH_4)_2SO_4$. Different proteins give different solubility curves and so conditions can be selected under which some proteins will remain in solution and others will precipitate. This is a convenient bulk method of fractionating solutions that contain large numbers of proteins, such as tissue extracts. Alterations in ionic composition can also be used to decrease protein solubility and thus promote crystal growth. [Based on figure 3.1 from T. L. Blundell and L. N. Johnson, 1976, *Protein crystallography*, Academic Press, London.]

and so the reactants will be stabilised relative to the transition state, leading to a decreased rate.

FURTHER READING

R.G. Compton and G.H.W. Sanders, 1996, 'Electrode potentials', Oxford University Press—*a good chapter on non-ideality in ionic solutions and the Debye-Hückel theory.*

E.B. Smith, 1990, 'Basic chemical thermodynamics', Oxford University Press—*a useful introduction to the concept of activity.*

PROBLEMS

1. Explain what is meant by an ideal and a non-ideal solution with reference to the following:

(a) 1:1 mixture of ethanol and propanol;

(b) 1 mM solution of ethanol in water;

(c) 1 mM solution of NaCl in water;

(d) 1 mM solution of lysozyme in water.

Which solution would you predict to deviate most from ideal behaviour?

2. At 310 K the vapour pressure of a 60% w/w solution of glycerol in water (the solution contains 60% glycerol by weight and 40% water by weight) is 0·0434 atm. If the vapour pressure of pure water is 0·0620 atm at this temperature, calculate the activity and the activity coefficient for water in the glycerol solution.

3. The following osmotic pressure data were obtained at 298 K for a solution of bovine serum albumin.

Protein concentration/g dm^{-3}	18	30	50	56
Osmotic pressure/kPa	0·75	1·36	2·56	2·96

First attempt to use the van't Hoff equation to determine the molecular weight of the protein; and then in the light of the evident non-ideality of the solution propose an alternative graphical solution that will give a more accurate estimate of the molecular weight.

4. Calculate the ionic strength of

(a) a 0·1 M $MgCl_2$ solution;

(b) a 5 mM Na_2SO_4 solution;

(c) a 0·1 mM solution of the sodium salt of a protein with a charge of −10;

(d) a 0·1 M solution of acetic acid ($pK_a = 4{\cdot}75$). [Hint: the degree of acid dissociation needs to be calculated to determine the solution concentrations of any ionic species.]

5. Assuming that the solutions (a) to (d) in the previous question follow the simple Debye–Hückel equation, calculate the activity coefficients of the ions present. Comment on the plausibility of the results that you obtain.

6. The solubility product of AgCl is $1{\cdot}6 \times 10^{-10}$ (M^2) at 298 K. Calculate the solution concentration of Ag^+ ions in the presence of (a) 0·01 M $NaNO_3$, (b) 0·01 M NaCl and (c) 0·01 M glucose. How would these solubilities change if $NaNO_3$, NaCl and glucose were added to much higher concentration?

7. The apparent pK_a of acetic acid (defined in terms of concentrations) was measured by titration of a 1 mM solution of acetic acid with NaOH in the presence of varying concentrations of NaCl, with the following results.

NaCl/mM	10	25	50	100	200
Apparent pK_a	4·65	4·59	4·53	4·48	4·45

Determine the true pK_a of acetic acid.

8. The hydrolysis of ATP is given by the following equilibrium

$$ATP + H_2O \rightleftharpoons ADP + P_i + H^+$$

Assuming that the charges on the species are ATP = − 4, ADP = − 3 and P_i = − 2, what effect would changing the ionic strength have on the apparent value of the equilibrium constant?

THE RATES OF CHEMICAL REACTIONS

9

Basic chemical kinetics and single-step reactions

KEY POINTS

- The rate of a chemical reaction increases with the concentration of the reactants in accordance with the *rate equation*.
- The *order* of a reaction with respect to each reactant is the power to which the concentration of that reactant is raised in the rate equation. The *overall order* is the sum of the orders of all the reactants.
- The rate of a single-step chemical reaction increases with temperature. This is given quantitatively by the Arrhenius equation, which relates the *rate constant* to the *temperature* and the *activation energy*.
- Reaction rates vary with ionic strength and with isotopic substitution. These effects can be used to give insight into the mechanism of the reaction.
- Rates for single-step reactions can be understood either in terms of *collision theory*, which emphasises the molecular nature of the reaction, or *transition state theory*, which emphasises the energy required to make the reaction occur.

Kinetics and thermodynamics

Thermodynamics enables us to predict which processes can occur spontaneously and at what stage they will reach equilibrium. We do this by comparing the work energy in the reactants and products, the state with the less work energy being the more stable. However, this does not tell us anything about the rate at which the process will occur. We can distinguish two different sorts of chemical stability.

- Thermodynamic stability—the chemicals present have less work energy than the products of any possible reaction. No reaction can occur spontaneously.
- Kinetic stability—the chemicals present have more work energy than the products of a possible reaction so the reaction can occur spontaneously.

However, the reaction is so slow that we cannot observe it happening and no change appears to occur over the time period of interest.

For example, we have seen that ATP is thermodynamically unstable with respect to ADP and P_i. For the hydrolysis of ATP, $\Delta G^\circ = -30{\cdot}5\,kJ\,mol^{-1}$ at pH 7.0 and 298 K. However, a solution of ATP at room temperature and pH 7.0 does not immediately decompose into ADP and P_i. ATP is only a useful store of chemical energy within cells because its hydrolysis is (i) thermodynamically favourable, (ii) occurs at a negligible rate in free solution, and (iii) is speeded up by enzymes that can then harness the available free energy (see Chapter 13). In fact, most biochemical and biological systems are thermodynamically unstable and kinetically stable. In order to understand this, we need to look at the distribution of energy as the reaction proceeds, not just to compare the reactants and products.

Energy profiles

Figure 9.1 illustrates a typical energy profile for a multistep reaction. The *reaction coordinate* is plotted as the x-axis and can be thought of as a measure of how far an individual reactant molecule has changed on its way to becoming a product molecule. If we simply compare the reactants and the products, the products contain less work energy than the reactants and so the reaction will proceed spontaneously. The line indicates the amount of energy each molecule needs as it is converted from a reactant into a product. In this case, a lot of energy has to be put into each reactant molecule to start the process of converting it into a product molecule. If the energy profile represented ATP hydrolysis, this might be the energy needed to break the P–O bond. The point of maximum energy on the curve is called the *transition state*. If we need to put a lot of energy (called the *activation energy*, $\Delta G^{\ddagger}$) into the reactants to reach the transition state, then the reaction will be slow, even if the products are much

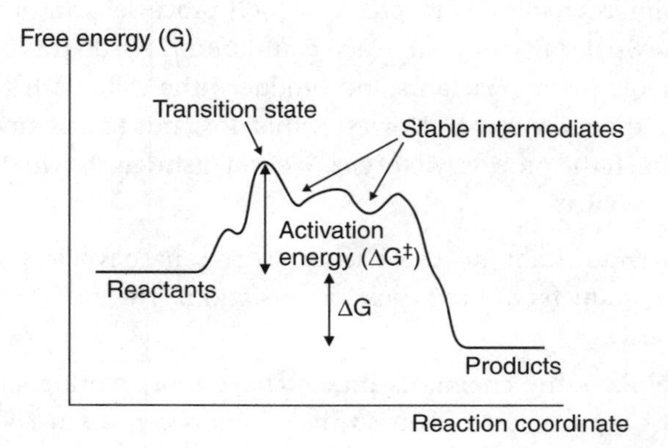

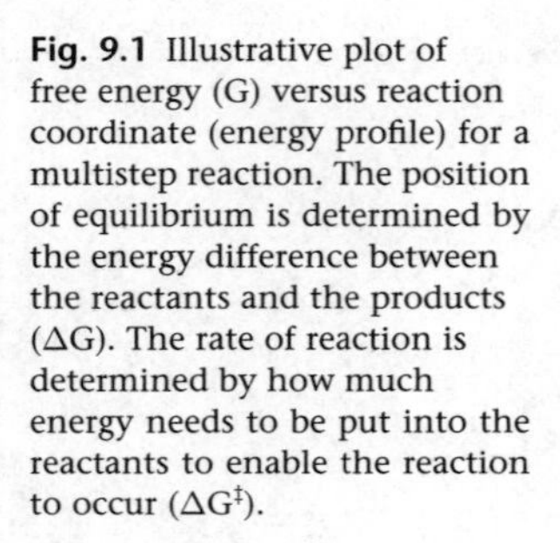
Fig. 9.1 Illustrative plot of free energy (G) versus reaction coordinate (energy profile) for a multistep reaction. The position of equilibrium is determined by the energy difference between the reactants and the products (ΔG). The rate of reaction is determined by how much energy needs to be put into the reactants to enable the reaction to occur ($\Delta G^{\ddagger}$).

more stable than the reactants. It is also possible to have stable intermediates formed during the process. This corresponds to a local minimum in the energy profile.

WORKED EXAMPLE

The energy profiles for three single-step reactions, A, B and C, are given in figure 9.2. Put these reactions in order according to:
(1) how far they will proceed before reaching equilibrium;
(2) how fast they will go.

SOLUTION: (1) How far a reaction will proceed before reaching equilibrium is determined by comparing the reactant and product energies. For B, the products are much more stable than the reactants and so the reaction will proceed a long way to give high product concentrations and low reactant concentrations. For C, the reactants are more stable than the products and so the reaction will only proceed a very short way. The final order is then $B > A > C$.

(2) How fast a reaction will proceed depends on the difference in energy between the reactants and the transition state. For A, the transition state is only slightly higher than the reactants and so the reaction will proceed very quickly. For B, the transition state is very much higher than the reactants and so the reaction will be very slow. The final order is A C B.

COMMENT: Although both position of equilibrium and rate of reaction depend on energy, there is no simple correlation between the two.

One of the problems with studying reaction rates is that usually we cannot investigate the transition states and intermediates directly because they are too short lived. Often, the only thing we can do is measure the concentrations of the reactants or products with time and hence calculate the reaction rate, then vary experimental conditions such as reactant concentrations or temperature and repeat the observations.

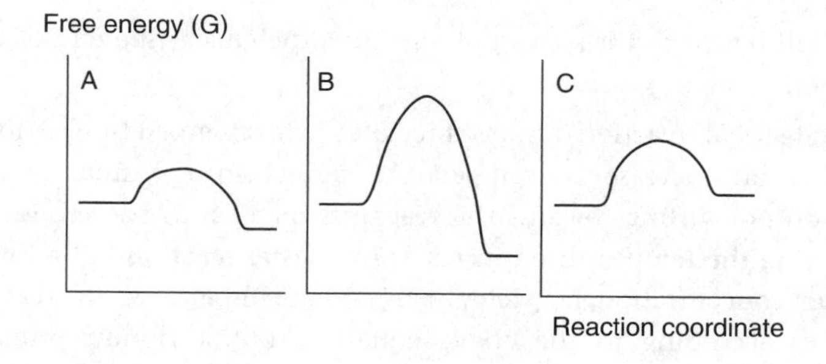

Fig. 9.2 Three possible energy profiles for single-step reactions.

Empirical observations

The rate of a reaction is defined as either the rate of decrease of reactant concentration or the rate of increase of product concentration.

$$\text{Rate} = -\frac{d[\text{Reactants}]}{dt} \quad \text{or} \quad \text{Rate} = \frac{d[\text{Products}]}{dt}$$

The rate of a reaction can be determined by measuring the product or reactant concentration as a function of time, plotting concentration versus time and measuring the gradient at the time of interest (figure 9.3a).

Two generalisations emerge from the measurements of reaction rates.

- The rate of a reaction usually decreases if we decrease the concentrations of the reactants. As reactants are used up as a reaction proceeds, this means that reactions get slower with time (figure 9.3b). The rate of the reaction can usually be fitted to an equation of the general form

$$\frac{d[\text{Products}]}{dt} = k[A]^x[B]^y \ldots$$

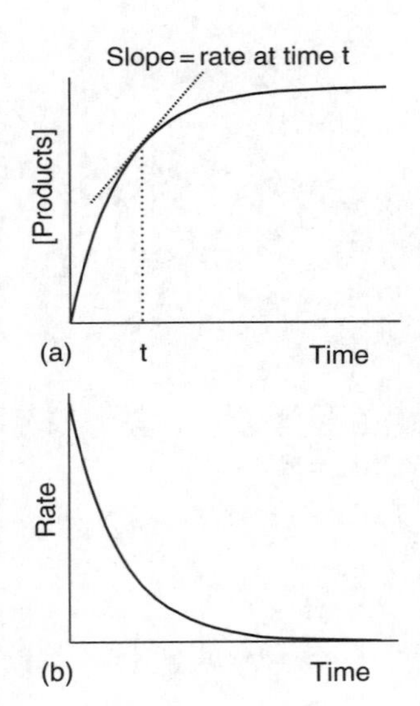

Fig. 9.3 (a) Plot of product concentration versus time for a reaction. The rate at any given time is determined by measuring the slope of the curve at that time. (b) Plot of rate versus time, determined from the [Products] versus time plot in (a). As the reactants are converted to products, the rate decreases.

k is a constant, called the *rate constant*, [A], [B], etc., are the reactant concentrations (or, more properly, activities) and x, y, etc., are constants, called the *order* of the reaction with respect to A, B, etc., respectively. Some reactions give very simple rate equations of the form above, for example

$$H_2 + I_2 \rightarrow 2HI \qquad \frac{d[HI]}{dt} = k[H_2][I_2]$$

whereas other, apparently similar, reactions can give much more complicated rate equations, for example

$$H_2 + Br_2 \rightarrow 2HBr \qquad \frac{d[HBr]}{dt} = \frac{2k_2[H_2]\sqrt{\frac{k_1}{k_{-1}}}[Br_2]}{1 + \frac{k_{-2}[HBr]}{k_3[Br_2]}}$$

We shall return to these examples in the problems at the end of the next chapter.

- The rate of a reaction always increases with temperature. This occurs whether an increase in temperature results in the final position of equilibrium shifting towards the reactants or towards the products. Thus increasing the temperature can result in a faster reaction but a lower final product concentration, i.e. yield. At any given temperature, the reaction still behaves according to the above equation, but k is now found to be temperature dependent and can be fitted to

$$k = \mathcal{A}\exp\left(\frac{-E_a}{RT}\right)$$

$\mathcal{A}$ and E_a can be treated as constants over small temperature ranges and their meaning will be discussed later. This is called the *Arrhenius equation.*

Ideally, what we would like to be able to do is to use this type of experimental information to give an insight into the nature of the reaction, the precise mechanism, the structures of the transition state and any intermediates, and the type of energy profile. In order to do this, we need to have some sort of theory that will enable us to understand the above equations and interpret their results

Reaction rate theories

There are two different theories to relate reactant concentrations to reaction rates, collision theory and transition state theory. At first glance, these theories are very different but both explain the observations above and give a final equation of the same general form. As they give alternative insights into what might be going on, we shall describe both of them here. We shall take a very simple reaction

$$A + B \rightarrow \text{Products}$$

that happens in a single step without intermediates.

Collision theory

Collision theory concentrates on the molecular nature of the reaction and so gives more insight into what individual molecules may be doing. It assumes that a molecule of A and a molecule of B have to collide before a reaction can occur. They also have to collide with enough energy to overcome the activation energy barrier. The rate of the reaction is then given by

$$\text{Rate} = \text{Rate of collision} \times \text{Probability that the reactants have enough energy}$$

The rate of collision is determined by the concentrations of A and B (the more molecules that are present the greater the chance of a collision), how large they are (the bigger the molecules the greater the chance of a collision) and how fast they are going (the faster the molecules move the shorter the time between collisions). This is given by the following equation

$$\text{Rate of collision} = \pi r^2 v_R[A][B]$$

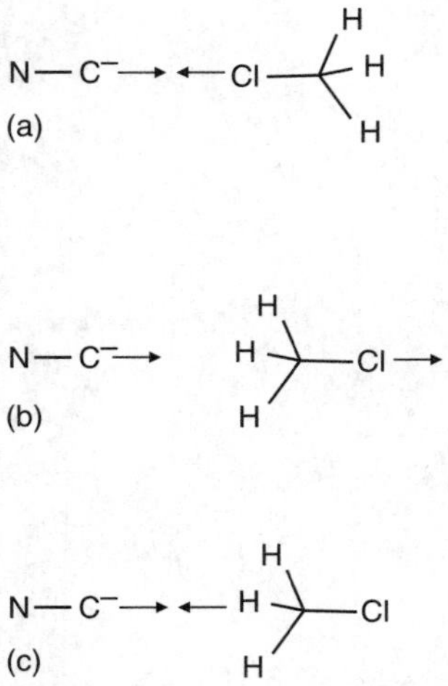

where r is the sum of the radii of A and B and v_R is their average relative speed. The probability that the reactants will have enough energy (E_a) during the collision is given by the Boltzmann distribution (see Chapter 14).

$$\text{Probability that the reactants have enough energy} = \exp\left(\frac{-E_a}{RT}\right)$$

Thus, the reaction rate is given by

$$\text{Rate} = \pi r^2 v_R \exp\left(\frac{-E_a}{RT}\right)[A][B]$$

This can be rewritten as

$$\text{Rate} = k[A][B] \quad \text{where} \quad k = \mathcal{A}\exp\left(\frac{-E_a}{RT}\right) \quad \text{and} \quad \mathcal{A} = \pi r^2 v_R$$

$\mathcal{A}$ is not a true constant but varies with temperature because v_R is temperature dependent.

These equations have exactly the same form as those obtained by empirical fitting of experimental data. However, the quantitative fit to experimental data is usually rather poor. This is because the assumption that molecules just need to collide in order to react is incorrect. The reactants usually need to collide in the correct orientation to give a reaction (figure 9.4) and so some collisions will not lead to a reaction. For charged reactants, there may be electrostatic attractions or repulsions that will alter the probability of collision. Extra terms can be added to the expression for $\mathcal{A}$ to allow for these factors, but these will vary on a case by case basis. Thus, while the concept is simple and the theory gives considerable insight into what happens to molecules during a reaction, the quantitative implementation of collision theory is usually very complex.

Fig. 9.4 Three of the possible collisions that could occur between CH_3Cl and CN^-. The reaction that occurs between CH_3Cl and CN^- is the displacement of the Cl^- by CN^- to give CH_3CN. This process involves the CN^- approaching from the opposite side to Cl to form a five-coordinate intermediate. For the reaction to occur, the reactants have to collide in the correct orientation and with sufficient energy. (a) This approach of the CN^- towards the Cl will not result in a reaction because the orientation is wrong. (b) This approach of the CN^- towards the C could result in a reaction, but if the molecules are moving in the same direction (small relative velocity) the collision is likely to be of low energy. (c) This collision with the correct orientation at high relative velocity will result in the reaction occurring.

Transition state theory

Transition state theory concentrates on the distribution of energy during the reaction rather than the molecular process (figure 9.5). This makes it easier to relate the kinetic and equilibrium thermodynamic properties of the system. It assumes that the reactants and the transition state are in equilibrium and there is then irreversible breakdown to give the products.

$$A + B \overset{K^{\ddagger}}{\rightleftharpoons} AB^{\ddagger} \overset{k_2}{\rightarrow} \text{Products}$$

The overall rate is given by the concentration of the transition state, $AB^{\ddagger}$, times the rate of breakdown.

$$\text{Rate} = k_2[AB^{\ddagger}]$$

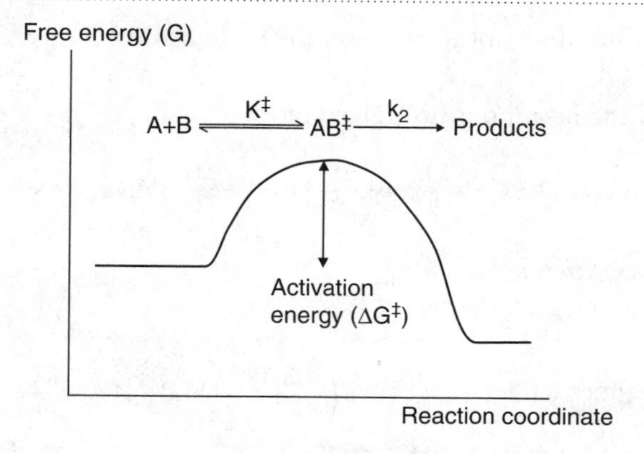

Fig. 9.5 Energy profile used in transition state theory.

The first step can be treated by equilibrium thermodynamics. As it is an equilibrium, there is an equilibrium constant given by

$$K^{\ddagger} = \frac{[AB^{\ddagger}]}{[A][B]}$$

$K^{\ddagger}$ can be converted into the free energy of activation ($\Delta G^{\ddagger}$) or the enthalpy ($\Delta H^{\ddagger}$) and entropy ($\Delta S^{\ddagger}$) of activation by standard thermodynamic equations

$$K^{\ddagger} = \exp\left(\frac{-\Delta G^{\ddagger}}{RT}\right) = \exp\left(\frac{\Delta S^{\ddagger}}{R}\right)\exp\left(\frac{-\Delta H^{\ddagger}}{RT}\right)$$

Thus

$$[AB]^{\ddagger} = K^{\ddagger}[A][B] = \exp\left(\frac{\Delta S^{\ddagger}}{R}\right)\exp\left(\frac{-\Delta H^{\ddagger}}{RT}\right)[A][B]$$

We cannot use equilibrium thermodynamics to evaluate k_2. In order to do this, we need to consider how energy is distributed within the transition state (called statistical thermodynamics). Much of the additional energy of the transition state compared to the reactants is present as vibrational energy. The transition state is inherently unstable and so these vibrations tend to lead to the transition state 'shaking itself apart'. Most possible vibrations result in the transition state breaking down to reactants, but one vibration will correspond to the transition state distorting itself into products. k_2 depends on how much energy is stored in that critical vibration and the probability that the critical vibration will lead to products instead of reactants. The energy stored in each vibration is given by kT/h where k is the Boltzmann constant and h is the Planck constant. The probability that this critical vibration leads to the formation of products rather than reactants is called the transmission coefficient, κ. Thus

$$k_2 = \kappa\frac{kT}{h}$$

κ is a constant for any given reaction and can have values between 0 and 1. In practice, κ is very difficult to determine and so it is frequently assumed that κ

equals 1, i.e. that the critical vibration in the transition state always leads to products.

The rate of the reaction is now given by

$$\text{Rate} = \kappa \frac{kT}{h} \exp\left(\frac{\Delta S^{\ddagger}}{R}\right) \exp\left(\frac{-\Delta H^{\ddagger}}{RT}\right) [A][B]$$

This can be rewritten as

$$\text{Rate} = k[A][B] \quad \text{where} \quad k = \mathcal{A} \exp\left(\frac{-\Delta H^{\ddagger}}{RT}\right) \quad \text{and} \quad \mathcal{A} = \kappa \frac{kT}{h} \exp\left(\frac{\Delta S^{\ddagger}}{R}\right)$$

This has an identical form to the equation derived from collision theory, although the two models are very different. In both cases, the rate at constant temperature is proportional to the concentrations of both reactants. In collision theory this is because higher concentrations lead to more collisions, in transition state theory this is because higher concentrations lead to a higher concentration of the transition state. Collision theory gives more insight into the molecular mechanism of a reaction but is very difficult to use quantitatively. Transition state theory gives more insight into the distribution of energy during a reaction and is easier to use and interpret quantitatively.

Order and molecularity

As we have seen above, for a simple single-step reaction

$$A + B \rightarrow \text{Products}$$

the theoretical rate of the reaction is given by

$$-\frac{d[A]}{dt} = -\frac{d[B]}{dt} = \frac{d[\text{Products}]}{dt} = k[A][B]$$

where k is called the rate constant. A more general form of the rate equation, as we have seen, is

$$-\frac{d[A]}{dt} = k[A]^x[B]^y[C]^z \cdots$$

where A, B, C, etc., are the reactant concentrations and x, y, z, etc., are the powers to which the reactant concentration terms are raised. These are called the orders of the reaction with respect to each reactant (the order with respect to A is x, etc.). The overall order of the reaction is $(x + y + z + \cdots)$. These orders can be any value. For simple reactions, they are usually integers (0, 1, 2, etc.)

and equal to the stoichiometry coefficients in the equation, but for more complex mechanisms they can be non-integral or even negative (as with the $H_2 + Br_2$ reaction above).

Order should not be confused with *molecularity*. The molecularity of a reaction is the minimum number of species involved in the *rate-determining step* of the reaction (often taken as synonymous with the slowest step of a multistep reaction, but see the next chapter for a discussion of this). This must be an integral value and is determined by the molecular mechanism of the 'slowest' step. In contrast, the order of the reaction depends on all the steps in the reaction mechanism and it is calculated experimentally from concentration versus time measurements. For example, consider the following reaction mechanism

$$A + B + C \xrightarrow{\text{fast}} AB + C \xrightarrow{\text{slow}} \text{Products}$$

with the rate equation

$$\frac{d[\text{Products}]}{dt} = k[A][B][C]$$

The overall order of the reaction is three, the order is one with respect to each reactant, and the molecularity is two, i.e. two species are involved in the slowest step.

WORKED EXAMPLE

For a reaction $A + B \rightarrow$ Products, calculate the shapes of the reactant concentration versus time plots if the reaction is overall (1) zeroth order, (2) first order for one reactant and zeroth order for the other reactant and (3) second order with first order for each reactant.

SOLUTION: We need to construct the appropriate rate equation and then integrate it to give an expression for [A] as a function of t.

(1) Zeroth order overall. In this case the rate of the reaction is independent of the concentration of either reactant.

$$A + B \xrightarrow{k} \text{Products} \qquad -\frac{d[A]}{dt} = k$$

In order to obtain [A] versus t we need to integrate the rate equation

$$\int_{[A]_o}^{[A]_t} d[A] = \int_0^t -k\, dt$$

where $[A]_o$ is the initial concentration of A at $t = 0$ and $[A]_t$ is the concentration of A at time t. This gives

$$[A]_t - [A]_o = -kt$$

Thus a plot of $[A]_t$ versus t is a straight line with slope of $-k$ (figure 9.6). Zeroth-order processes are very uncommon; however, enzyme-catalysed reactions under certain conditions do show zeroth-order behaviour (see Chapter 11).

(2) First order for one reactant and first order overall. In this case the reaction rate depends linearly on the concentration of a single reactant.

$$A + B \xrightarrow{k} \text{Products} \qquad -\frac{d[A]}{dt} = k[A]$$

Again, we need to integrate this equation to give [A] versus t.

$$\int_{[A]_o}^{[A]_t} \frac{1}{[A]} d[A] = \int_0^t -kdt$$

which gives

$$\log_e[A]_t - \log_e[A]_o = -kt \quad \text{or} \quad [A]_t = [A]_o \exp(-kt)$$

Thus a plot of $[A]_t$ versus t gives an exponential decay curve (figure 9.6) and a plot of $\log_e[A]_t$ versus t gives a straight line of slope $-k$. Some reactions are truly first order, for instance those that involve the breakdown of a single reactant such as radioactive decay. Under certain conditions some reactions which are actually second order may appear to be first order, called *pseudo-first-order* reactions. For example, if the concentration of B is much larger than A then $[B]_t$ will remain approximately constant during the reaction and the rate will only change as $[A]_t$ changes. This is the case for hydrolysis of ATP where the second reactant, water, is in vast excess.

(3) First order for two reactants and second order overall. In this case the reaction rate depends linearly on both reactants.

$$A + B \xrightarrow{k} \text{Products} \qquad -\frac{d[A]}{dt} = k[A][B]$$

If we assume that x moles of A and B have been used at time t, then

$$[A]_t = [A]_o - x \quad \text{and} \quad [B]_t = [B]_o - x$$

Differentiating the equation for $[A]_t$ with respect to x gives

$$\frac{d[A]_t}{dx} = -1$$

Therefore

$$-\frac{d[A]}{dt} = -\frac{d[A]}{dx}\frac{dx}{dt} = \frac{dx}{dt}$$

The rate equation can then be written as

$$\frac{dx}{dt} = k([A]_o - x)([B]_o - x)$$

Integration gives

$$\frac{1}{[A]_o - [B]_o}\log_e\left(\frac{[B]_o([A]_o - x)}{[A]_o([B]_o - x)}\right) = kt$$

This can be rearranged to give

$$\log_e\left(\frac{[A]_t}{[B]_t}\right) = ([A]_o - [B]_o)kt - \log_e\left(\frac{[B]_o}{[A]_o}\right)$$

In this case, the shape of the plot of $[A]_t$ versus t is complex and depends on the relative values of $[A]_o$ and $[B]_o$. More usefully, a plot of $\log_e([A]_t/[B]_t)$ versus t will give a straight line of slope $k([A]_o - [B]_o)$. If $[A]_o = [B]_o$ then we cannot use the above equation. However, as [A] must equal [B] at any subsequent time as well, the rate is now given by

$$-\frac{d[A]}{dt} = k[A]^2$$

which gives on integration

$$\frac{1}{[A]_t} - \frac{1}{[A]_o} = kt$$

In this case, the shape of [A] versus t is just a reciprocal curve (figure 9.6)

COMMENT: The curves in figure 9.6 are similar at very short time and so data points need to be collected over a large percentage of the total reaction to decide reaction orders unequivocally.

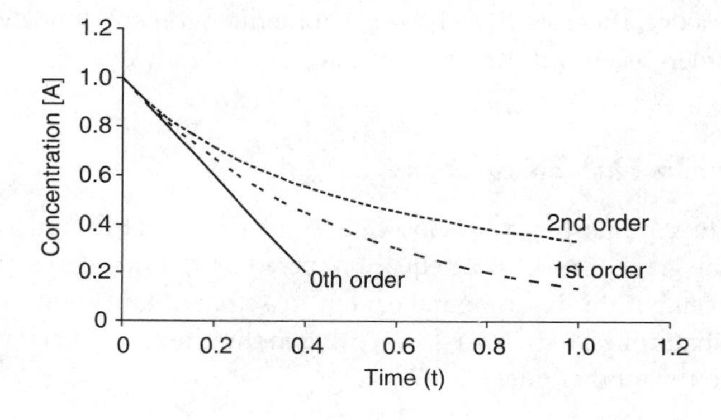

Fig. 9.6 Plots of reactant concentration, expressed relative to the initial concentration, versus time, in arbitrary units, for zeroth-, first- and second-order reactions of the form $d[A]/dt = -k[A]^x$. The curves show similar behaviour at short times, only differing significantly at longer times.

Reaction half-times

The half-time ($t_{\frac{1}{2}}$) for a reaction is the time taken for the concentration of a reactant to fall to half its initial value, i.e. $t_{\frac{1}{2}}$ is the point at which $[A]_t = \frac{1}{2}[A]_o$. In some cases, the use of consecutive half-times, i.e. the times taken for the concentration to fall from $[A]_o$ to $\frac{1}{2}[A]_o$, from $\frac{1}{2}[A]_o$ to $\frac{1}{4}[A]_o$, etc., can simplify the determination of reaction orders.

WORKED EXAMPLE

Calculate the first two consecutive half-times for a reaction if it is (1) zeroth order, (2) first order or (3) second order with equal initial concentrations of reactants.

SOLUTION: We have already obtained equations for $[A]_t$ versus t for these three cases above. For the first half-time all we need to do is set $[A]_t = \frac{1}{2}[A]_o$. For the second half-time, we need to replace $[A]_o$ with $\frac{1}{2}[A]_o$ and set $[A]_t = \frac{1}{4}[A]_o$

	First $t_{\frac{1}{2}}$	Second $t_{\frac{1}{2}}$
(1) Zeroth order	$[A]_o/2k$	$[A]_o/4k$
(2) First order	$\log_e 2/k$	$\log_e 2/k$
(3) Second order	$1/k[A]_o$	$2/k[A]_o$

For a first-order reaction the half-time is independent of the starting concentration, it takes the same time for the concentration to fall from $[A]_o$ to $\frac{1}{2}[A]_o$ as from $\frac{1}{2}[A]_o$ to $\frac{1}{4}[A]_o$. For a second-order reaction, each subsequent half-time is double the previous one (figure 9.7).

Experimental determination of reaction orders and rate constants

Reaction orders are obtained from experimental concentration versus time data. In practice, there are three basic experimental protocols for determining reaction orders, each with its own advantages and disadvantages.

Fit the data to a full rate equation

This involves proposing a specific rate equation, working out a suitable straight-line graph, or half-time equation, based on the integrated rate equation and seeing if the experimental data fit to a straight line, or fit the correct pattern of half-times. If the data do not fit, then the rate equation is wrong and we need to try another one.

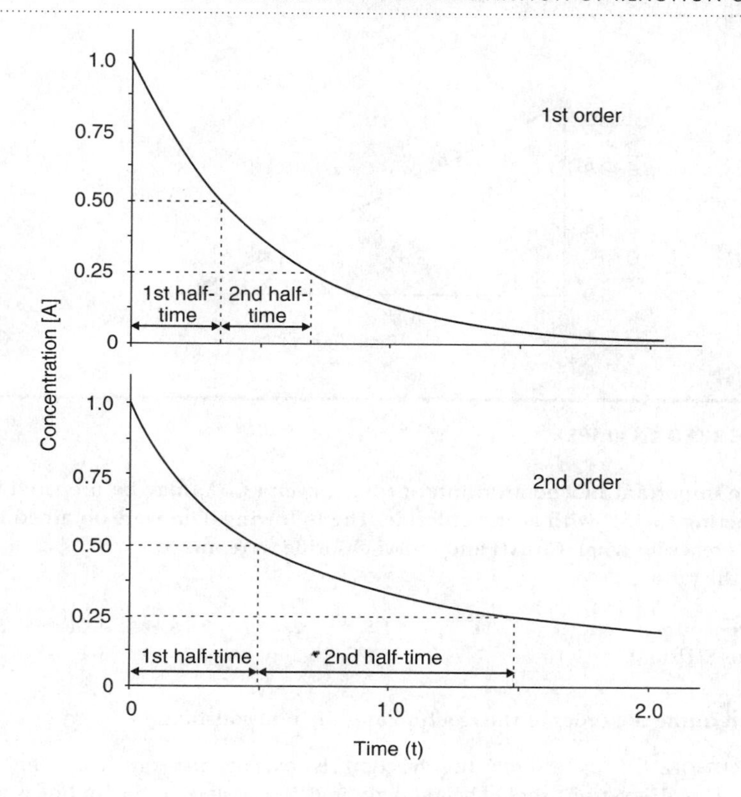

Fig. 9.7 For a first-order reaction, each successive half-time is the same. For a second-order reaction, each successive half-time is double the previous one.

WORKED EXAMPLE

The following data were obtained for the radioactive decay of ^{24}Na.

Time/hours	0	4	8	12	16	20	24
Activity/disintegrations per min	478	395	329	272	226	187	155

Show that radioactive decay follows first-order kinetics and calculate the corresponding rate constant.

SOLUTION: The equation relating concentration to time for a first-order reaction is

$$\log_e[A]_t - \log_e[A]_o = -kt$$

For the data above, a plot of $\log_e[\text{activity}]_t$ versus t is linear (figure 9.8), showing that the reaction is indeed first order. The gradient of the line is $-0{\cdot}0469$, thus

$$k = 0{\cdot}0469\ h^{-1}$$

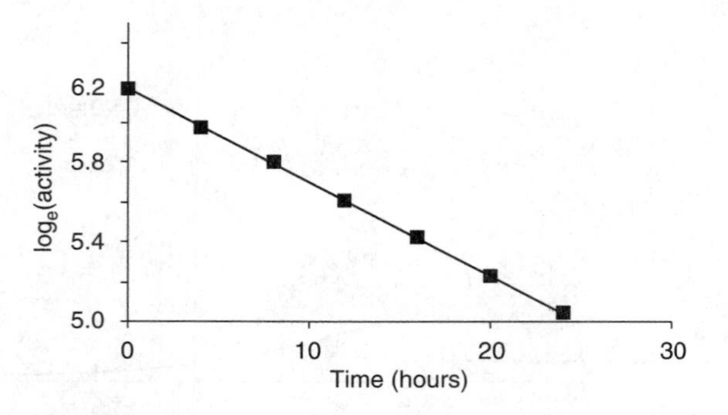

Fig. 9.8 A plot of $\log_e$(activity) versus time gives a straight line, proving that radioactive decay is first order.

WORKED EXAMPLE

The important biochemical intermediate, acetyl CoA, may be prepared by reacting CoASH with acetyl chloride. The following data were obtained for this reaction when CoASH and acetyl chloride were mixed in the ratio 1:1.

Time/min	0	1	1·5	2	2·75	4·41	5
[CoASH]/mM	10	6·7	5·8	5·2	4·2	3·1	3·0

Determine the order of the reaction and the rate constant.

SOLUTION: Let us assume the reaction is overall first order. A plot of $\log_e[\text{CoASH}]_t$ versus t should be a straight line. In this case, a straight line is not obtained (figure 9.9a) and so the reaction is not first order.

Let us assume the reaction is overall second order. As the initial concentrations of both reactants are equal, the equation relating concentration to time is

$$\frac{1}{[\text{CoASH}]_t} - \frac{1}{[\text{CoASH}]_o} = kt$$

In this case, a plot of $1/[\text{CoASH}]_t$ versus t is linear (figure 9.9b), showing that the reaction is overall second order. The gradient of the line is 0·0483, thus

$$k = 0{\cdot}0483\,\text{mM}^{-1}\ \text{min}^{-1}$$

COMMENT: This shows that the reaction is overall second order. However, as [CoASH] = [acetyl chloride] this is consistent with any of the following three rate laws (assuming that the orders with respect to the individual reactants are integral)

$$-d[\text{CoASH}]/dt = k[\text{CoASH}][\text{acetyl chloride}]$$

$$-d[\text{CoASH}]/dt = k[\text{CoASH}]^2$$

$$-d[\text{CoASH}]/dt = k[\text{acetyl chloride}]^2$$

The only way to distinguish between these is to measure concentrations as a function of time with $[\text{CoASH}]_o \neq [\text{acetyl chloride}]_o$.

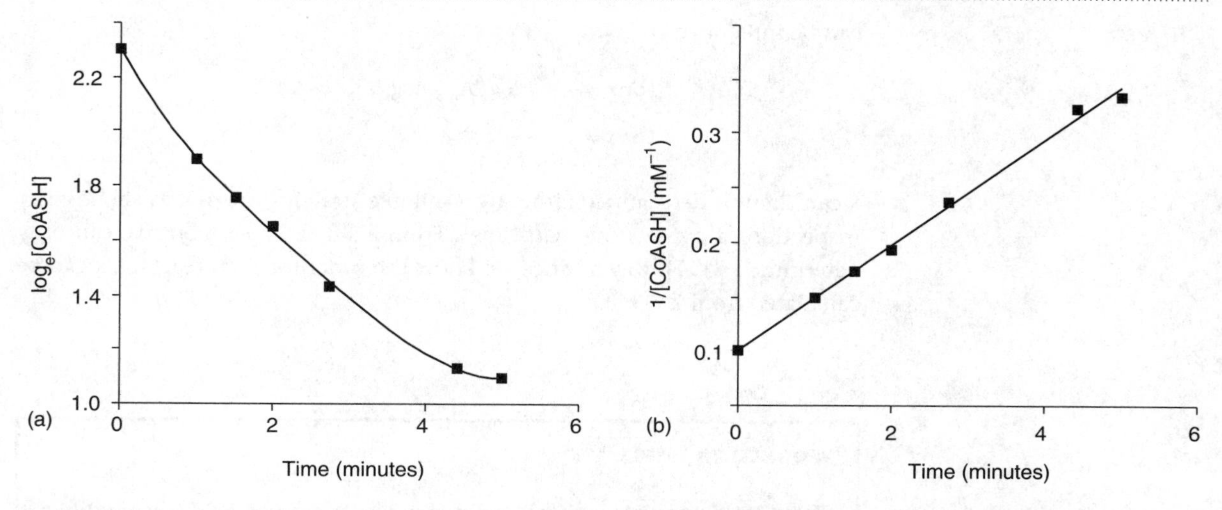

Fig. 9.9 (a) A plot of $\log_e$[concentration] versus time does not give a straight line whereas (b) a plot of 1/[concentration] versus time does, proving that the reaction between CoASH and acetyl chloride is second order.

The advantages of this method are: (i) that once we get a straight line, we know we have the answer; (ii) that we only need one set of experimental data; and (iii) that the same graph will also give us the rate constant. The disadvantage is that it may take many attempts to guess the right rate equation and, in very complex cases, we may never guess the correct form. This is a particular problem with non-integral orders. It is also important to have very good quality data to ensure that we have a good fit to a straight line.

Have all reactants except one in large excess

Consider a reaction

$$A + B \rightarrow \text{Products}$$

with a rate equation of the form

$$-\frac{d[A]}{dt} = k[A]^x[B]^y$$

If B is in large excess, then its concentration is effectively constant during the reaction (so little of B is used up) and so the rate equation simplifies to

$$-\frac{d[A]}{dt} = k^{app}[A]^x$$

where k^{app} is the apparent rate constant, $k^{app} = k[B]^y$. A plot of $\log_e$(rate) versus $\log_e[A]_t$ will then give a straight line of gradient x. Alternatively, this equation can be integrated to give

$$\text{For } x = 1 \qquad \log_e[A]_t - \log_e[A]_o = -k^{app}t$$

$$\text{For } x \neq 1 \qquad [A]_o^{(1-x)} - [A]_t^{(1-x)} = (1-x)k^{app}t$$

x can then be determined either by a suitable straight-line plot as above or by inspection of consecutive half-times. From a single experiment we can only determine k^{app}. However, once we know the order for each reactant k can be calculated from k^{app}.

WORKED EXAMPLE

Sheep liver pyruvate carboxylase at a concentration of 1 μM was modified with cyanate and the extent of modification was followed by measuring the loss of the acetyl-CoA-dependent activity. From the data below, verify that the rate equation is

$$-\frac{d[\text{enzyme}]}{dt} = k[\text{enzyme}][\text{cyanate}]$$

and determine the pseudo-first-order rate constants (k^{app}) and the second-order rate constant (k) for the modification.

Time / sec	(% initial activity)			
	[cyanate]/mM			
	50	100	200	300
0	100	100	100	100
600	97	94	89	84
1200	94	89	79	71
2400	89	79	63	50
3600	84	71	50	35
4800	79	63	40	25
6000	75	56	31	18
9000	64	42	18	7.5
12000	56	31	10	3

SOLUTION: We shall assume that the rate of enzyme deactivation by cyanate is

$$-\frac{d[\text{enzyme}]}{dt} = k[\text{enzyme}][\text{cyanate}]$$

Because the cyanate concentration is many orders of magnitude greater than the enzyme concentration, the cyanate concentration remains effectively constant during the reaction. The rate expression for the reaction can be written as

$$-\frac{d[\text{enzyme}]}{dt} = k^{app}[\text{enzyme}]$$

where k^{app} is the pseudo-first-order rate constant ($k^{app} = k[\text{cyanate}]$). This expression can be integrated to give

$$\log_e\left(\frac{[\text{enzyme}]}{[\text{enzyme}]_o}\right) = -k^{app}t$$

k^{app} may be obtained by plotting $\log_e([\text{enzyme}]/[\text{enzyme}]_o)$ against t (figure 9.10a). This gives a straight line for each value of [cyanate], confirming that the reaction is first order with respect to [enzyme]. k^{app} values are obtained from the slopes of the plots, giving

[Cyanate]/mM	k^{app}/s^{-1}
50	$0{\cdot}486 \times 10^{-4}$
100	$0{\cdot}971 \times 10^{-4}$
200	$1{\cdot}914 \times 10^{-4}$
300	$2{\cdot}906 \times 10^{-4}$

The second-order rate constant, k, is given by plotting k^{app} against [cyanate] (figure 9.10b)

$$k^{app} = k[\text{cyanate}]$$

Again, this gives a straight line confirming that the reaction is first order with respect to [cyanate]. k is obtained from the slope of the plot, giving

$$k = 9{\cdot}66 \times 10^{-7}\,\text{mM}^{-1}\,\text{s}^{-1}$$
$$= 9{\cdot}66 \times 10^{-4}\,\text{M}^{-1}\,\text{s}^{-1}$$

The advantages of this method are: (i) that data interpretation is very easy; and (ii) that non-integral orders can be easily determined. The disadvantages are: (i) that we need to repeat the experiment once per reactant; and (ii) that we cannot always use physiological conditions because most reactants have to be present at very high concentrations. We cannot be sure that the same mechanism applies at very high reactant concentrations, for instance we may have reactant inhibition that would dominate at a large excess of that reactant.

Measure initial reaction rates keeping the concentrations of all reactants except one constant

Consider a reaction

$$A + B \rightarrow \text{Products}$$

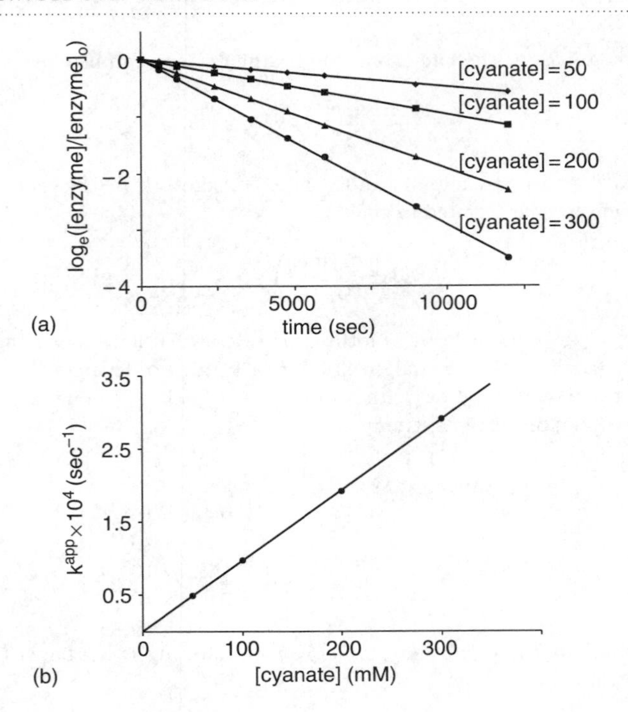

Fig. 9.10 (a) Pseudo-first-order plots for the reaction between sheep liver pyruvate carboxylase and cyanate give straight lines, proving that the reaction is first order with respect to [enzyme]. (b) Plotting the pseudo-first-order rate constant versus [cyanate] proves that the reaction is first order with respect to [cyanate], the gradient giving the true second-order rate constant.

with an initial rate equation of the form

$$-\frac{d[A]_o}{dt} = k[A]_o^x[B]_o^y$$

If $[B]_o$ is kept constant, the rate equation simplifies to

$$-\frac{d[A]_o}{dt} = k^{app}[A]_o^x$$

where $k^{app} = k[B]_o^y$. A series of experiments with different $[A]_o$ can be performed and the initial rates calculated. The data can then be analysed in an identical fashion to the situation where all reactants bar one are in large excess.

The advantages of this method are: (i) that data interpretation is very easy; (ii) that non-integral orders can easily be determined; and (iii) that it is not necessary to use a large excess of any of the reagents. The disadvantages are: (i) that we need to repeat the experiment many times per reactant with different initial concentrations; and (ii) that as we are only looking at initial rates we shall miss any processes that occur later in the reaction, such as product inhibition, or further reactions between reactants and products. This can also be an advantage as a useful simplifying strategy for very complex reactions.

Effect of temperature on the rate of a reaction

As we have already seen, reaction rates are usually strongly temperature dependent. The temperature dependence of the rate constant is given by the Arrhenius equation

$$k = \mathcal{A}\exp\left(\frac{-E_a}{RT}\right)$$

Both collision theory and transition state theory give the same general equation although the interpretation of the constants is a little different.

- Collision theory — $\mathcal{A}$ is a probability term including the speed, size and orientation of the molecules during the collision; E_a is the activation energy
- Transition state theory — $\mathcal{A}$ depends on the entropy of activation, $\Delta S^{\ddagger}$; E_a is the enthalpy of activation, $\Delta H^{\ddagger}$

In many textbooks, activation energy and activation enthalpy are used interchangeably.

If $\Delta H^{\ddagger}$ is assumed to be independent of temperature, then the Arrhenius equation can be differentiated and reintegrated (see problems below) to give

$$\log_e\left(\frac{k_{T_1}}{k_{T_2}}\right) = -\frac{\Delta H^{\ddagger}}{R}\left(\frac{1}{T_1} - \frac{1}{T_2}\right)$$

where k_{T_1} is the rate constant at temperature T_1 and k_{T_2} is the rate constant at temperature T_2.

WORKED EXAMPLE

A reaction rate doubles on going from 293 K to 303 K. What is the activation enthalpy of the reaction? The rate constant for the reaction is 105.5 s^{-1} at 293 K. What is the activation entropy of the reaction? Comment on this value.

SOLUTION: The relationship between k and T at two temperatures is given by

$$\log_e\left(\frac{k_{T_1}}{k_{T_2}}\right) = -\frac{\Delta H^{\ddagger}}{R}\left(\frac{1}{T_1} - \frac{1}{T_2}\right)$$

In this case, $T_1 = 293\,K$, $T_2 = 303\,K$ and $k_{T_1}/k_{T_2} = \frac{1}{2}$. Substituting these values gives

$$\Delta H^{\ddagger} = 51{\cdot}4\,kJ\,mol^{-1}$$

The rate constant is given by

$$k = \mathcal{A}\exp\left(\frac{-\Delta H^{\ddagger}}{RT}\right)$$

In this case, $k = 105.5\,s^{-1}$, $\Delta H^{\ddagger} = 51\,400\,J\,mol^{-1}$ and $T = 293\,K$. Thus

$$\mathcal{A} = 1{\cdot}538 \times 10^{11}\,s^{-1}$$

From transition state theory, assuming that the transmission coefficient is one, $\mathcal{A}$ is given by

$$\mathcal{A} = \frac{kT}{h}\exp\left(\frac{\Delta S^{\ddagger}}{R}\right)$$

and so

$$\Delta S^{\ddagger} = -30{\cdot}6\,J\,K^{-1}\,mol^{-1}$$

COMMENT: A negative $\Delta S^{\ddagger}$ indicates that the transition state is more ordered than the reactants. This is common for bimolecular reactions where two reactant molecules have to come together to form the transition state.

Effect of ionic strength on the rate of a reaction

The ionic strength of a solution can affect the rate of a reaction as well as the position of equilibrium (Chapter 8). These effects are called *salt effects* and they arise because changes in the ionic strength change the activity coefficients of charged reactants and the transition state. An increase in ionic strength stabilises all charged species in solution. If the reactants and the transition state have the same total charge both would be stabilised equally and so the activation energy will be unchanged. If the transition state has a greater total charge than the reactants, it will be stabilised more and so the activation energy will be reduced. The observation of a salt effect implies that there is a difference in total charge between the reactants and the transition state and so can give mechanistic information about a reaction. An increase in rate with increasing ionic strength implies that the transition state has a higher charge than the reactants, while a decrease in rate implies that the transition state has a lower charge than the reactants.

Salt effects can be dealt with quantitatively within transition state theory. In the expression for the equilibrium between the reactants and the transition state

$$K^{\ddagger} = \frac{[AB]^{\ddagger}}{[A][B]}$$

we have to use activities rather than concentrations. The relationship between activity coefficients and ionic strength is given by the Debye–Hückel theory. For aqueous solutions of low ionic strength, the final equation is

$$\log_{10}\frac{k'}{k_o} = z_A z_B \sqrt{I}$$

where k' is the rate constant at an ionic strength I, k_o is the rate constant at zero ionic strength, and z_A and z_B are the charges with signs of the ions taking part in the reaction $A + B \rightarrow$ products (see Problems below for the derivation). This effect is known as the *primary salt effect* (see Problems for an example of this effect). If one of the reactants is uncharged, the product $z_A \times z_B$ is zero and the rate is independent of the ionic strength.

A second type of salt effect is observed in pH-dependent reactions (see next chapter) when changes in the ionic strength can affect the degree of dissociation of weak acids and bases and hence the concentrations of H^+ or OH^-. This indirect effect on the reaction rate is known as the *secondary salt effect.*

Effect of isotopic substitution on the rate of a reaction

The rate of a reaction may change if one of the atoms involved in bond breaking or bond making is replaced by a different isotope. Consider the replacement of hydrogen in a C–H bond by deuterium (^2H or D). Because of the heavier mass of deuterium, the zero-point energy of the C–D bond is less than that of C–H (see Chapter 14 for a discussion of zero-point energy). The dissociation energy of the C–D bond is approximately 5.0 kJ mol^{-1} greater than that of the C–H bond (figure 9.11). As it takes more energy to break the C–D bond compared to the C–H bond, if bond breaking occurs in the rate-determining step then the activation energy will be higher and hence the reaction rate will be slower.

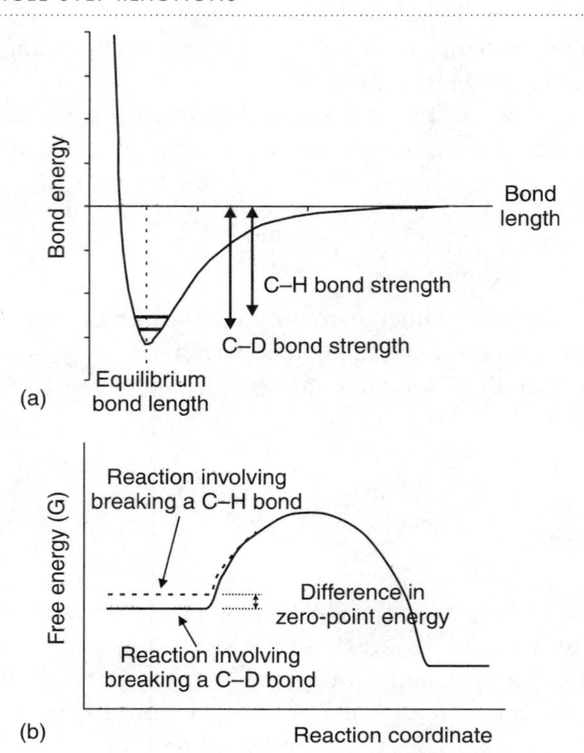

Fig. 9.11 (a) The C–D bond is slightly stronger than the C–H bond because the former has a slightly higher mass and hence a slightly lower zero-point energy. (b) This results in the activation energy for breaking a C–D being slightly higher than for breaking a C–H bond.

WORKED EXAMPLE

Assuming that a C–H bond is broken completely in the rate-determining step of a reaction, calculate the decrease in the rate at 300 K if the hydrogen is replaced by deuterium.

SOLUTION: The rate constant for the reaction that involves breaking a C–H bond is given by

$$k_H = \mathcal{A}\exp\left(\frac{-\Delta H^{\ddagger}_{C-H}}{RT}\right)$$

where the activation enthalpy, $\Delta H^{\ddagger}_{C-H}$ includes the C–H bond strength. If the hydrogen is replaced by a deuterium, then the rate constant is given by

$$k_D = \mathcal{A}\exp\left(\frac{-\Delta H^{\ddagger}_{C-D}}{RT}\right)$$

As the C–D bond strength is 5.0 kJ mol^{-1} greater than that of the C–H bond,

$$\Delta H^{\ddagger}_{C-D} = \Delta H^{\ddagger}_{C-H} + 5000\,J\,mol^{-1}$$

The ratio of the rate constants is given by

$$\frac{k_H}{k_D} = \frac{A\exp\left(\frac{-\Delta H^{\ddagger}_{C-H}}{RT}\right)}{A\exp\left(\frac{-(\Delta H^{\ddagger}_{C-H} + 5000)}{RT}\right)} = \frac{A\exp\left(\frac{-\Delta H^{\ddagger}_{C-H}}{RT}\right)}{A\exp\left(\frac{-\Delta H^{\ddagger}_{C-H}}{RT}\right)\exp\left(\frac{-5000}{RT}\right)} = 7{\cdot}4$$

Thus replacing hydrogen with deuterium reduces the reaction rate by a factor of 7·4, if the C–H bond is broken completely in the rate-determining step.

WORKED EXAMPLE

Alcohol dehydrogenases catalyse the following class of reactions

$$RCH_2OH + NAD^+ \rightleftharpoons RCHO + NADH + H^+$$

Replacing RCH_2OH with RCD_2OH leads to a reduction in the reaction rate by a factor of three to five for the yeast enzyme, whereas replacing RCH_2OH with RCH_2OD leads to no reduction in rate. Replacing RCH_2OH with RCD_2OH gives no reduction in rate for the horse liver enzyme. Comment on these observations.

SOLUTION: The reaction involves both proton transfer and hydride transfer from the alcohol (figure 9.12). The observation of a significant kinetic isotope effect for the yeast enzyme with RCD_2OH indicates that hydride transfer occurs in the rate-determining step. The kinetic isotope effect of three to five is less than the theoretical maximum of 7.4 calculated above. This suggests that the C–D bond is not fully broken in the transition state but that partial hydride transfer has occurred (figure 9.12). The lack of a kinetic isotope effect with RCH_2OD indicates that proton transfer does not occur in the rate-determining step but is a separate fast step.

The lack of isotope effect with RCD_2OH for the horse liver enzyme indicates that hydride transfer does not take place during the rate-determining step. In this case, the overall mechanism is similar to the yeast enzyme but the rate-determining step is dissociation of the enzyme–NADH complex.

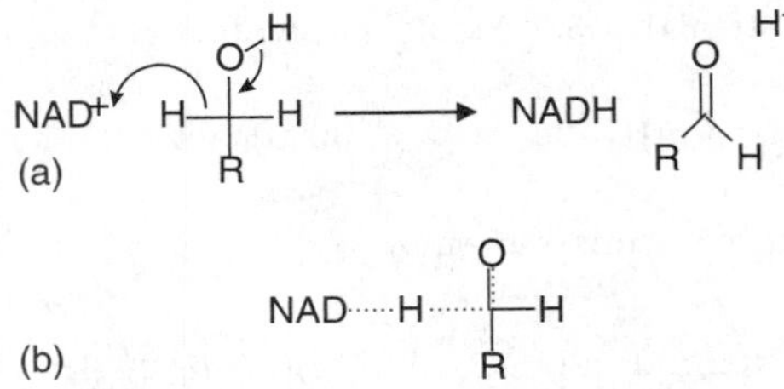

Fig. 9.12 (a) Alcohol dehydrogenase catalyses the conversion of an alcohol to an aldehyde. This involves loss of a proton (H^+) from the –OH group and transfer of hydride (H^-) from carbon to NAD^+. (b) Schematic representation of the transition state of the rate-determining step for yeast alcohol dehydrogenase, consistent with the kinetic isotope effect studies. The dotted lines represent partial bonds. The proton from the –OH group has already been lost. The H^- is being transferred from C to NAD, with the C–H bond having been partially broken and the NAD–H bond partially formed.

FURTHER READING

S.R. Logan, 1996, 'Fundamentals of chemical kinetics', Longman—*a comprehensive introduction to kinetics from a chemical perspective.*

M.J. Pilling and P.W. Seakins, 1995, 'Reaction kinetics', Oxford University Press—*a more advanced account of chemical kinetics, including material outside the interests of most biochemists.*

PROBLEMS

1. Distinguish between the molecularity and order of a chemical reaction. An organic acid decarboxylates spontaneously at 298 K and pH 6. The evolution of CO_2 from 10 cm^3 of a 10 mM solution was measured at various times as follows

Time/min	25	50	75	100	125	150	200
CO_2 evolved/cm^3	0·64	1·10	1·45	1·70	1·90	2·05	2·23

Determine the order and rate constant for this reaction. Assume that no CO_2 remains in solution.

2. The stoichiometry of the reaction between *N*-acetylcysteine and iodoacetamide is 1:1. Use the following results for the reaction of 1 mM *N*-acetylcysteine with 1 mM iodoacetamide to determine the overall order of the reaction and the rate constant.

N-acetylcysteine concentration/mM	0·770	0·580	0·410	0·315	0·210	0·155
Time/s	10	20	40	60	100	150

The results below were obtained for the reaction of 1 mM *N*-acetylcysteine with 2 mM iodoacetamide under different conditions.

N-acetylcysteine concentration/mM	0·74	0·58	0·33	0·21	0·12	0·09
Time/s	5	10	25	35	50	60

Determine the order with respect to each reactant and calculate the rate constant under these conditions.

3. The reaction between two compounds, A and B, in solution is first order in B. The following results were obtained for an initial concentration of B of 1·0 M at 300 K.

Concentration of A/mM	1·000	0·692	0·478	0·29	0·158	0·110
Time/s	0	20	40	70	100	120

Determine the order in A and calculate the rate constant for the reaction. Would the half-time of A be different if the initial concentration of B = 0·5 M?

4. By differentiation and reintegration of the Arrhenius equation, derive the equation

$$\log_e\left(\frac{k_{T_1}}{k_{T_2}}\right) = -\frac{\Delta H^{\ddagger}}{R}\left(\frac{1}{T_1} - \frac{1}{T_2}\right)$$

5. The rate constant of a reaction is 15 M^{-1} min^{-1} at 298 K and 37 M^{-1} min^{-1} at 308 K. What is the activation energy for the reaction and the rate constant at 283 K?

6. The rate constant for the association of an inhibitor with carbonic anhydrase was studied as a function of temperature, with the results shown below. What is the activation energy for this reaction?

T/K	289·0	293·5	298·1	303·2	308·0	313·5
$k \times 10^{-6}/M^{-1}\,s^{-1}$	1·04	1·34	1·53	1·89	2·29	2·84

7. The first-order decomposition of diacetyl ($CH_3COCOCH_3$) has values of $\Delta G^{\circ\ddagger}$ and $\Delta H^{\circ\ddagger}$ of 231.5 kJ mol^{-1} and 264.2 kJ mol^{-1} respectively at 558 K. Calculate the value of $\Delta S^{\circ\ddagger}$ and comment on its magnitude.

8. By applying the Debye–Hückel theory for dilute solutions (Chapter 8)

$$\log_{10}\gamma_i = -\tfrac{1}{2}z_i^2\sqrt{I}$$

to the equilibrium between the reactants A and B and the activated complex AB

$$A + B \overset{K^{\ddagger}}{\rightleftharpoons} AB^{\ddagger} \overset{k_2}{\rightarrow} \text{Products}$$

where

$$K^{\ddagger} = \frac{a_{AB^{\ddagger}}}{a_A a_B} = \frac{\gamma_{AB^{\ddagger}}}{\gamma_A\gamma_B}\frac{[AB^{\ddagger}]}{[A][B]} \quad \text{and} \quad \text{Rate} = k_2[AB^{\ddagger}]$$

use transition state theory to derive the equation relating the change in rate constant to the ionic strength of the solution

$$\log_{10}\frac{k'}{k_o} = z_A z_B\sqrt{I}$$

k' = rate constant at ionic strength I, k_O = rate constant at zero ionic strength, z_A and z_B are the charges (with signs) on the ionic species involved, and I = ionic strength.
For the reaction

$$Cr(H_2O)_6^{3+} + SCN^- \rightleftharpoons Cr(H_2O)_5SCN^{2+} + H_2O$$

the following data were obtained

k/k_O	0·87	0·81	0·76	0·71	0·62	0·50
$I \times 10^3$/M	0·4	0·9	1·6	2·5	4·9	10·0

Are these data consistent with this equation? What is the dependence of the rate of the reverse reaction on ionic strength?

9. Given that the bond strength of the C–^{3}H bond is 6.3 kJ mol^{-1} greater than that of C–^{1}H bond, calculate the change in rate that would be expected on replacing ^{1}H by ^{3}H in a C–H bond that is fully broken in the rate-determining step of a reaction at 300 K. Would you expect this value to be observed in practice?

10

Applications of chemical kinetics to multistep reactions

KEY POINTS

- The rate equation for a multistep reaction can be derived from the rate equations for the individual steps.
- The *rate-determining step* of a reaction is the step with the highest activation energy. It is often used to indicate the slowest step in a process but this idea can be very misleading.
- The *steady-state approximation* can be used when the concentration of a particular intermediate remains constant. This approximation is usually valid for short-lived intermediates that are present at low concentrations, but not at the very start of the reaction.
- The *steady-state approximation* can be used to greatly simplify the rate equations for complex reactions.

The previous chapter dealt with the kinetic behaviour of single-step reactions, e.g. $A+B \rightarrow$ Products. Most biochemical reactions are much more complex than this, involving several discrete steps, for instance $A+B \rightarrow$ Intermediates $\rightarrow$ Products, and the reactants and products frequently take part in other reactions as well, as in metabolic pathways. Each step in these more complicated reaction sequences can be treated in the same way as a single-step reaction with a simple rate equation, but the overall rate equation generally becomes more complex.

Parallel reactions

In this case, a compound A can give rise to two distinct products, B and C,

$$A \xrightarrow{k_1} B$$

$$A \xrightarrow{k_2} C$$

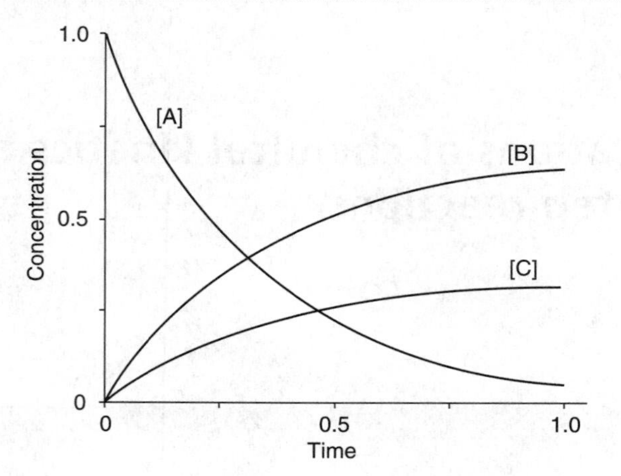

Fig. 10.1 Plots of concentration versus time for two parallel reactions, $A \xrightarrow{k_1} B$ and $A \xrightarrow{k_2} C$, where $k_1 = 2k_2$. All three curves are first order.

with respective rate constants k_1 and k_2. The rates of production of B and C are given by the equations introduced in the previous chapter

$$\frac{d[B]}{dt} = k_1[A] \quad \text{and} \quad \frac{d[C]}{dt} = k_2[A]$$

The rate of loss of A is simply given by the rate of loss of A to B plus the rate of loss of A to C

$$-\frac{d[A]}{dt} = k_1[A] + k_2[A] = (k_1 + k_2)[A]$$

The disappearance of A is still first order, but the overall rate constant is now the sum of the rate constants for the individual processes (figure 10.1)

WORKED EXAMPLE

Erythrocytes were labelled with ^{59}Fe, which decays with a half-time of 45 days. The erythrocytes were harvested, washed and counted at set intervals.

Time/hours	0	12	24	48	96	192	384
Activity/counts	532	525	518	504	478	429	347

What is the biological half-time (rate at which Fe exchanges with the environment) for Fe in erythrocytes?

SOLUTION: The activity of ^{59}Fe is lost by two processes, radioactive decay and loss to the environment. Both of these are first-order processes, thus

$$-\frac{d[Fe]}{dt} = (k_1 + k_2)[Fe] = k^{app}[Fe]$$

where k_1 is the rate of radioactive decay and k_2 is the rate of loss to the environment. The overall first-order rate constant, k^{app}, can be determined from the experimental data by plotting $\log_e$(activity) versus time to give

$$k^{app} = 0{\cdot}0011\,h^{-1}$$

The rate constant for radioactive decay is given by

$$k_1 = \frac{\log_e(2)}{t_{\frac{1}{2}}} = 0{\cdot}0006\,h^{-1}$$

Thus,

$$k_2 = k^{app} - k_1 = 0{\cdot}0005\,h^{-1}$$

giving the biological half-time as

$$t_{\frac{1}{2}} = \frac{\log_e(2)}{k_2} = 57{\cdot}8 \text{ days}$$

COMMENT: Although Fe is an essential component of many metalloproteins, the concentration of free Fe in solution within cells is very low because it is highly toxic. Thus Fe within cells is bound to macromolecules and scavenger proteins exist to bind any free Fe. It is these systems for holding Fe that lead to its very slow exchange with the extracellular environment.

Reversible reactions

In principle, all reactions are reversible but in many cases we can ignore the back reaction because it is much slower than the forward reaction. If the rate of the back reaction becomes significant, then the rate laws are more complex. Consider the reaction

$$A \underset{k_{-1}}{\overset{k_1}{\rightleftharpoons}} B$$

where k_1 and k_{-1} are the rate constants for the forward and back reactions. The rate equation for this reaction is

$$-\frac{d[A]}{dt} = \frac{d[B]}{dt} = k_1[A] - k_{-1}[B]$$

Because the reaction is reversible, at some time it will reach equilibrium with no net formation of A or B, i.e. the reaction rates in both directions will be equal. At this point $d[A]/dt = d[B]/dt = 0$. If $[A]_{eq}$ and $[B]_{eq}$ are the equilibrium concentrations of A and B then

$$k_1[A]_{eq} = k_{-1}[B]_{eq} \quad \text{or} \quad \frac{k_1}{k_{-1}} = \frac{[B]_{eq}}{[A]_{eq}} = K$$

Thus, the equilibrium constant is the ratio of the forward and backward rate constants (see Chapter 3 for a more detailed discussion). This can be used as an alternative definition of an equilibrium constant and provides another direct link between kinetics and thermodynamics.

Returning to the rate equation, if the initial concentration of A is $[A]_o$ and of B is zero then

$$[A] = [A]_o - [B] \qquad \text{and} \qquad [A]_{eq} = [A]_o - [B]_{eq}$$

and

$$\frac{d[B]}{dt} = k_1([A]_o - [B]) - k_{-1}[B]$$

Using $k_1([A]_o - [B]_{eq}) = k_{-1}[B]_{eq}$ to replace k_{-1} gives

$$\frac{d[B]}{dt} = k_1(A_o - [B]) - \frac{k_1([A]_o - [B]_{eq})}{[B]_{eq}}[B]$$

which can be rearranged to give

$$\frac{d[B]}{dt} = \frac{k_1[A]_o}{[B]_{eq}}([B]_{eq} - [B])$$

This can be expressed in an alternative form, again by using $k_1([A]_o - [B]_{eq}) = k_{-1}[B]_{eq}$, to give

$$\frac{d[B]}{dt} = (k_1 + k_{-1})\left([B]_{eq} - [B]\right)$$

This shows that the rate constant for the approach to equilibrium is equal to the sum of the first-order rate constants k_1 and k_{-1}. As k_1, k_{-1} and $[B]_{eq}$ are constants, this expression may be integrated (noting that $[B] = 0$ when $t = 0$) to give

$$\log_e \frac{[B]_{eq}}{[B]_{eq} - [B]} = (k_1 + k_{-1})t$$

This equation is sometimes written in the form

$$\log_e \frac{\Delta B_o}{\Delta B_t} = (k_1 + k_{-1})t$$

where ΔB_o is the difference between the initial and equilibrium concentrations of B, and ΔB_t is the difference between the concentration of B at time t and at equilibrium.

$$\Delta B_o = [B]_o - [B]_{eq} \qquad \text{and} \qquad \Delta B_t = [B] - [B]_{eq}$$

In this form, the equation will apply regardless of whether there is any B present at the start and of whether $[B]_o$ is smaller or larger that $[B]_{eq}$.

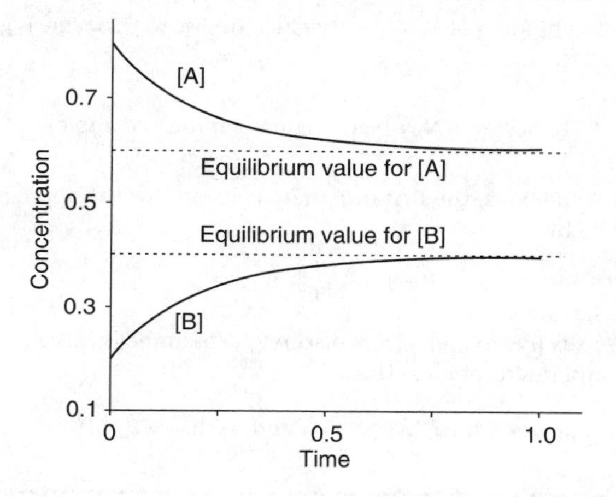

Fig. 10.2 Plots of concentration versus time for the reversible reaction $A \underset{k_{-1}}{\overset{k_1}{\rightleftharpoons}} B$ where $k_1 = \frac{2}{3}k_{-1}$. The dotted lines give the equilibrium concentrations.

It will also work for A. Both [A] and [B] approach equilibrium at the same rate (figure 10.2).

The treatment of second-order reactions is rather more complex but can be simplified if the perturbation from equilibrium is very small. We shall just state the result for one such reaction here.

$$A + B \underset{k_{-1}}{\overset{k_1}{\rightleftharpoons}} C \qquad -\frac{d(\Delta C_t)}{dt} = \{k_1([A]_{eq} + [B]_{eq}) + k_{-1}\}\Delta C_t$$

where ΔC_t is the difference in concentration of C from its equilibrium value and $[A]_{eq}$ and $[B]_{eq}$ are the equilibrium concentrations of A and B.

This analysis is the basis of the *relaxation approach* for the kinetic study of reversible reactions. If a reaction is at equilibrium and the equilibrium is suddenly altered (e.g. by altering the temperature or pressure) then the equilibrium will 'relax' to a new state, with a time course defined by equations of the kind given above. This approach has great advantages for studying *rapid reactions*, such as the reversible binding of a substrate to an enzyme, because it avoids the need to start the reaction by mixing the reactants, a process that can take a significant time of the order of 1 ms. It will only give the sum of k_1 and k_{-1} for opposed first-order reactions, but as the equilibrium constant equals the ratio of k_1 and k_{-1}, both can be calculated.

WORKED EXAMPLE

The binding of *N,N*-biacetylglucosamine to lysozyme was studied using a temperature-jump experiment to perturb the equilibrium. The first-order rate constants for the approach to equilibrium are

[lysozyme] + [*N,N*-biacetylglucosamine]/M	2×10^{-4}	4×10^{-4}	6×10^{-4}
k/s^{-1}	1.8×10^{3}	2.7×10^{3}	3.6×10^{3}

Calculate the rate constants for formation and dissociation of the complex.

SOLUTION: The binding of *N,N*-biacetylglucosamine to lysozyme is given by the equation

$$\text{lysozyme} + N,N\text{-biacetylglucosamine} \underset{k_{-1}}{\overset{k_1}{\rightleftharpoons}} \text{complex}$$

As we have seen above, the first-order rate constant for the approach to equilibrium is given by

$$k = k_1([A]_{eq} + [B]_{eq}) + k_{-1}$$

Plotting k versus [lysozyme] + [*N,N*-biacetylglucosamine] gives a straight line of gradient k_1 and intercept k_{-1}. Thus,

$$k_1 = 4{\cdot}5 \times 10^6\ M^{-1}\ s^{-1} \qquad \text{and} \qquad k_{-1} = 9 \times 10^2\ s^{-1}$$

Consecutive reactions

An important class of reactions comprises those in which the product of a reaction is subsequently converted to a second product. Consider the process

$$A \xrightarrow{k_1} B \xrightarrow{k_2} C$$

with rate constants k_1 and k_2 for the two reactions respectively. B could be a stable species in its own right, such as a component in a metabolic pathway, or B could be a reactive intermediate in a multistep reaction mechanism. The rate equations for this sequence of reactions are

$$\frac{d[A]}{dt} = -k_1[A]$$

$$\frac{d[B]}{dt} = k_1[A] - k_2[B]$$

$$\frac{d[C]}{dt} = k_2[B]$$

The solution to these equations is relatively complex and so we will simply state them here. If the initial concentration of A is $[A]_o$ and the initial concentrations of B and C are zero, then

$$[A] = [A]_o \exp(-k_1 t)$$

$$[B] = \frac{[A]_o k_1}{k_2 - k_1}\{\exp(-k_1 t) - \exp(-k_2 t)\}$$

$$[C] = [A]_o - \frac{[A]_o k_2}{k_2 - k_1}\exp(-k_1 t) + \frac{[A]_o k_1}{k_2 - k_1}\exp(-k_2 t)$$

As these equations give little simple insight into the behaviour of the system, we shall consider two limiting cases, the first where B is a relatively stable

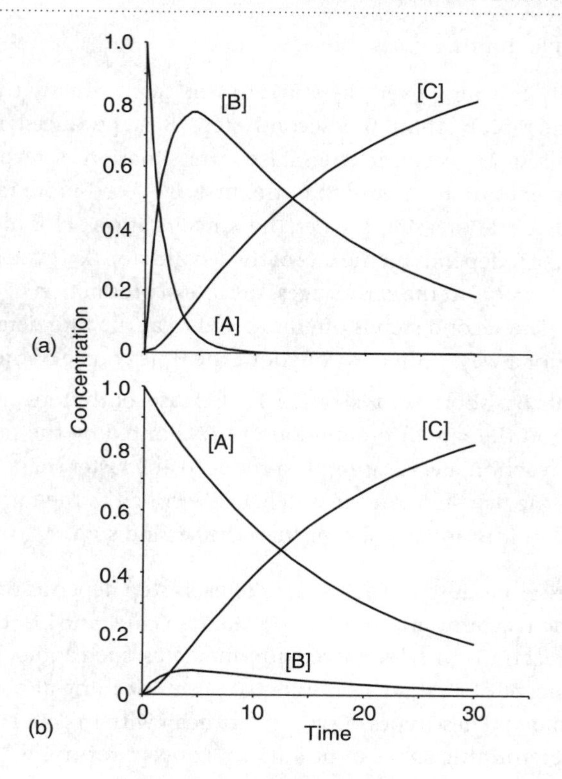

Fig. 10.3 Plots of concentration versus time for the sequential reactions $A \xrightarrow{k_1} B \xrightarrow{k_2} C$ where (a) $k_1 = 10k_2$ and (b) $k_1 = 0{\cdot}1k_2$. B is a long-lived intermediate in (a) and a short-lived intermediate in (b).

species in its own right, as in many biochemical pathways, and the second where B is a highly reactive intermediate, as in a multistep reaction.

- Intermediate B is long lived ($k_2 \ll k_1$). Initially B is formed faster than it is converted into C and so the concentration of B rises as the concentration of A falls. As [A] falls and [B] rises, the rate of the first step decreases and the rate of the second step increases until the second step becomes faster than the first. At this point [B] starts to fall again. This is shown schematically in figure 10.3a.
- Intermediate B is short lived ($k_2 \gg k_1$). As B is formed it is rapidly converted into C. The concentration of B never gets very high and, except for very early in the reaction, is approximately constant (shown schematically in figure 10.3b).

The rate-determining step in consecutive reactions

In consecutive reactions, one reaction is often referred to as the *rate-determining step*. This is defined as the step that governs the rate of the overall process and it is often interpreted as the slowest step. However, this concept has to be used with *extreme* caution and only in selected cases, as we shall see if we take the two very simple limiting cases above.

- Intermediate B is long lived ($k_2 \ll k_1$). In this case, initially the first step is going more rapidly than the second step, B is produced more rapidly than it is used. However at later stages the situation is reversed and the second step goes more rapidly than the first, B is used more rapidly than it is produced. At the later stages, when the concentration of B is high, the rate of production of C depends on the rate of the second step (k_2) even though this is now the faster step. At the early stages, the rate of production of C depends on both steps. The second step is often referred to as the rate-determining step but this is not always correct throughout the time course of the reaction.
- Intermediate B is short lived ($k_2 \gg k_1$). In this case, both steps are going at the same rate and the rate of production of C is limited by the rate of the first step of the reaction, i.e. C cannot be produced any faster than A is converted to B. Thus, the first step can accurately be described as the rate-determining step although it is not any slower than the second step.

Confusion arises because the actual rate of each step depends on the concentrations of the reactants and not just on the rate constants. In the context of consecutive reactions with long-lived intermediates, such as biochemical pathways, the concept of a rate-determining step is misleading at best and should always be avoided. These types of systems are dealt with in detail in Chapter 13.

The rate-determining step can be a useful concept for trying to understand the relationship between a multistep reaction mechanism with short-lived intermediates and its kinetics. In this context, it is useful to define the rate-determining step as the step with the highest activation energy. However, the rate constants of the other steps appear frequently in the rate equations. This is the only situation in which the rate-determining step is used in this book.

The steady-state approximation in consecutive reactions

The case where an intermediate is short lived leads to the *steady-state approximation*, which states that the concentration of an intermediate reactive species can be considered to remain constant, i.e. the rate of its formation equals the rate of its breakdown. For the scheme

$$A \xrightarrow{k_1} B \xrightarrow{k_2} C$$

where B is a short-lived intermediate

$$\frac{d[B]}{dt} = k_1[A] - k_2[B] = 0 \qquad \text{or} \qquad k_1[A] = k_2[B]$$

The time during which this is true is known as the *steady-state period* (figure 10.4). The initial time in which the concentration of B rises rapidly is known as the *pre-steady-state period*.

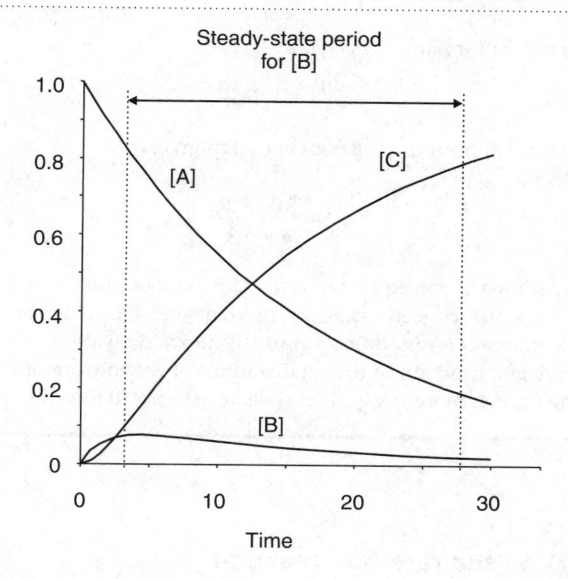

Fig. 10.4 For two sequential reactions $A \xrightarrow{k_1} B \xrightarrow{k_2} C$ where B is a short-lived intermediate, the concentration of B can be assumed to be constant for most of the reaction. This is called the steady-state approximation.

The importance of the steady-state approximation is that it greatly simplifies the solution of the rate equations in complex consecutive reaction pathways. The steady-state approximation allows us to set up equations for each reactive intermediate in a reaction. These equations can then be solved to find the steady-state concentrations of these intermediates. Obtaining the overall rate of the reaction is then a simpler task.

WORKED EXAMPLE

The enzyme-catalysed reaction of a substrate (S) to give products (P) involves initial reversible binding to the enzyme (E) followed by a subsequent reaction.

$$E + S \underset{k_{-1}}{\overset{k_1}{\rightleftharpoons}} ES \xrightarrow{k_2} E + P$$

By applying the steady-state approximation to the complex ES, obtain an equation for the rate of formation of product.

SOLUTION: According to the steady-state approximation the rate of production of ES equals the rate of its breakdown.

$$\text{Rate of production of ES} = k_1[E][S]$$

$$\text{Rate of breakdown of ES} = k_{-1}[ES] + k_2[ES]$$

Thus

$$k_1[E][S] = k_{-1}[ES] + k_2[ES] \quad \text{or} \quad [ES] = \frac{k_1[E][S]}{k_{-1} + k_2}$$

The rate of product formation is given by

$$\frac{d[P]}{dt} = k_2[ES]$$

Substituting for [ES] from the previous equation gives

$$\frac{d[P]}{dt} = \frac{k_1 k_2 [E][S]}{k_{-1} + k_2}$$

COMMENT: Although the equation gives the rate of this enzyme-catalysed reaction, it uses the concentration of free enzyme [E]. This can be almost impossible to measure. A modification of the above derivation can be used to obtain the rate as a function of the total amount of enzyme present, which can be determined much more easily. This is dealt with in Chapter 11.

Effect of pH on the rate of a reaction

Reactions where the rate depends on the pH form a fairly small subset, but because of the importance of these reactions in biological systems we shall deal with them explicitly here. We can divide this section into two principal categories, those reactions in which the ionisation of groups directly affects the reactivity of the reactants, and reactions subject to acid or base catalysis.

The effect of reactant ionising groups

Many examples can be found in which a compound is reactive in either the protonated or the deprotonated form but not in both. For example, the basic form of an amine can be a powerful nucleophile, whereas its conjugate acid is unreactive.

$$\underset{\text{Unreactive}}{RNH_3^+} + H_2O \rightleftharpoons \underset{\text{Reactive}}{RNH_2} + H_3O^+$$

Clearly as the pH is raised, $[RNH_2]$ is increased and hence the reaction rate increases.

For the general case

$$\underset{\text{Unreactive}}{HA} + H_2O \overset{k_a}{\rightleftharpoons} \underset{\text{Reactive}}{A^-} + H_3O^+$$

$$A^- + B \xrightarrow{k_2} \text{products}$$

the rate of reaction is given by

$$\frac{d[\text{products}]}{dt} = k_2[A^-][B]$$

The concentration of A^- can be determined from the acid dissociation constant

$$K_a = \frac{[A^-][H_3O^+]}{[HA]}$$

The concentration of the protonated acid, HA, is given by

$$[HA] = [A]_{tot} - [A^-]$$

where $[A]_{tot}$ is the concentration of all forms of the acid, protonated and deprotonated. Thus

$$K_a = \frac{[A^-][H_3O^+]}{[A]_{tot} - [A^-]}$$

which can be rearranged to give

$$[A^-] = \frac{K_a}{K_a + [H_3O^+]}[A]_{tot}$$

The rate of the reaction is then given by

$$\frac{d[\text{products}]}{dt} = k_2 \frac{K_a}{K_a + [H_3O^+]}[A]_{tot}[B]$$

This can be written as

$$\frac{d[\text{products}]}{dt} = k^{app}[A]_{tot}[B] \quad \text{and} \quad k^{app} = k_2 \frac{K_a}{K_a + [H_3O^+]}$$

k^{app} is the apparent rate constant measured as a function of the total amount of A present and is dependent on the pH. k^{app} is often easier to measure than k_2 because it is easier to measure $[A]_{tot}$ (the amount of reactant we added at the start) than $[A^-]$.

An alternative approach to the derivation of the rate equation, and one that we shall use in the chapters on enzyme kinetics, is to write the rate as

$$\frac{d[\text{products}]}{dt} = k_2\theta[A]_{tot}[B]$$

where θ is the fraction of $[A]_{tot}$ in the reactive form. If A^- is the reactive component then

$$\theta = \frac{[A^-]}{[A]_{tot}} = \frac{[A^-]}{[A^-] + [HA]}$$

We saw above that $[A^-] = \{K_a/(K_a + [H_3O^+])\}[A]_{tot}$ and so

$$\theta = \frac{[A^-]}{[A]_{tot}} = \frac{K_a}{K_a + [H_3O^+]}$$

The apparent rate constant k^{app} is then given by

$$k^{app} = k_2\theta = k_2 \frac{K_a}{K_a + [H_3O^+]}$$

We can consider two limiting cases of this equation. The first is when $[H_3O^+] \gg K_a$, i.e. when $pH < pK_a$. In this case

$$k^{app} = k_2 \frac{K_a}{[H_3O^+]}$$
$$\log_{10} k^{app} = \log_{10} k_2 - pK_a + pH$$

so that a plot of $\log_{10} k^{app}$ versus pH will be linear with a slope of unity. This is observed provided that $pH < pK_a - 1{\cdot}5$.

The second limiting case is when $[H_3O^+] \ll K_a$, i.e. when $pH > pK_a$. In this case

$$k^{app} = k_2$$

The overall plot of $\log_{10} k^{app}$ versus pH is shown in figure 10.5. If the two straight lines at high and low pH are extrapolated, the intersection point is at $pH = pK_a$. Thus, this method can be used to find the pK_a of an ionising group involved in a reaction.

In the case of the protonated form (HA) being active and the deprotonated form inactive, the plot gives a horizontal line at low pH and a line of gradient −1 at high pH.

Acid–base catalysis

This is a very large subject and it will only be treated in outline here. Acid–base catalysis can be subdivided into *general* or *specific* catalysis.

- General acid catalysis—all the acidic species present in the solution (e.g. H^+, CH_3CO_2H, H_2O, etc.) contribute to the rate. The overall rate equation includes terms representing each of these species.
- Specific acid catalysis—only the term in H^+ (H_3O^+ in aqueous solution) is important, the undissociated acids do not contribute.

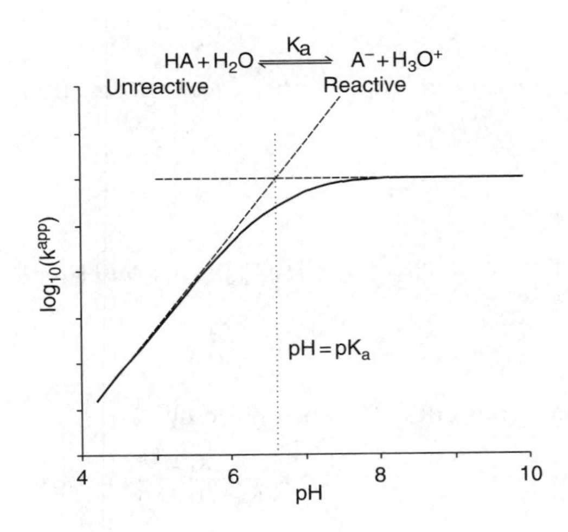

Fig. 10.5 Plot of $\log_{10}(k^{app})$ versus pH for a reaction in which one of the reactants is only active in the deprotonated form. k^{app} is an apparent rate constant that equals the actual rate constant, k, times the fraction of the reactant in the active form. The curve is linear at very low and very high pH values. The intersection point of the extrapolated straight lines from these two regions occurs at $pH = pK_a$.

An experimental distinction between the two types can be made by varying the concentration of undissociated acid at constant pH, i.e. by varying the concentration of the conjugate base. As the concentration of the undissociated acid is increased, the rate of a reaction subject to general acid catalysis will rise, whereas the reaction subject to specific acid catalysis will be unaffected.

In order to understand more clearly why some reactions are subject to specific and others to general acid–base catalysis, we must examine the various steps involved. Consider the following general scheme involving initial proton transfer followed by a second step to give products

$$A + HX \underset{k_{-1}}{\overset{k_1}{\rightleftharpoons}} AH^+ + X^-$$

$$AH^+ + B \xrightarrow{k_2} \text{Products}$$

The rate of formation of products is given by

$$\frac{d[\text{products}]}{dt} = k_2[AH^+][B]$$

Treating AH^+ as a reactive intermediate and using the steady-state approximation gives

$$\frac{d[AH^+]}{dt} = k_1[A][HX] - k_{-1}[AH^+][X^-] - k_2[AH^+][B] = 0$$

which can be solved for $[AH^+]$ to give

$$[AH^+] = \frac{k_1[A][HX]}{k_{-1}[X^-] + k_2[B]}$$

and thus

$$\frac{d[\text{products}]}{dt} = \frac{k_1 k_2[A][HX][B]}{k_{-1}[X^-] + k_2[B]}$$

We can consider two limiting cases of this equation. The first is when $k_{-1}[X] \gg k_2[B]$. This corresponds to the first equilibrium being rapidly established. The initial proton transfer is fast, and the equilibrium is only slightly disturbed by the slow second step (figure 10.6a). The rate equation then simplifies to

$$\frac{d[\text{products}]}{dt} = \frac{k_1 k_2[A][HX][B]}{k_{-1}[X^-]}$$

This equation can be rewritten using the equilibrium constant for the association of H^+ and X^-

$$H^+ + X^- \overset{K_{HX}}{\rightleftharpoons} HX \qquad K_{HX} = \frac{[HX]}{[H^+][X^-]} \quad \text{or} \quad \frac{[HX]}{[X^-]} = K_{HX}[H^+]$$

which gives

$$\frac{d[\text{products}]}{dt} = \frac{K_{HX} k_1 k_2}{k_{-1}}[A][H^+][B]$$

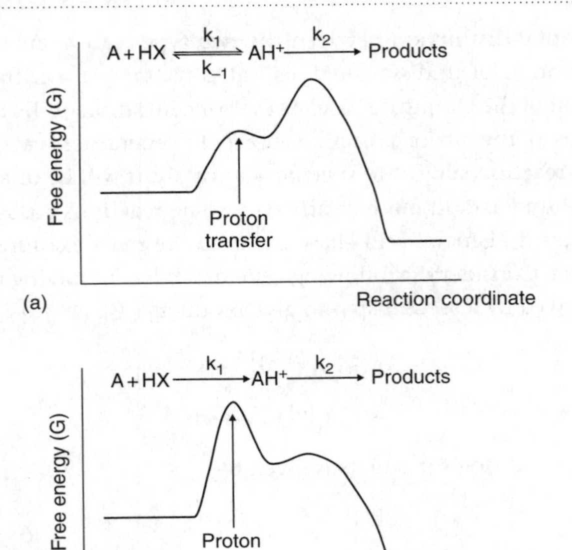

Fig. 10.6 (a) Energy profile for an acid-catalysed reaction in which rapid proton transfer is followed by a slow final step, giving specific acid catalysis. (b) Energy profile for an acid-catalysed reaction in which slow proton transfer is followed by a fast final step, giving general acid catalysis.

This rate equation corresponds to specific acid catalysis. Although the reaction rate does not vary explicitly with the concentration of the acid HX, the nature of the acid does affect the rate because K_{HX} appears in the rate equation.

The second limiting case is when $k_2[B] \gg k_{-1}[X^-]$. The initial proton transfer is slow and so the rate-determining step is the forward reaction of the first step (figure 10.6b). The rate equation then simplifies to

$$\frac{d[\text{products}]}{dt} = \frac{k_1 k_2 [A][HX][B]}{k_2[B]} = k_1[A][HX]$$

If other acids (HX′, HX″, etc.) are present the rate would be

$$\frac{d[\text{products}]}{dt} = k_1[A][HX] + k_1'[A][HX'] + k_1''[A][HX''] + \cdots$$

This situation corresponds to general acid catalysis.

Thus, the difference between general and specific acid catalysis is whether the proton transfer step is rate determining or not.

FURTHER READING

As for chapter 9:

S.R. Logan, 1996, 'Fundamentals of chemical kinetics', Longman—*a comprehensive introduction to kinetics from a chemical perspective.*

M.J. Pilling and P.W. Seakins, 1995, 'Reaction kinetics', Oxford University Press—*a more advanced account of chemical kinetics, including material outside the interests of most biochemists.*

PROBLEMS

1. The compound A can give two alternative products B and C.

$$A \rightarrow B$$
$$A \rightarrow C$$

The first-order rate constants are 0·15 min^{-1} and 0·06 min^{-1} respectively. What is the half-life of A? If the initial concentration of A is 0·1 M, at what time does the concentration of B = 0·05 M? What is the maximum concentration of B?

2. ^{209}Tl decays by β-emission to ^{209}Pb with a half-life of 2 minutes. ^{209}Pb decays to ^{209}Bi, also by β-emission, with a half-life of 3 hours. Assuming that only ^{209}Tl is present initially, calculate the relative proportions of these three species after (a) 5 minutes and (b) 2 hours.

3. The enzyme substitution, or ping-pong, mechanism involves one substrate reacting with an enzyme E to give a modified enzyme E′ plus the first product and the second substrate reacting with E′ to regenerate E and give the second product. If this mechanism involves the following steps

$$E + A \underset{k_{-1}}{\overset{k_1}{\rightleftharpoons}} E' + P$$

$$E' + B \xrightarrow{k_2} E + Q$$

obtain an equation for the rate of formation of the products P and Q in terms of the solution concentrations of the enzyme, reactants and products. [Note: a discussion of this mechanism, together with the rate in terms of the total enzyme concentration, is given in Chapter 12.]

4. Assuming that the reaction between H_2 and Br_2 follows the radical chain reaction mechanism below

$Br_2 \xrightarrow{k_1} 2Br$	Initiation
$Br + H_2 \xrightarrow{k_2} HBr + H$	Propagation
$H + Br_2 \xrightarrow{k_3} HBr + Br$	Propagation
$H + HBr \xrightarrow{k_{-2}} H_2 + Br$	Retardation
$2Br \xrightarrow{k_{-1}} Br_2$	Termination

use the steady-state approximation for [H] and [Br] to show that the rate of the reaction is given by the equation

$$H_2 + Br_2 \rightarrow 2HBr \qquad \frac{d[HBr]}{dt} = \frac{2k_2[H_2]\sqrt{\dfrac{k_1}{k_{-1}}[Br_2]}}{1 + \dfrac{k_{-2}[HBr]}{k_3[Br_2]}}$$

Comment on the mechanistic implications for the reaction between H_2 and I_2, given that the rate equation for this reaction is

$$H_2 + I_2 \rightarrow 2HI \qquad \frac{d[HI]}{dt} = k[H_2][I_2]$$

5. The rate of the reaction between an amino acid and trinitrobenzene sulphonate was studied in buffered solutions as a function of the pH of the solution.

pH	6·5	7·0	7·5	8·0	8·5	9·0	9·5	10·0	10·5	11·0	11·5
$k_{relative}$	0·074	0·25	0·8	2·4	7·5	20	56·6	72	90	97	100

Deduce what you can about the species involved in the reaction. The pK_a of the sulphonic acid group is so low that it remains as the anion throughout the pH range.

$NH_3^+CHRCO_2^-$ + 2,4,6-(NO_2)$_3$$C_6H_2$$SO_3^-$ → 2,4,6-(NO_2)$_3$$C_6H_2$$NHCHRCO_2^-$ + HSO_3^- + H^+

6. Nitramide (NH_2NO_2) decomposes in aqueous solution according to the equation

$$NH_2NO_2 \rightarrow N_2O + H_2O$$

The rate of the reaction, as monitored by the release of N_2O, was studied in acetic acid/acetate buffers of varying composition. The following data were obtained

pH	**Total concentration of acetic acid + acetate/mM**	$\mathbf{k \times 10^3/min^{-1}}$
4·09	18·2	2·12
4·11	20·3	2·46
4·40	20·3	3·82
4·46	26·5	5·26
4·77	28·4	8·00
5·01	20·3	7·26

Given that the pK_a of acetic acid is 4·7, deduce whether the rate constant depends on the concentration of H^+, acetate, or acetic acid.

11

Catalysis and enzyme kinetics

KEY POINTS

- Catalysts act by reducing the activation energy for the reaction.
- Single-substrate enzymes can do this (a) by using binding energy to stabilise the transition state relative to the reactants and (b) by providing an alternative pathway for the reaction involving reactive groups in the active site.
- The rate of a single-substrate enzyme-catalysed reaction is usually given by the *Michaelis–Menten equation*. In this equation, enzymes are characterised by two parameters, K_m and V_{max}.
- Under certain circumstances single-substrate enzyme-catalysed reactions do not follow Michaelis–Menten kinetics. This indicates that the simple kinetic scheme and/or the steady-state assumption used to derive the equation are incorrect.
- The rates of enzyme-catalysed reactions vary frequently with both temperature and pH, reflecting the influence of these factors on the enzyme structure and the groups in the active site.

Catalysis of chemical reactions

Catalysts are defined as species that increase the rate of a reaction but do not alter the final position of equilibrium. A catalyst may play a direct part in the reaction mechanism but it is regenerated at the end of the reaction. For example, an uncatalysed reaction between A and B may be

$$A + B \rightarrow \text{Products}$$

The addition of a catalyst, C, may then give

$$A + C \rightarrow I$$

$$I + B \rightarrow J$$

$$J \rightarrow \text{Products} + C$$

giving an overall reaction of

$$A + B + C \rightarrow \text{Products} + C$$

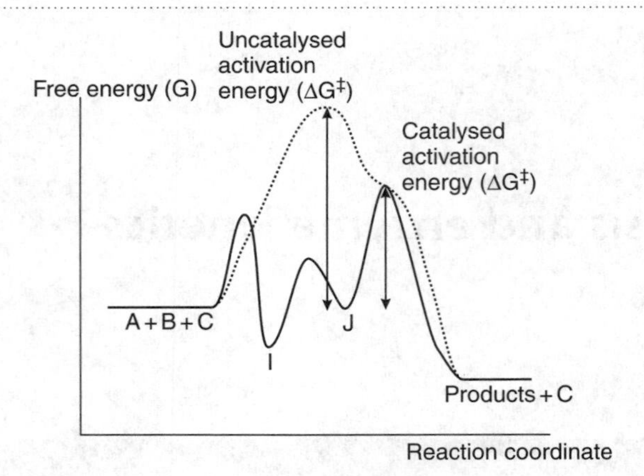

Fig. 11.1 Energy profiles for an uncatalysed (dotted line) and a catalysed (solid line) reaction. In this case, the uncatalysed reaction A + B → Products occurs in a single step with a high activation energy. The catalysed reaction occurs in three steps forming two short-lived intermediates, I and J, and re-forming the catalyst C at the end. Each of these steps has a lower activation energy than the uncatalysed reaction.

Although C is altered during the reaction it is re-formed at the end. Because of this, a very small amount of C can have a very large effect on the reaction rate as each molecule of C can catalyse many reactions between A and B molecules.

We can see how a catalyst works by going back to our energy profile for a reaction. By changing the mechanism of the reaction (in the example above we have three easy steps rather than one difficult step) we can lower the activation energy and thus increase the reaction rate (figure 11.1).

Enzymes

Enzymes are much more efficient catalysts than any man-made catalysts yet devised. They work in a similar fashion to all other catalysts but are remarkably good at their job, giving both very large rate enhancements and a very high degree of selectivity and specificity.

- Speed — Enzymes give typical rate enhancements of 10^6–10^{14} compared to the uncatalysed rate. A few enzymes, such as triosephosphate isomerase, perform their reactions so quickly that the limiting step is how fast the substrate can reach the active site. These reactions are called diffusion limited, the diffusion rate being approximately 10^8 M^{-1} s^{-1}.
- Selectivity — Enzymes are very good at selecting substrates. For instance, a glucosidase will hydrolyse glucose from an oligosaccharide but not galactose, although they only differ at one chiral centre on the opposite side of the ring from the point of hydrolysis.
- Specificity — Enzymes will often only catalyse a single reaction and carry out that reaction stereoselectively. For instance, a specific glucosyl transferase will only add glucose on to the 2 position of another glucose and not the 3, 4 or 6 positions.

In addition, enzymes have a range of properties not normally associated with catalysts, such as their ability to respond to external signals or to their environment. This enables the activities of many enzymes to be altered, a necessary prerequisite for control of a complex biological system in which many competing reactions occur.

Enzymes are classified according to the reaction that they catalyse (see Table 11.1). Lysases and isomerases only have a single substrate. Kinetically, hydrolases can also be treated as single-substrate enzymes because the second substrate, water, is almost always present in large excess and so its concentration does not vary during the reaction. This chapter will deal with the kinetics of single-substrate enzyme reactions. The kinetics of enzyme reactions in the other categories (oxidoreductases, transferases and ligases), with two or more substrates, will be dealt with in the next chapter.

Use of binding energy in catalysing single-substrate reactions

The rate of a reaction at a given temperature is determined by the free-energy difference between the reactants and the transition state. For any catalyst to work, it must reduce this free-energy difference. The only source of free energy available to an enzyme to do this comes from the binding of the reactants to the enzyme.

> Enzymes work by using ΔG of binding to change ΔG of activation for the reaction.

TABLE 11.1

Class of enzyme	Type of reaction
Oxidoreductases	Catalyse redox reactions in which one substrate is reduced at the expense of a second that is oxidised
Transferases	Catalyse reactions in which a group is transferred from one substrate to another
Hydrolases	Catalyse reactions in which a substrate is hydrolysed
Lyases	Catalyse reactions in which a group is eliminated from a substrate to form a double bond
Isomerases	Catalyse isomerisation reactions
Ligases	Catalyse the joining together of two molecules at the expense of ATP, or some other free-energy source

For single-substrate reactions, there are two basic ways in which an enzyme can do this.

Using binding energy to destabilise the substrate/stabilise the transition state

Ligand binding to proteins usually has a large negative ΔG. However, for an enzyme this would be unfavourable because it would stabilise the substrate in the enzyme complex and so increase the activation energy for the reaction. Instead, this free energy of binding can be used to favour distortion of the substrate, e.g. through changes in bond angles or bond lengths or through changes in electron distribution. This is achieved by the enzyme-binding site matching the distorted substrate better than the undistorted substrate so that maximum binding is only obtained after the substrate has become distorted. For efficient catalysis, the enzyme needs to favour a distortion of the substrate towards the transition state of the reaction. The catalysed reaction will have a similar energy profile to the uncatalysed reaction but because the transition state has been stabilised by more than the reactants, the activation energy is reduced (figure 11.2).

One example of this is lysozyme, an enzyme that catalyses the hydrolysis of bacterial cell wall polysaccharides (figure 11.3). This enzyme has a binding site that binds six residues of the polysaccharide, in subsites that are labelled A to F,

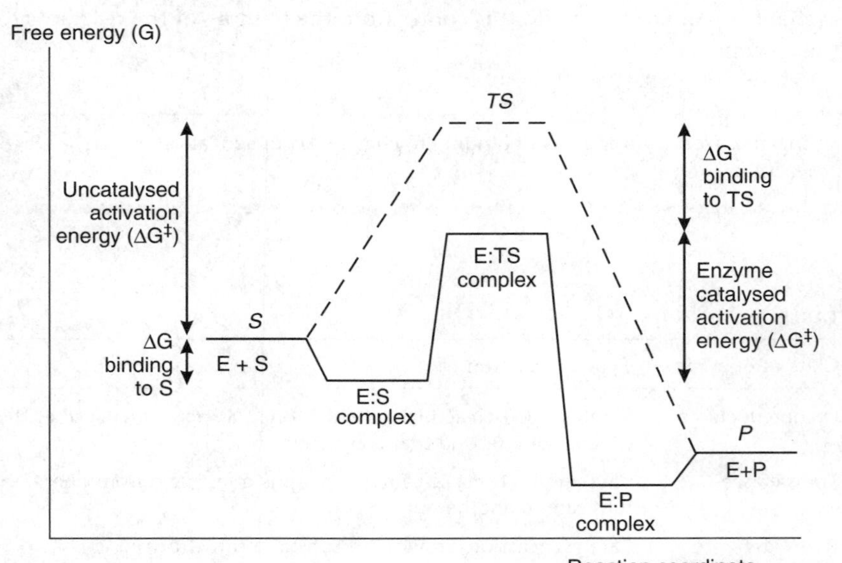

Fig. 11.2 Energy profile for an uncatalysed (dotted line) and enzyme-catalysed reaction (solid line) in which binding energy is used to reduce the activation energy. In general, the binding of a substrate to an enzyme is a favourable process ($\Delta G < 0$) and so will stabilise the substrate. In order to decrease the activation energy, the enzyme must bind to the transition state with a higher affinity than to the substrate. E, enzyme; S, substrate; TS, transition state; P, products.

and cleaves the linkage between the fourth and fifth residues (sites D and E). Sites A to C, E and F can bind with high affinity to undistorted saccharide residues. Site D is the wrong shape to bind with high affinity to an undistorted saccharide ring (shown schematically in figure 11.4) but can bind with higher

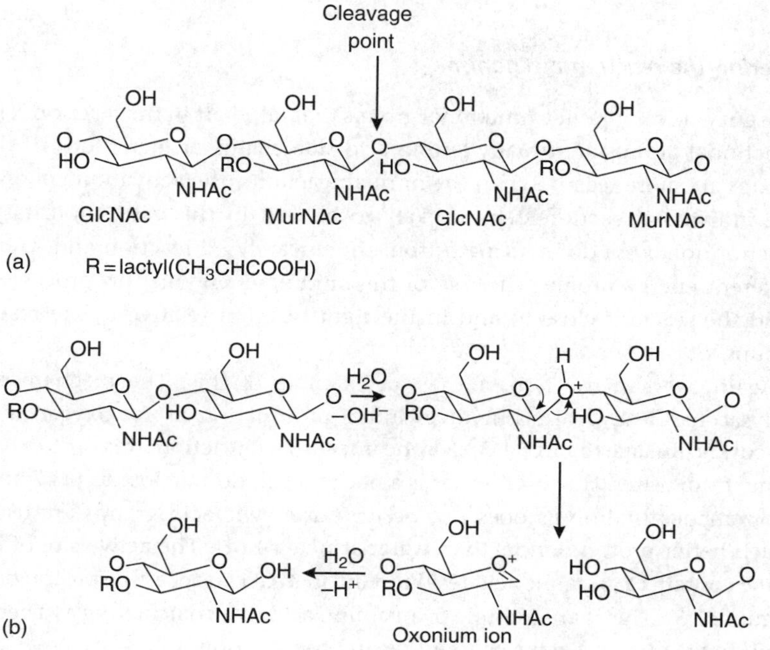

Fig. 11.3 (a) The bacterial cell wall (MurNAcβ1 $\rightarrow$ 4GlcNAcβ1 $\rightarrow$ 4)$_n$ polysaccharide showing the cleavage point for lysozyme. (b) Polysaccharide hydrolysis in aqueous solution involves an initial proton transfer from water to the glycosidic linkage oxygen, followed by bond breakage to give an oxonium ion intermediate. The MurNAc and GlcNAc rings are normally in a puckered conformation, called the chair conformation, whereas the ring of the oxonium ion is flattened by the presence of the double bond.

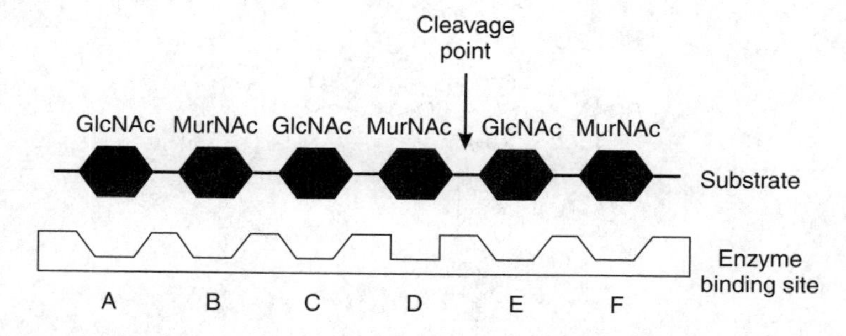

Fig. 11.4 Schematic representation of the binding site of lysozyme. The enzyme binds to six residues of the (MurNAcβl $\rightarrow$ 4GlcNAcβl $\rightarrow$ 4)$_n$ polysaccharide, cleaving between the fourth and fifth. The first three and last two residues fit in their binding sites in the chair conformation. The binding site for the fourth residue matches the conformation of the oxonium ion intermediate better than the normal chair conformation of the substrate, promoting distortion of the substrate towards the transition state at residue four.

affinity to a flattened saccharide ring, as is formed in the oxonium ion transition state. Site D also contains a negatively charged amino acid, Asp 52, that interacts with the positively charged oxonium ion more strongly than with the uncharged substrate. Thus, the transition state binds with higher affinity than the substrate resulting in a reduced activation energy.

Altering the reaction mechanism

The enzyme can provide functional groups that take part in the reaction. These functional groups are usually amino acid side chains or metal ions. If these groups are more reactive than the normal reactants, the activation energy for the individual reaction steps is likely to be less. In this case, the catalysed reaction follows a different path from the uncatalysed reaction and so has a different energy profile. The role of the binding energy in this process is to hold the reactant close to and in the right position relative to the reactive groups.

Again, we can use lysozyme as an example of this. The mechanism of polysaccharide hydrolysis involves initial protonation of the oxygen in the glycosidic linkage (figure 11.3). For the uncatalysed reaction, this proton has to come from water. However, water is a poor proton donor close to pH 7 and so spontaneous hydrolysis does not occur. Carboxylic acid groups are usually much better proton donors than water at neutral pH. The active site of lysozyme contains a glutamic acid residue, Glu 35, that can act as the proton donor (figure 11.5). This is an example of an amino acid side chain acting as a general acid catalyst (see previous chapter) during the reaction.

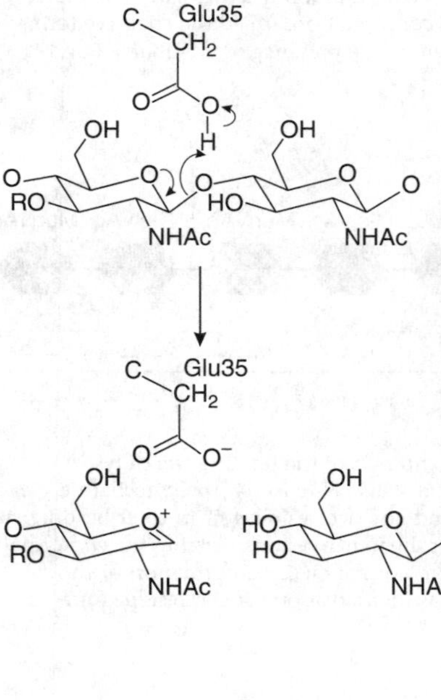

Fig. 11.5 The mechanism of hydrolysis of the MurNAcβ1→4GlcNAcβ linkage by lysozyme involves initial proton transfer to the glycosidic linkage oxygen from the side chain of Glu 35. This group is a much better proton donor than water (figure 11.3). Other residues in the active site are involved in later steps.

Kinetics of single-substrate enzyme reactions

There are two ways of deriving the rate equation for a single-substrate enzyme-catalysed reaction. Both are based on the following kinetic scheme

$$E + S \underset{k_{-1}}{\overset{k_1}{\rightleftharpoons}} ES \overset{k_2}{\rightarrow} E + \text{Products}$$

and they differ in the assumptions that are made. The first method uses the steady-state approximation applied to the ES intermediate. The second assumes that ES is in equilibrium with E + S and that the equilibrium is only slightly perturbed by the breakdown to products.

Steady-state approximation

Consider the following kinetic scheme for a single-substrate enzyme reaction.

$$E + S \underset{k_{-1}}{\overset{k_1}{\rightleftharpoons}} ES \overset{k_2}{\rightarrow} E + \text{Products}$$

If we assume that we can apply the steady-state approximation to ES, then

$$\frac{d[ES]}{dt} = 0 = k_1[E][S] - k_{-1}[ES] - k_2[ES]$$

The concentration of free enzyme [E] is given by the total amount of enzyme present $[E]_{total}$ minus the amount present as the complex [ES]. As no enzyme is consumed during the reaction, $[E]_{total}$ is the concentration of enzyme present at the start of the reaction.

$$\frac{d[ES]}{dt} = 0 = k_1([E]_{total} - [ES])[S] - k_{-1}[ES] - k_2[ES]$$

This can be rearranged to give

$$[ES] = \frac{[E]_{total}[S]}{\frac{k_{-1} + k_2}{k_1} + [S]}$$

The rate of the reaction, v, is given by

$$\text{Rate} = v = k_2[ES] = \frac{k_2[E]_{total}[S]}{\frac{k_{-1} + k_2}{k_1} + [S]}$$

This is normally rewritten as

$$v = \frac{V_{max}[S]}{K_m + [S]}$$

where

$$K_m = \frac{k_{-1} + k_2}{k_1} \quad \text{and} \quad V_{max} = k_2[E]_{total}$$

This is called the Michaelis–Menten equation. The interpretations of the constants K_m and V_{max} are discussed in the next section. [S] refers to the concentration of substrate not bound to the enzyme. One further simplifying approximation can be made when the substrate is present at much higher concentration than the enzyme. In this case, the concentration of the complex ES must be very small compared to the concentration of the unbound substrate so that

$$[S] = [S]_{total} - [ES] \approx [S]_{total}$$

where $[S]_{total}$ is the total amount of substrate present. If initial rate measurements are made, then $[S]_{total}$ is the concentration of substrate at the start of the reaction.

Equilibrium approximation

An alternative way of deriving the Michaelis–Menten equation is to assume that the enzyme complex ES is in equilibrium with E and S. This is a less rigorous assumption than the steady-state approximation but it gives a simpler derivation. The form of the final equation is identical, but the meaning of K_m is slightly different (see next section). For preference, you should always use the results of the steady-state approximation. We are only going to go through this derivation here because we shall use a similar approach in the next chapter to deal with enzyme inhibition, where the simplifications provided are much more useful.

$$E + S \overset{K_m}{\rightleftharpoons} ES \overset{k_2}{\rightarrow} E + \text{Products}$$

If E + S are in equilibrium with ES and K_m is the dissociation constant of ES, then

$$K_m = \frac{[E][S]}{[ES]} \quad \text{or} \quad [ES] = \frac{[E][S]}{K_m}$$

The fraction θ of the enzyme in the active form is given by

$$\theta = \frac{[ES]}{[E]_{total}} = \frac{[ES]}{[E] + [ES]} = \frac{\frac{[E][S]}{K_m}}{[E] + \frac{[E][S]}{K_m}}$$

Dividing by [E] and multiplying by K_m gives

$$\theta = \frac{[S]}{K_m + [S]}$$

The rate of the reaction, v, is given by the limiting rate, V_{max}, when all the enzyme is present as ES multiplied by the fraction of the enzyme in the form ES.

$$v = V_{max} \times \theta = \frac{V_{max}[S]}{K_m + [S]}$$

Again, the further simplification can be made that $[S] = [S]_{total}$ as long as the substrate is present in large excess compared to the enzyme.

WORKED EXAMPLE

For a single-substrate enzyme-catalysed reaction, calculate the change in substrate concentration required to increase the rate of the reaction from 10% to 90% of the limiting rate. What further change would be required to increase the rate to 95% of the limiting rate?

SOLUTION: At $v = 0{\cdot}1\,V_{max}$, [S] is given by

$$v = \frac{V_{max}}{10} = \frac{V_{max}[S]}{K_m + [S]}$$

which gives

$$[S] = \frac{K_m}{9}$$

At $v = 0{\cdot}9\,V_{max}$, $[S] = 9\,K_m$. Thus, [S] has to increase by a factor of 81.
At $v = 0{\cdot}95\ V_{max}$, $[S] = 19\,K_m$. Thus, [S] has to increase by a further factor of 2·11.

Discussion of the Michaelis–Menten equation

The dependence of v on [S] for a single-substrate reaction that follows the Michaelis–Menten equation is shown in figure 11.6. At low substrate

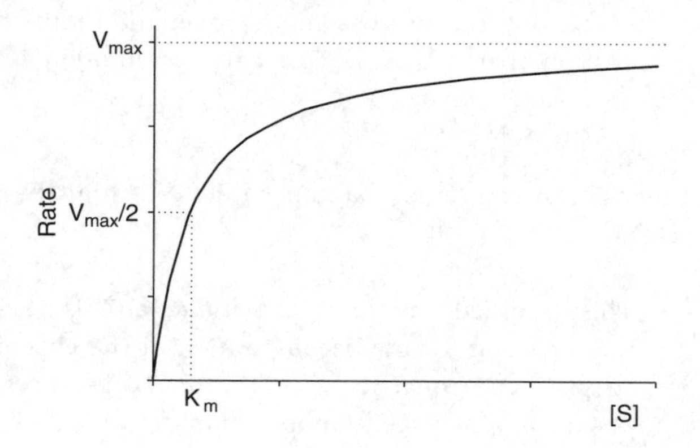

Fig. 11.6 Plot of rate (v) versus [S] for a single-substrate enzyme reaction that follows Michaelis-Menten kinetics. At low substrate concentration, the rate is linearly dependent on [S]. At high substrate concentration, the rate becomes independent of [S] and approaches the limiting rate, V_{max}.

concentrations, when $[S] \ll K_m$, the rate depends linearly on [S] and the concentration of the substrate is rate limiting. At high substrate concentrations, when $[S] \gg K_m$, the rate is independent of [S], the enzyme is saturated by substrate and so the concentration of enzyme is rate limiting. The rate, v, is linearly dependent on $[E]_{total}$ at any given value of [S].

There are two constants in the Michaelis–Menten equation.

$V_{max} \equiv k_2[E]_{total}$

This is the *limiting velocity* of the reaction at a given concentration of enzyme. It is the rate at saturating levels of substrate when all the enzyme is present as the complex. It is sometimes referred to as the *maximum velocity* for the reaction. However, this term is slightly misleading because the velocity can always be increased by adding more enzyme.
Units: strictly speaking this is a reaction rate ($M\,s^{-1}$). However, often it is quoted as moles of substrate used per weight of enzyme per unit time.
V_{max} has the same units as v.

$K_m \equiv \frac{k_{-1} + k_2}{k_1}$ (steady state approximation)

$\equiv \frac{k_{-1}}{k_1}$ (equilibrium approximation)

This is called the *Michaelis constant* for the reaction. This has a specific meaning in terms of the rate equation. When $[S] = K_m$ then $v = \frac{1}{2}V_{max}$, and so K_m is the substrate concentration at half the limiting velocity.
The precise interpretation of K_m depends on which derivation of the rate equation is used, i.e. on the assumptions made during the derivation. For the equilibrium approximation, K_m is the dissociation constant of the ES complex. For the steady-state approximation, K_m can be interpreted as a measure of the lifetime of the ES complex, the ratio of the rate of breakdown by any route to the rate of formation.
If $k_2 \ll k_{-1}$ then both derivations give the same value for K_m. Thus, the equilibrium approximation is equivalent to making the steady-state approximation plus the assumption that $k_2 \ll k_{-1}$. This latter assumption is frequently invalid.
Units: M.

There are other useful constants that can be derived from the Michaelis–Menten equation.

$k_{cat} \equiv \frac{V_{max}}{[E]_{total}}$

This is called the *catalytic constant* and is the first-order rate constant for the decomposition of the enzyme complex to products (equal to k_2 in the mechanistic schemes used above, but may be a more complex term if formation of the products from the complex involves several steps). It is also the first-order rate constant for the reaction at high substrate concentration. If $[S] \gg K_m$

$$v = \frac{V_{max}[S]}{[S]} = k_{cat}[E]_{total}$$

Units: s^{-1}.

$k_A \equiv \frac{k_{cat}}{K_m}$

This is called the *specificity constant* or *catalytic efficiency*. It is also the second-order rate constant for the reaction at low substrate concentration. If $[S] \ll K_m$

$$v = \frac{V_{max}[S]}{K_m} = \frac{k_{cat}[E]_{total}[S]}{K_m} = k_A[E]_{total}[S]$$

The maximum value of the specificity constant is the diffusion rate constant, when diffusion of the reactant to the active site is the slowest step in the mechanism. Comparison of the specificity constants for two different substrates gives a measure of the degree of selectivity of the enzyme for one over the other.
Units: $M^{-1}\ s^{-1}$.

Enzyme activities

The activity of an enzyme, or an impure enzyme preparation, is defined by the rate at which the catalysed reaction proceeds rather than by the amount of enzyme present. It is thus a kinetic, or functional, definition. There are several ways in which enzyme activity can be expressed (see Table 11.2).

TABLE 11.2

Quantity	Units	Description
Enzyme activity	katal	An amount of protein which, under the specified conditions, catalyses the production of 1 mole of product (or the consumption of 1 mole of substrate) per second. Thus x katals will catalyse the production of x moles of product per second. In fact, the katal is not commonly used and activities are often expressed in the units of $\mu mol\ min^{-1}$.
Specific enzyme activity	katal (kg protein)$^{-1}$	Moles of substrate consumed or product formed per kg protein per second. The specific activity can be quite low for an impure enzyme preparation because only a small fraction of the total protein is active enzyme. The specific activity increases as the enzyme purity increases. These units can be used for quoting v or V_{max}. Again, the more commonly used unit for specific activity is μmol (mg protein)$^{-1}$ min^{-1}.
Turnover number	s^{-1}	Moles of substrate consumed or product formed per mole enzyme per unit time. The turnover number is given by the specific activity of the pure enzyme multiplied by its molecular weight and is the same as k_{cat}.

WORKED EXAMPLE

Under certain conditions the maximum specific activity of a purified preparation of triosephosphate isomerase was quoted as 10 000 μmoles glyceraldehyde 3-phosphate transformed per mg protein per minute. Express this in terms of katal kg^{-1} and evaluate the turnover number of the enzyme given that its molecular weight per active site is 26 500 Da.

SOLUTION:

$$10\,000\,\mu mol\,mg^{-1}\,min^{-1} \equiv \frac{10\,000}{60}\,\mu mol\,mg^{-1}\,s^{-1}$$

$$\equiv \frac{10\,000}{60} \times 10^{6}\,\mu mol\,kg^{-1}\,s^{-1}$$

$$\equiv \frac{10\,000}{60} \times 10^{6} \times 10^{-6}\,mol\,kg^{-1}\,s^{-1}$$

$$\text{Specific activity} = 166{\cdot}7\ \text{katal kg}^{-1}$$

The specific activity gives the number of moles of substrate converted per kg of enzyme per second. To convert to moles of substrate converted per mole of enzyme per second, the specific activity is multiplied by the number of kg enzyme per mole.

$$\text{Turnover number} = 166{\cdot}7 \times 26{\cdot}5 = 4420\,s^{-1}$$

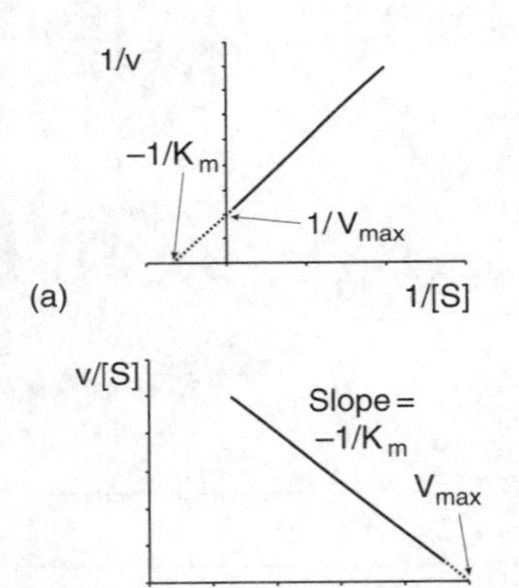

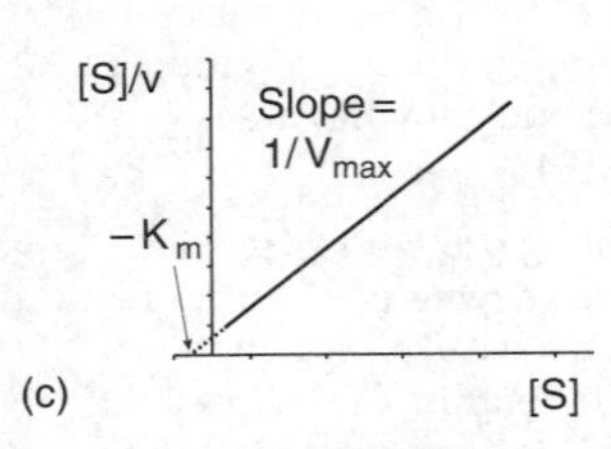

Fig. 11.7 Plots of rearranged versions of the Michaelis–Menten equation that give straight lines. (a) Lineweaver–Burk plot, (b) Eadie–Hofstee plot, (c) Hanes plot.

Analysis of kinetic data

Values of K_m and V_{max} are usually obtained by fitting the experimental data to the Michaelis–Menten equation on the basis of a non-linear regression analysis of the plot of v versus [S]. However, it is also possible to rearrange the basic rate equation to obtain straight-line plots from which K_m and V_{max} can be extracted. The three best-known versions are given in Table 11.3 and typical plots shown in figure 11.7.

TABLE 11.3

Equation	Plot	K_m	V_{max}
Lineweaver–Burk plot			
$\frac{1}{v} = \frac{K_m}{V_{max}}\frac{1}{[S]} + \frac{1}{V_{max}}$	$\frac{1}{v}$ versus $\frac{1}{[S]}$	x-intercept $= -\frac{1}{K_m}$	y-intercept $= \frac{1}{V_{max}}$
Eadie–Hofstee plot			
$\frac{v}{[S]} = \frac{V_{max}}{K_m} - \frac{v}{K_m}$	$\frac{v}{[S]}$ versus v	gradient $= -\frac{1}{K_m}$	x-intercept $= V_{max}$
Hanes plot			
$\frac{[S]}{v} = \frac{[S]}{V_{max}} + \frac{K_m}{V_{max}}$	$\frac{[S]}{v}$ versus [S]	x-intercept $= -K_m$	gradient $= \frac{1}{V_{max}}$

WORKED EXAMPLE

The activity of the enzyme urease, which catalyses the reaction

$$CO(NH_2)_2 + H_2O \rightleftharpoons CO_2 + 2NH_3$$

was studied as a function of urea concentration with the following results.

Urea concentration/mM	30	60	100	150	250	400
Velocity/mmol urea consumed (mg enzyme)$^{-1}$ min^{-1}	3·37	5·53	7·42	8·94	10·70	12·04

What are the values of K_m and V_{max} for this reaction?

SOLUTION: This is a hydrolysis reaction and so we would expect it to follow single-substrate kinetics. The data are rearranged into an appropriate form for a linear plot (i.e. 1/v versus 1/[S], v/[S] versus v, [S]/v versus [S]). These three plots are shown in figure 11.8. All three plots are linear, confirming that the enzyme follows Michaelis–Menten kinetics, and give

$$K_m = 105\,\text{mM}$$

$$V_{max} = 15{\cdot}2\,\text{mmol urea consumed}\,\text{mg}^{-1}\,\text{min}^{-1}$$

COMMENT: We would not normally do all three plots. This was done here merely for illustrative purposes. For perfect data all three methods will give the same result. It might be noted that the distribution of points is markedly different in the three plots. The three plots also give different distributions of errors. In particular, the Lineweaver–Burk plot gives a very non-uniform distribution of errors and so is a much less accurate method of analysis than the other two. In addition, extrapolating straight lines to determine intercepts will always accentuate any errors and give a less reliable result than measuring gradients. It is for these reasons that non-linear regression analysis is preferred.

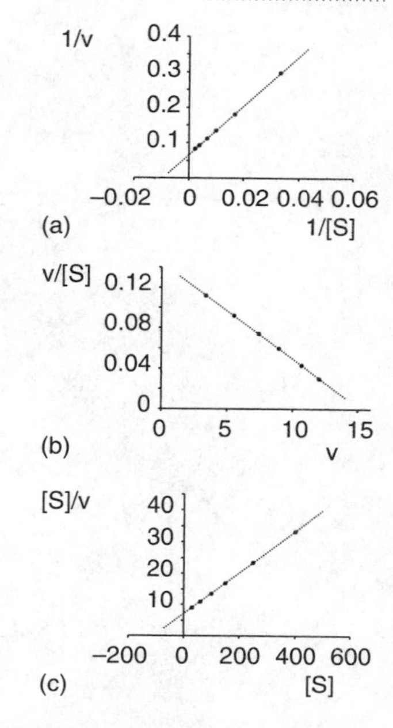

Fig. 11.8 (a) Lineweaver-Burk plot, (b) Eadie-Hofstee plot, (c) Hanes plot.

Complications to the basic rate equation

Pre-steady-state period

The basic rate equation is derived using the steady-state approximation. At the very early stages of a reaction this will not apply. This is likely to be particularly apparent for multistep reactions. In the steady-state period, all the individual steps will have the same rate (this is the definition of the steady state) but before this the different steps may go at different rates. Studies during the pre-steady-state period may enable the rate constants of the individual steps to be determined and so can be much more informative than steady-state kinetics; however, the rate equations are much more complex.

Approach to equilibrium

The kinetic scheme used to derive the basic rate equation assumes that the forward reaction from complex (ES) to products is irreversible. In practice, all enzyme reactions are reversible if the product concentration becomes high enough. Under these conditions the following kinetic scheme has to be used.

$$\mathrm{E + S \underset{k_{-1}}{\overset{k_1}{\rightleftharpoons}} ES \underset{k_{-2}}{\overset{k_2}{\rightleftharpoons}} E + P}$$

Product inhibition

As well as being involved in the back reaction, products may also act as inhibitors of the enzyme (see next chapter for a more detailed discussion of enzyme inhibition). This may occur because the product can bind to the active site and act as a competitive inhibitor of the substrate. Some enzymes have secondary sites for binding their products as a mechanism to reduce the enzyme activity when the product concentration goes up (see the worked example on enzyme inhibition in the next chapter).

Substrate inhibition

Some enzymes are inhibited by a high substrate concentration (figure 11.9). This usually implies a secondary binding site for the substrate that reduces the enzyme activity, the substrate acting as a non-competitive or mixed inhibitor. One example of this is insect acetylcholinesterase, which catalyses the reaction

$$\text{acetylcholine} + \text{water} \rightleftharpoons \text{acetate} + \text{choline}$$

where marked inhibition occurs at acetylcholine concentrations above about 1 mM. We shall come back to this example in the next chapter.

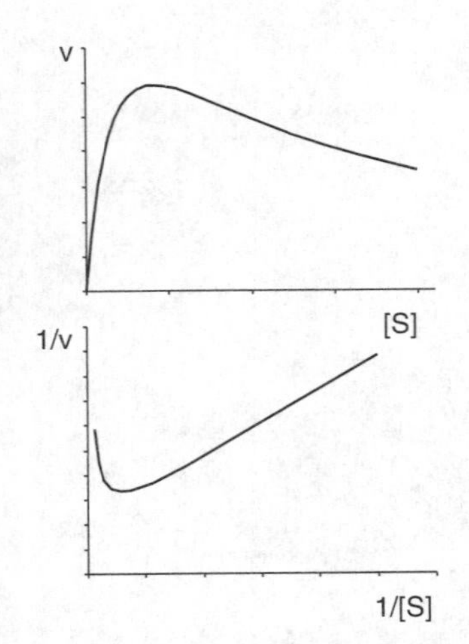

Fig. 11.9 Typical plots of v versus [S] and 1/v versus 1/[S] (Lineweaver–Burk plot) for a single-substrate enzyme reaction when substrate inhibition occurs.

Enzymes with more than one active site

Complex kinetics are often observed when an enzyme is composed of multiple subunits and therefore possesses a number of active sites. We have already discussed this type of system in connection with ligand binding (Chapter 4). The effects of interactions between subunits, which can be manifested as either positive or negative cooperativity, are often of considerable importance in the alteration of enzyme activity, since they render the enzyme activity more or less sensitive to changes in [S]. An example of positive cooperativity is aspartate transcarbamylase, which shows a sigmoidal v versus [S] curve for aspartate. An example of negative cooperativity is seen with the substrate NAD^+ for beef-liver glutamate dehydrogenase.

Effect of temperature on enzyme-catalysed reactions

Most chemical reactions follow the Arrhenius equation for the dependence of reaction rate on the temperature. However, there are several possible additional complications that may arise for enzyme-catalysed reactions.

- Above a certain temperature, the enzyme molecule will become unfolded, so that the three-dimensional integrity of the catalytic site will become lost. The temperature at which this process occurs varies from enzyme to enzyme, and it is also generally dependent upon the pH, the concentration of substrates or other ligands, ionic strength, etc. Above this critical temperature the reaction reverts to the uncatalysed rate, which in many cases is unmeasurably small and so the reaction appears to stop (figure 11.10).
- In a multistep reaction, different steps may be rate determining at different temperatures. The rate of a step with a high positive $\Delta H^{\ddagger}$ and a low negative $\Delta S^{\ddagger}$ will be very sensitive to temperature, being slow at low temperatures but much faster at high temperatures. Conversely, the rate of a step with a low positive $\Delta H^{\ddagger}$ and a higher negative $\Delta S^{\ddagger}$ will be relatively insensitive to temperature and can become the rate-determining step at higher temperatures. This will result in the slope of the Arrhenius plot changing at a certain temperature (figure 11.11).
- The enzyme may exist in two or more interconvertible, active forms that possess different activation energies. We would then expect a break in the Arrhenius plot around the temperature where the change-over between the two forms becomes significant (figure 11.12). Two examples of this type of behaviour are seen with the ATPase enzymes that are involved with the transport of (a) Na^+ and K^+ and (b) Ca^{2+}. In these cases, the transition probably arises from structural changes within the tightly bound phospholipid molecules that are associated with the enzymes.

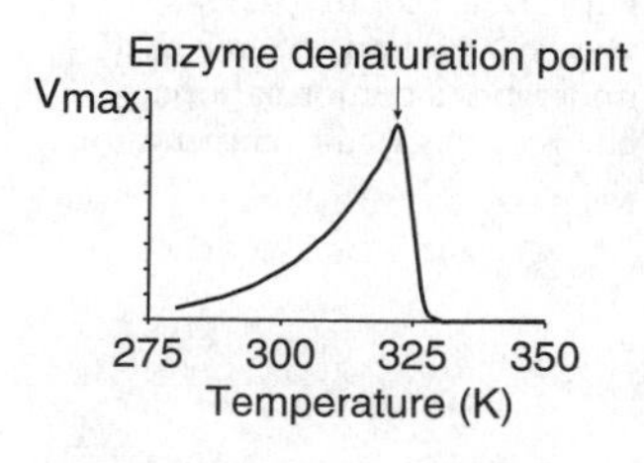

Fig. 11.10 Typical plot of V_{max} versus temperature for an enzyme-catalysed reaction. Below the denaturation point of the enzyme, Arrhenius behaviour is observed. Above the denaturation point, the rate drops very rapidly to the uncatalysed rate. Note that the denaturation point varies substantially between different enzymes.

Despite these factors, we still get linear behaviour over reasonable temperature ranges for many enzymes and so activation enthalpies and entropies for

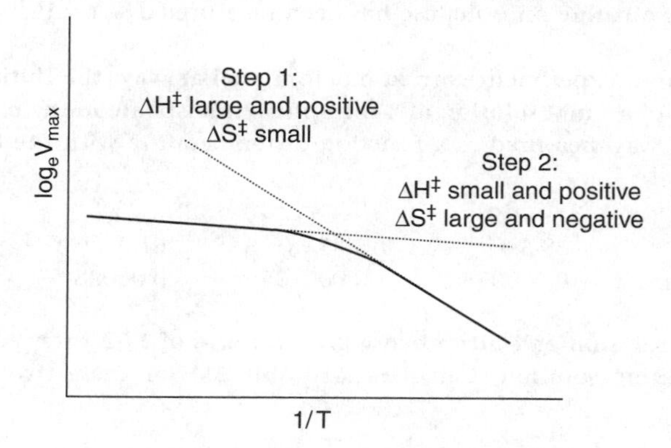

Fig. 11.11 Plot of $\log_e V_{max}$ versus $1/T$ for a two-step reaction in which step 1 is rate determining at low temperature and step 2 is rate determining at high temperature.

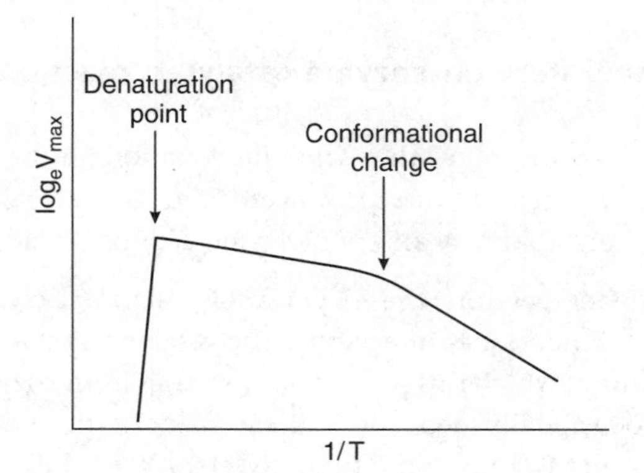

Fig. 11.12 Plot of $\log_e V_{max}$ versus 1/T for an enzyme that undergoes a conformational change at a specific temperature.

enzyme-catalysed reactions can be obtained and compared to those for the uncatalysed reactions. This comparison can provide insight into the mechanism of catalysis.

WORKED EXAMPLE

The hydrolysis of 4-nitrophenylphosphate, which can be catalysed either by OH^- ions or by the enzyme alkaline phosphatase, can be followed by monitoring the change in absorption of the solution at 400 nm (ΔA_{400}) due to the formation of 4-nitrophenol. In practice, this is done by taking samples from the reaction mixture at known times and quenching the reaction by addition of an equal volume of strong alkali.

0·02 mg of alkaline phosphatase was added to 2 mM 4-nitrophenylphosphate in 50 mM Na^+/glycine buffer at pH 9·5 and the initial rate of hydrolysis measured as a function of temperature, with the following results

T/°C	1·4	11·0	22·0	31·0	39·5	50·2	61·0	70·2
ΔA_{400}/min	0·0344	0·0592	0·107	0·164	0·228	0·278	0·205	0·096

k_{cat}/K_m for alkaline phosphatase has been measured as $2{\cdot}6 \times 10^4\ M^{-1}\ s^{-1}$ at 30°C.

In a second experiment carried out in a similar way, the initial rate of hydrolysis of a 1 mM solution of 4-nitrophenylphosphate in the presence of 1 M NaOH was measured as a function of temperature with the following results

T/°C	39·5	50·5	61·0	70·2
ΔA_{400}/min	0·000035	0·000105	0·00038	0·00133

A 1 mM solution of 4-nitrophenol gives an A_{400} of 11·2, after addition of the quenching solution. Calculate $\Delta H^\ddagger$ and $\Delta S^\ddagger$ for these two reactions at 30°C.

SOLUTION: An Arrhenius plot of $\log_e(\Delta A_{400})$ versus 1/T, remembering to convert the temperature from °C to K, should be a straight line of slope $-\Delta H^{\ddagger}/R$. $\Delta S^{\ddagger}$ can then be obtained from the second-order rate constant using the equation obtained from transition state theory.

The alkaline phosphatase-catalysed hydrolysis of 4-nitrophenylphosphate gives a straight-line Arrhenius plot over the temperature range 0°C to 40°C but the line then starts to curve (figure 11.13a). The slope of the line in the range 0°C to 40°C is $-4435\ K^{-1}$, giving $\Delta H^{\ddagger} = 37\ kJ\ mol^{-1}$.

The second-order rate constant for the reaction at 30°C is $k = 2{\cdot}6 \times 10^{4}\ M^{-1}\ s^{-1}$. From transition state theory, assuming that the transmission coefficient is one,

$$k = \frac{\mathit{k}T}{h} \exp\left(\frac{\Delta S^{\ddagger}}{R}\right) \exp\left(\frac{-\Delta H^{\ddagger}}{RT}\right)$$

Substituting in for k, $\Delta H^{\ddagger}$, T and the various constants gives $\Delta S^{\ddagger} = -39\ J\ K^{-1}\ mol^{-1}$.

The OH^--catalysed hydrolysis of 4-nitrophenylphosphate gives a straight-line Arrhenius plot over the whole temperature range studied (figure 11.13b). The slope of the line is $-12\,639\ K^{-1}$, giving $\Delta H^{\ddagger} = 105\ kJ\ mol^{-1}$.

By extrapolation of the Arrhenius plot, the rate of the reaction at 30°C is $\Delta A_{400} = 8{\cdot}66 \times 10^{-6}\ min^{-1}$. This can be converted into a rate of production of 4-nitrophenol by dividing by 11·2 and converting from $mM\ min^{-1}$ to $M\ s^{-1}$.

$$\text{rate} = \frac{8{\cdot}66 \times 10^{-6}}{11{\cdot}2 \times 1000 \times 60} = 1{\cdot}29 \times 10^{-11}\ M\,s^{-1}$$

The rate equation for this reaction is

$$\text{rate} = k[\text{4-nitrophenylphosphate}][OH^-]$$

$[OH^-] = 1\ M$ and [4-nitrophenylphosphate] = 1 mM and so the second-order rate constant at 30°C is $1{\cdot}29 \times 10^{-8}\ M^{-1}\ s^{-1}$. Substitution into the transition state theory equation as before gives $\Delta S^{\ddagger} = -48\ J\ K^{-1}\ mol^{-1}$.

These results are summarised as follows

	$\Delta H^{\ddagger}$/kJ mol^{-1}	$\Delta S^{\ddagger}$/J K^{-1} mol^{-1}
Enzyme-catalysed hydrolysis	37	−39
OH^--catalysed hydrolysis	105	−48

COMMENT: The major effect of alkaline phosphatase on this reaction is to reduce $\Delta H^{\ddagger}$, although there is also a small reduction in $\Delta S^{\ddagger}$. In this case, the enzyme provides an alternative pathway for the reaction involving initial rapid transfer of the phosphate group to the enzyme, followed by the slower release of phosphate from the enzyme:phosphate complex. Above 50°C the activity of the enzyme decreases, which reflects the onset of the loss of the three-dimensional structure required for activity. A number of studies have shown that the active site of an enzyme is more prone to loss of structure than the rest of the molecule.

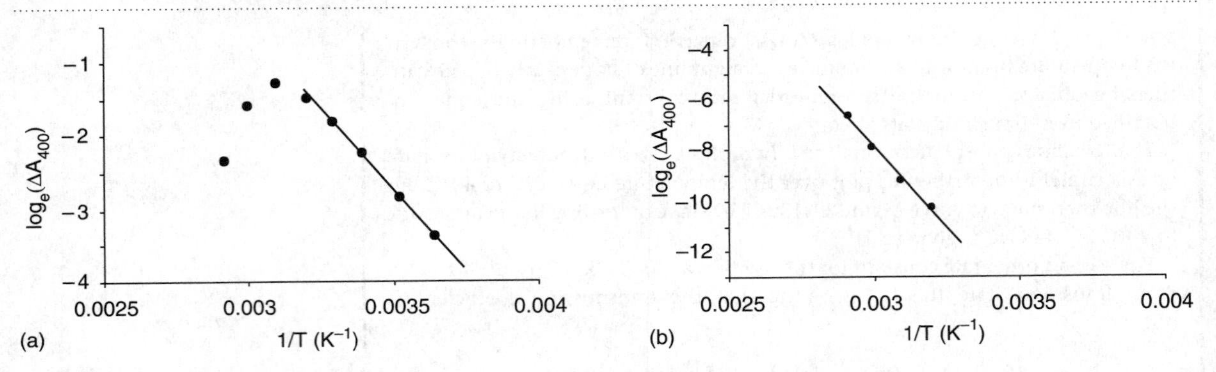

Fig. 11.13 (a) Plot of $\log_e(\Delta A_{400})$ versus 1/T for the hydrolysis of 4-nitrophenylphosphate catalysed by alkaline phosphatase. (b) Plot of $\log_e(\Delta A_{400})$ versus 1/T for the hydrolysis of 4-nitrophenylphosphate catalysed by 1 M NaOH.

Effect of pH on enzyme-catalysed reactions

Changes in pH can have a number of effects on the rate of enzyme-catalysed reactions.

- Direct effects — $[H^+]$ or $[OH^-]$ appear in the rate equation.
- Effects on the substrate — Changes in the ionisation state of the substrate leading to additional acid/base catalysis (as we have seen in Chapter 10).
  Changes in the ionisation state of the substrate leading to altered binding to the enzyme and hence changing K_m.
- Effects on the enzyme — Unfolding of the enzyme outside a certain pH range leading to its complete inactivation.
  Change in the ionisation state of amino acid side chains in the enzyme leading to changes in K_m if the side chain is involved in binding to the substrate or in V_{max} if the side chain takes part in the reaction.

We shall only consider the effects on V_{max} here as these are usually associated with effects on side chains in the enzyme at the active site and hence provide useful mechanistic information.

First, we shall consider an enzyme active site containing a titratable group that is inactive in the protonated form and active in the deprotonated form.

$$\underset{\text{Inactive}}{EH^+{:}S} \overset{K_a}{\rightleftharpoons} \underset{\text{Active}}{E{:}S} + H^+$$

At low pH ($pH \ll pK_a$) the enzyme will be inactive, at high pH the enzyme will be active. In this case H^+ is acting as a non-competitive inhibitor (see next chapter for a discussion of inhibitors), deactivating the enzyme without changing K_m. Enzyme inhibition is dealt with in detail in the next chapter and so all we will do here is quote the result. At any given concentration of H^+, V_{max} is given by

$$V_{max} = \frac{\bar{V}_{max}}{\left(1 + \frac{[H^+]}{K_a}\right)}$$

where $\bar{V}_{max}$ is the limiting velocity for the enzyme in the fully active form. A plot of V_{max} versus pH will follow a typical sigmoidal titration curve. A more useful plot is $\log_{10}(V_{max})$ versus pH. At very low pH, where $[H^+]/K_a \gg 1$, this will give straight line of slope one, and at very high pH, where $[H^+]/K_a \ll 1$, this will give a horizontal line. These two straight lines intersect at $pH = pK_a$ (figure 11.14).

If there are two or more titratable groups in the active site then the behaviour becomes more complex. Consider the case where there are two groups with different pK_a values, one of which needs to be protonated and one of which does not.

$$\underset{\text{Inactive}}{\overset{HX \quad YH}{\diagdown\!\diagup}\atop E} \underset{+H^+}{\overset{-H^+}{\rightleftharpoons}} \underset{\text{Active}}{\overset{HX \quad Y^-}{\diagdown\!\diagup}\atop E} \underset{+H^+}{\overset{-H^+}{\rightleftharpoons}} \underset{\text{Inactive}}{\overset{X^- \quad Y^-}{\diagdown\!\diagup}\atop E}$$

(first equilibrium: pK_{a1}; second equilibrium: pK_{a2})

In this case, the enzyme will be inactive at very low pH and very high pH. It will only be active in a well-defined pH range between the two pK_a values. In this case, V_{max} is given by

$$V_{max} = \frac{\bar{V}_{max}}{\left(1 + \frac{[H^+]}{K_{a1}} + \frac{K_{a2}}{[H^+]}\right)}$$

Again, the pK_a values can be found from the appropriate linear portions of a plot of $\log_{10}(V_{max})$ versus pH (figure 11.15). However, this only works if the two pK_a values are well separated and the curve reaches a maximum with the enzyme in its ideal form (corresponding to $V_{max} = \bar{V}_{max}$). If the pK_a values are too close in value, then one group will not be fully deprotonated before the other group starts becoming deprotonated as well and the ideal arrangement of one group fully deprotonated and the other fully protonated is never achieved. In this case, V_{max} never reaches $\bar{V}_{max}$. A plot of $\log_{10}(V_{max})$ versus pH still gives three linear regions (figure 11.16), but these can no longer be extrapolated to give the pK_a values and instead the data have to be fitted to the above equation directly.

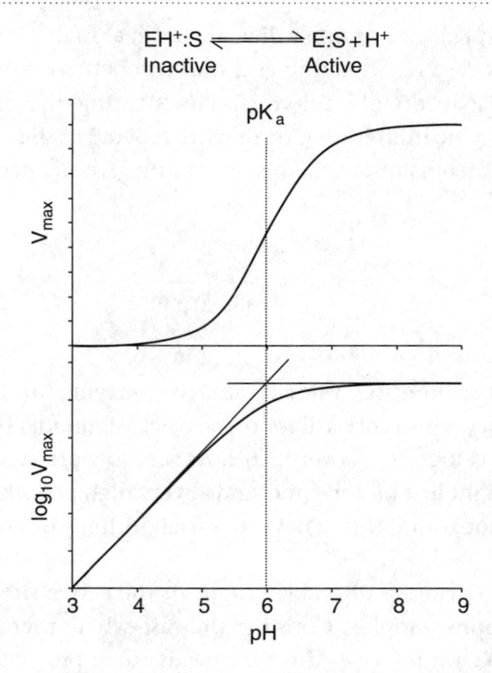

Fig. 11.14 Typical plots of V_{max} and $\log_{10}V_{max}$ versus pH for an enzyme with one titratable group in the active site. The enzyme is only active when the titratable group is deprotonated. The pK_a of the group can be determined by extrapolating from the linear regions of the $\log_{10}V_{max}$ versus pH plot.

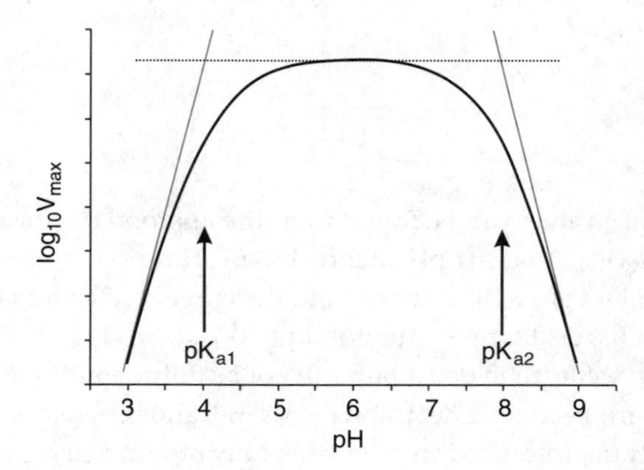

Fig. 11.15 Typical plot of $\log_{10} V_{max}$ versus pH for an enzyme with two titratable groups in the active site. The enzyme is only active when one group is deprotonated and the other group protonated. This gives a small pH range over which the enzyme is fully active. The pK_a values of the groups can be determined by extrapolating from the linear regions of the plot as long as the two pK_a values are sufficiently different.

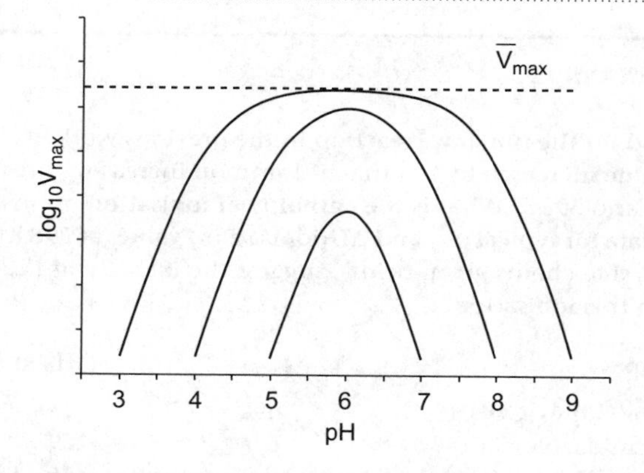

Fig. 11.16 Plots of $\log_{10} V_{max}$ versus pH for an enzyme with two titratable groups in the active site. The enzyme is only active when one group is deprotonated and the other group protonated. The upper line corresponds to groups with pK_a values of 4 and 8, the middle line to pK_a values of 5 and 7, and the lower line to both groups having pK_a values of 6. $\bar{V}_{max}$ is the same for all three plots but only in the first plot does V_{max} approach the value of $\bar{V}_{max}$.

WORKED EXAMPLE

The data listed below show the pH dependence of V_{max} for the reaction catalysed by fumarase at 298 K.

pH	**5·2**	**5·5**	**6·0**	**6·5**	**6·75**	**7·0**	**7·5**	**8·0**	**8·5**
V_{max}/arbitrary units	**54**	**105**	**184**	**224**	**236**	**230**	**162**	**86**	**33**

Comment on the data and deduce pK_a values for the ionising groups involved.

SOLUTION: The plot of $\log_{10} V_{max}$ against pH (figure 11.17) suggests that two ionising groups are involved in the catalytic activity of fumarase. The curve appears to reach a plateau at the top, with three very similar data points, and so the pK_a values can be determined by extrapolating from the linear parts of the curve. The points of intersection give the pK_a values as 5·9 and 7·5 for pK_{a1} and pK_{a2} respectively.

COMMENT: The difference in pK_a values of the two groups is only 1·6. This is a little on the small side for this method to work reliably, even though the curve appears to reach a plateau at the top. In this case, accurate values can only be obtained by a least squares fit to the entire curve.

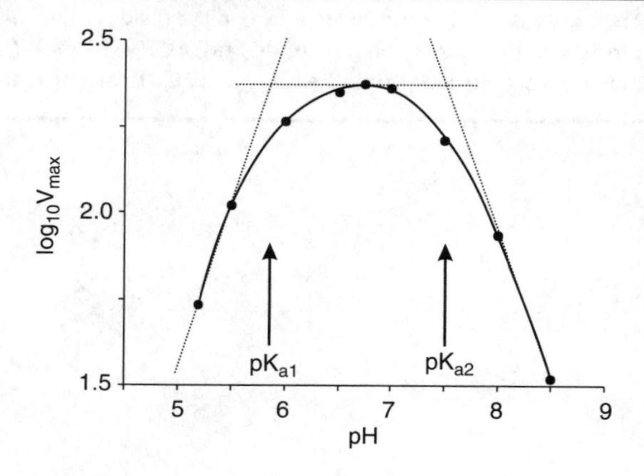

Fig. 11.17 A plot of $\log_{10}V_{max}$ versus pH for fumarase. The active site contains at least two titratable groups, one of which must be protonated and one deprotonated for full activity. In this case, the optimum pH range for the enzyme is relatively small.

WORKED EXAMPLE

It was found for the fumarase reaction in the previous worked example that the pK_{a1} value increased by less than 0·1 unit on increasing the temperature from 298 K and 308 K. What is the enthalpy of ionisation of this side chain? Using the data for typical pK_a and ΔH(ionisation) values at 298 K of some free amino acid side chains given below, suggest the identity of the amino acid involved in this ionisation.

Group	pK_a	ΔH_i/kJ mol^{-1}
Asp or Glu (– COOH)	4·4	$\sim\pm 4$
His (imidazole)	6·9	~ 29
Lys (– NH_2)	10·0	~ 46
Tyr (– OH)	10·0	~ 25

SOLUTION: Since the pK_a values at 298 and 308 K are less than 0·1 unit different

$$\log_{10}\left(\frac{K_{a308}}{K_{a298}}\right) \le 0{\cdot}1 \qquad \text{or} \qquad \log_e\left(\frac{K_{a308}}{K_{a298}}\right) \le 0{\cdot}2303$$

The variation in equilibrium constant with temperature is given by the van't Hoff isochore (Chapter 3)

$$\log_e\left(\frac{K_{T_2}}{K_{T_1}}\right) = -\frac{\Delta H^o}{R}\left(\frac{1}{T_2} - \frac{1}{T_1}\right)$$

Thus

$$-\frac{\Delta H^o}{R}\left(\frac{1}{308} - \frac{1}{298}\right) \le 0{\cdot}2303$$

giving

$$\Delta H^o \le 17{\cdot}5 \text{ kJ mol}^{-1}$$

The pK_a of 5·9 for this ionisable group is intermediate between Glu/Asp and His. We have already seen (Chapter 5) that the pK_a values of amino acid side chains can vary considerably in structured proteins from the values quoted above, but a value of 5·9 is highly unlikely to indicate a Lys or Tyr residue. The ΔH^o value for the ionisation is usually a more reliable guide and, as 17·5 kJ mol^{-1} is an upper limit, this suggests that the side chain belongs to aspartic or glutamic acid.

FURTHER READING

A. Cornish-Bowden, 1995, 'Fundamentals of enzyme kinetics', Portland Press—*a lucid and rigorous description of enzyme kinetics.*

A. Fersht, 1999, 'Structure and mechanism in protein science', Freeman—*a comprehensive account of the molecular mechanisms of enzyme action, including a good discussion of the use of binding energies in catalysis.*

N.C. Price and L. Stevens, 1999, 'Fundamentals of enzymology', Oxford University Press—*a user-friendly account of enzymes and enzyme kinetics.*

PROBLEMS

1. Derive the Michaelis-Menten equation for an enzyme-catalysed single-substrate reaction. In what circumstances will such a reaction not obey the Michaelis-Menten equation?

2. For an enzyme that obeys simple Michaelis-Menten kinetics, (i) what is V_{max} if the velocity when $[S] = K_m$ is QJ;45 μM min^{-1} and (ii) what is the K_m if the velocity equals 50 μM min^{-1} when $[S] = 3 \times 10^{-5}$ M?

3. The following results were obtained for the hydrolysis of a *p*-nitrophenyl derivative of a pentasaccharide derived from chitin (PNP-$GlcNAc_5$) by wild-type human lysozyme (WT) and three single-site mutants, Tyr 63 changed to Phe (Y63F), Tyr 63 changed to Leu (Y63L) and Asp 52 changed to Ala (D52A). The enzyme concentration was 10 μM in all cases.

[PNP-$GlcNAc_5$]/M		0·00001	0·00002	0·00005	0·00010	0·00025
Initial rate/μM s^{-1}	WT	0·020	0·035	0·063	0·084	0·107
	Y63F	0·011	0·020	0·039	0·056	0·076
	Y63L	0·000037	0·000075	0·00019	0·00037	0·00092
	D52A	0·00079	0·0014	0·0025	0·0033	0·0042

What can you deduce about the roles that Tyr 63 and Asp 52 play in the catalytic mechanism of lysozyme?

In a further experiment using *M. luteus* cells as the substrate, the single-site mutants N46A and the double mutant N46A/D52A gave the following values of k_{cat} relative to the wild-type protein

	k_{cat} relative to WT
WT	1
N46A	0·06
N46A/D52A	0·02

What further information can you deduce about the roles that Asn 46 and Asp 52 play in the catalytic mechanism of lysozyme?

4. The liver contains two enzymes that can convert glucose to glucose 6-phosphate at the expense of ATP. One of these, hexokinase, has a K_m for glucose of 40 μM and a maximum catalytic activity of 0·7 μmol glucose transformed min^{-1} (g tissue)$^{-1}$. The other, glucokinase, has a K_m for glucose of 10 mM and a maximum catalytic activity of 4·3 μmol min^{-1} (g tissue)$^{-1}$. [The data refer to the enzymes from rat liver.]

Use these results to calculate the proportion of glucose that is phosphorylated by glucokinase, as compared with hexokinase, in a fasting rat (glucose concentration = 3 mM) and in a well-fed rat (glucose concentration = 9·5 mM). Assume that ATP is present in saturating concentrations in each case.

5. The activity of a phosphatase as a function of temperature was determined at saturating substrate concentrations with the following results.

T/K	284	295	303	313
Activity/mmol min^{-1} (mg protein)$^{-1}$	0·022	0·058	0·136	0·160

Determine the activation enthalpy for the enzyme-catalysed reaction, stating any assumptions that you have made.

6. The activity of the enzyme nitrogenase, which catalyses the reduction of nitrogen to ammonia, from *Azotobacter* was studied as a function of temperature with the following results.

T/K	278	284	290	296	302	308	314	320	326	330
Velocity/arbitrary units	1·65	11·7	72·2	284	483	781	1180	1260	200	9·5

Comment on these data.

7. The trypsin-catalysed hydrolysis of *N*-benzoyl-L-arginine ethyl ester was studied as a function of pH at 298 K with the following results.

pH	4·0	4·5	5·0	5·5	6·0	6·5
V_{max}/katal kg^{-1}	0·0028	0·0087	0·027	0·076	0·180	0·320
pH	7·0	7·5	8·0	8·5	9·0	
V_{max}/katal kg^{-1}	0·425	0·473	0·491	0·497	0·499	

What is the pK_a of the ionising group in this process? At 308 K the pK_a of this group is 6·08. What is the enthalpy of ionisation of this group? Comment on the likely identity of this amino acid.

12

Multisubstrate enzyme kinetics and enzyme inhibition

KEY POINTS

- Multisubstrate enzymes can catalyse reactions in the same way as single-substrate enzymes, and they can also use binding energy to offset the activation enthalpy and entropy terms associated with bringing the reactants together.
- Different mechanisms for multisubstrate reactions give different steady-state rate equations. This can be used experimentally to distinguish between them.
- Enzyme inhibitors can be classified according to their effects on K_m and V_{max} for the enzyme.
- This can be used experimentally to identify the type of inhibitor, but caution is required in interpreting such kinetic data in terms of a specific molecular mechanism.

The previous chapter dealt with the kinetics of single-substrate enzyme reactions. The majority of enzymes have two or more substrates. Whilst this does not change any of the basic concepts of enzyme kinetics, it does make their application a little more complicated. The general procedure is to apply the steady-state approximation to evaluate the concentrations of the various enzyme-containing complexes in the reaction pathway and then the velocity of the overall reaction is set equal to the concentration of the complex preceding enzyme regeneration multiplied by the rate constant for the step that regenerates enzyme. We shall only deal here with two substrate reactions and limit ourselves to discussing the possible mechanisms and resulting kinetic equations of such reactions.

Use of binding energy in catalysing multisubstrate reactions

As in the case of single-substrate enzymes, multisubstrate enzymes catalyse reactions by using ΔG of binding to alter ΔG of activation. They can do this in the same ways that enzymes can catalyse single-substrate reactions, by destabilising the substrate on binding or by providing reactive groups in the active site that can take part in the reaction. There are also additional ways of reducing $\Delta H^{\ddagger}$ and $\Delta S^{\ddagger}$ specific to multisubstrate reactions.

Using binding energy to overcome reactant repulsion

Many biochemical reactions involve charged reactants. The reactants will repel each other if they carry the same type of charge, resulting in a large $\Delta H^{\ddagger}$ contribution to the activation energy. An enzyme can overcome this by binding to both reactants and neutralising the local charges on the reactants with amino acid side chains with complementary charges. Alternatively, the enzyme can react with one substrate and release the first product before binding the second substrate (see ping-pong mechanism below) so that the two reactants never have to come together.

Using binding energy to bring together multiple reactants

Reactions that involve three reactants are usually very slow indeed because the chances of all three molecules colliding at the same time are very small. This gives a large unfavourable $\Delta S^{\ddagger}$ contribution to the activation energy. An enzyme can overcome this by binding to each reactant sequentially and holding them in the active site until the other reactants arrive. Instead of all the reactants having to collide simultaneously, the enzyme can collect the reactants together over a longer period of time.

Using binding energy to bring together reactants in the correct orientation

As we have seen previously, reactants have to collide in the correct orientation to react (one of the contributions to $\Delta S^{\ddagger}$). Reactants bound to enzymes are held in a specific orientation and thus an enzyme can ensure that multiple reactants are always in the correct orientation with respect to each other.

WORKED EXAMPLE

How can the *in vivo* generation of antigen-specific antibodies be used to produce artificial enzymes?

SOLUTION: Antibodies, or immunoglobulins, are ligand-binding proteins (not enzymes) that usually show a very high degree of specificity for their ligand. When presented with a potential antigen, mammals produce specific antibodies that will recognise the antigen. If we present a mammal (such as a rabbit) with a stable molecule that has a structure very similar to the transition state of the reaction we are interested in, antibodies will be produced that will bind to that structure very tightly. If we now isolate these antibodies and add them to the reactants, they will bind the reactants in the correct orientations and the correct relative distances to form the transition state (figure 12.1). It is also likely that the antibody will bind to the transition state with higher affinity than to the reactants and so will also reduce the activation energy by stabilising the transition state. This is one approach to producing 'designer enzymes' in which most of the design work is done by the immune system of the rabbit.

COMMENT: One example of this is the generation of catalytic antibodies to carry out a Diels–Alder cyclisation reaction (figure 12.2). There are no natural enzymes that catalyse such reactions. While the antibodies generated in this case are very efficient at catalysing the reaction, the product binds to the antibody with greater affinity than the reactants leading to product inhibition. This can be circumvented by producing a product that undergoes a further reaction.

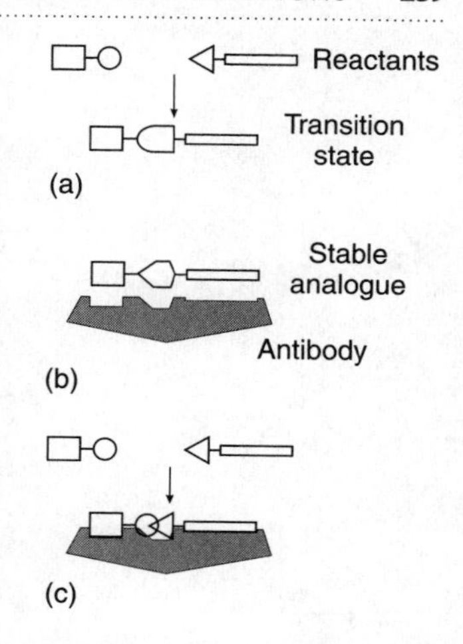

Fig. 12.1 Theory for producing catalytic antibodies. The reaction of interest between two reactants goes via a transition state (a). A stable analogue of the transition state is synthesised and used to produce an antibody response. The isolated antibody will have a binding site matched to the transition state analogue (b). On mixing the reactants with the antibody, the reactants will bind in the correct positions to form the transition state (c).

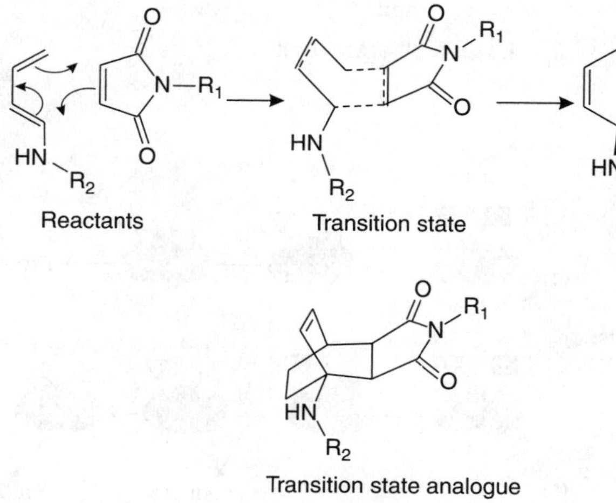

Reactants — Transition state — Product

Transition state analogue

Fig. 12.2 A Diels–Alder reaction goes through a transition state with a six-membered ring in a boat conformation. A stable transition state analogue with a similar conformation can be made using a bridged six-membered ring. Antibodies raised against this compound catalyse this Diels–Alder reaction. [B. S. Green, 1991, *Current Biology*, 2, 395.]

Kinetics and mechanisms of two-substrate enzyme reactions

Two-substrate reactions can be categorised according to whether or not both substrates bind to the enzyme simultaneously (figure 12.3).

Reactions involving a ternary complex

In these reactions, both substrates (A and B) bind simultaneously to the enzyme, where they are converted into products.

$$E + A + B \rightarrow EAB$$

$$EAB \rightarrow E + \text{products}$$

This category can be subdivided into

(i) Those reactions in which the ternary complex (EAB) is formed in an *ordered* manner,

$$E + A \rightarrow EA + B \rightarrow EAB$$
$$(\text{but not } E + B \rightarrow EB)$$

(ii) Those reactions in which the ternary complex is formed in a *random* manner,

$$E + A \rightarrow EA + B \rightarrow EAB$$
$$\text{and}$$
$$E + B \rightarrow EB + A \rightarrow EAB$$

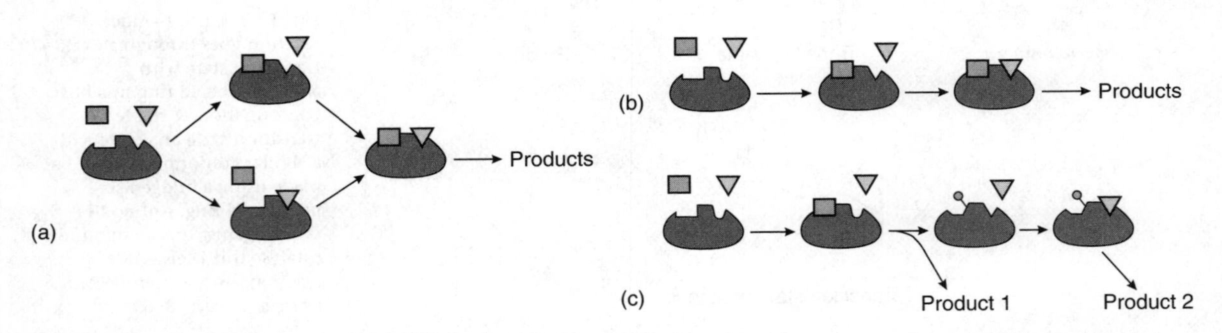

Fig. 12.3 The three most common reaction mechanisms for two-substrate enzyme reactions. (a) Random ternary complex: either substrate can bind first and both substrates bind at the same time. (b) Ordered ternary complex: one substrate cannot bind until the other is bound. (c) Ping-pong: the first substrate reacts to form products and a modified enzyme which then reacts with the second substrate. A ternary complex is not formed.

Both these mechanisms produce a steady-state kinetic equation of the form

$$v = \frac{V_{max}[A][B]}{K_m^{A/EA}K_m^{B/EAB} + K_m^{B/EAB}[A] + K_m^{A/EAB}[B] + [A][B]}$$

V_{max} is the limiting velocity of the reaction at given enzyme concentration and saturating levels of A and B. The constants $K_m^{A/EAB}$, $K_m^{B/EAB}$ and $K_m^{A/EA}$ are all combinations of rate constants of individual steps of the reaction. $K_m^{A/EAB}$ can be thought of as the Michaelis constant for A reacting with EB. This can be readily shown by dividing numerator and denominator of the above equation by [B] and then setting $[B] \to \infty$. The equation reduces to

$$v_{[B]\to\infty} = \frac{V_{max}[A]}{K_m^{A/EAB} + [A]}$$

Thus, $K_m^{A/EAB}$ is the Michaelis constant for A as the concentration of B tends to infinity. Similarly, dividing through by [A] shows that $K_m^{B/EAB}$ is the Michaelis constant for B as the concentration of A tends to infinity.

There is not such a general interpretation of $K_m^{A/EA}$. However, $K_m^{A/EAB}$, $K_m^{B/EAB}$ and $K_m^{A/EA}$ all have a relatively simple meaning in the random-order ternary complex mechanism if we make the further assumption that the enzyme complexes are in equilibrium with the free enzyme.

$$\begin{array}{ccccc} & & EA + B & & \\ & \nearrow\!\!\swarrow & & \searrow\!\!\nwarrow & \\ E + A + B & & & & EAB \rightarrow E + \text{Products} \\ & \searrow\!\!\nwarrow & & \nearrow\!\!\swarrow & \\ & & EB + A & & \end{array}$$

In this case $K_m^{A/EAB}$, $K_m^{B/EAB}$ and $K_m^{A/EA}$ are the dissociation constants of A from EAB, B from EAB and A from EA respectively. The dissociation constant of B from EB is not an independent variable but a function of the other three constants, $K_m^{B/EB} = K_m^{A/EA}K_m^{B/EAB}/K_m^{A/EAB}$.

Reactions not involving a ternary complex

In these reactions, the substrates do not bind simultaneously to the enzyme, the first product is formed before the second substrate is bound. These cases involve a modification of the enzyme and are known as *enzyme substitution* or *ping-pong* mechanisms.

$$E + A \rightarrow EA \rightarrow E' + P$$

$$E' + B \rightarrow E'B \rightarrow E + Q$$

The steady-state rate equation for this type of reaction is

$$v = \frac{V_{max}[A][B]}{K_m^{B/E'B}[A] + K_m^{A/EA}[B] + [A][B]}$$

where $K_m^{A/EA}$ is the Michaelis constant for A binding to E and $K_m^{B/E'B}$ is the Michaelis constant for B binding to E′. This equation is very similar to that for the ternary complex mechanisms except that the denominator lacks a constant term. This equation again reduces to the Michaelis–Menten equation in the limit of either $[A] \to \infty$ or $[B] \to \infty$.

WORKED EXAMPLE

The following data were derived from a study of the reaction

$$\text{ethanol} + NAD^+ \rightleftharpoons \text{ethanal} + NADH + H^+$$

catalysed by yeast alcohol dehydrogenase

[ethanol]/mM	Velocity/katal kg^{-1}			
	[NAD$^+$]/mM			
	0·05	0·1	0·25	1·0
10	0·30	0·51	0·89	1·43
20	0·44	0·75	1·32	2·11
40	0·57	0·99	1·72	2·76
200	0·76	1·31	2·29	3·67

Determine the kinetic parameters for this reaction. What information can be obtained on the reaction mechanism from these data?

SOLUTION: The rate equation for a two-substrate ternary complex reaction is

$$v = \frac{V_{max}[A][B]}{K_m^{A/EA}K_m^{B/EAB} + K_m^{B/EAB}[A] + K_m^{A/EAB}[B] + [A][B]}$$

The inverse of this is

$$\frac{1}{v} = \left(1 + \frac{K_m^{A/EAB}}{[A]} + \frac{K_m^{B/EAB}}{[B]} + \frac{K_m^{A/EA}K_m^{B/EAB}}{[A][B]}\right)\frac{1}{V_{max}}$$

A plot of 1/v versus 1/[A], at constant value of [B], will be linear with a slope of

$$\text{Gradient} = \left(K_m^{A/EAB} + \frac{K_m^{A/EA}K_m^{B/EAB}}{[B]}\right)\frac{1}{V_{max}}$$

and an intercept on the y-axis of

$$\text{y-intercept} = \left(1 + \frac{K_m^{B/EAB}}{[B]}\right)\frac{1}{V_{max}}$$

Plots of 1/v versus 1/[NAD^+] at various values of [ethanol] are shown in figure 12.4. From these plots we can tabulate the following data

[ethanol]/mM	10	20	40	200
Gradient	0·139	0·095	0·073	0·055
Y-intercept	0·55	0·38	0·28	0·22

The plots of gradient versus 1/[ethanol] and y-intercept versus 1/[ethanol] should also give straight lines and are shown in figure 12.5a and 12.5b. The y-intercept of figure 12.5b gives $1/V_{max}$. The gradient of figure 12.5b gives the value of $K_m^{B/EAB}/V_{max}$, allowing $K_m^{B/EAB}$ to be determined. The y-intercept of figure 12.5a gives the value of $K_m^{A/EAB}/V_{max}$, allowing $K_m^{A/EAB}$ to be determined. The gradient of figure 12.5a gives the value of $K_m^{A/EA}K_m^{B/EAB}/V_{max}$, allowing $K_m^{A/EA}$ to be determined.

This analysis gives the following kinetic parameters.

$$V_{max} = 5\,\text{katal kg}^{-1} \qquad K_m^{A/EAB} = 0{\cdot}25\,\text{mM}$$

$$K_m^{A/EA} = 0{\cdot}25\,\text{mM} \qquad K_m^{B/EAB} = 17{\cdot}5\,\text{mM}$$

where A = NAD^+, B = ethanol and E = alcohol dehydrogenase.

COMMENT: The plot of gradient versus 1/[ethanol] proves that the mechanism of the reaction is ternary complex rather than ping-pong. For a ping-pong reaction, where the $K_m^{A/EA}K_m^{B/EAB}$ term does not appear in the rate equation, the gradient of the 1/v versus 1/[A] plot is given by

$$\text{Gradient} = \frac{K_m^{A/EAB}}{V_{max}}$$

which is independent of [B] and so a plot of gradient versus 1/[ethanol] would give a straight line with a slope of zero. For NAD^+ binding to alcohol dehydrogenase, $K_m^{A/EAB} = K_m^{A/EA}$. If we treat $K_m^{A/EAB}$ and $K_m^{A/EA}$ as equilibrium constants (see above), then the binding of NAD^+ to the enzyme appears to be unaffected by the presence or absence of ethanol. This strongly suggests, but does not prove, that the reaction is random ternary complex rather than ordered ternary complex.

As with single-substrate enzyme reactions, the study of pre-steady-state kinetics is much more complex but can also be much more informative because individual steps in the mechanism can often be identified and their rate constants determined.

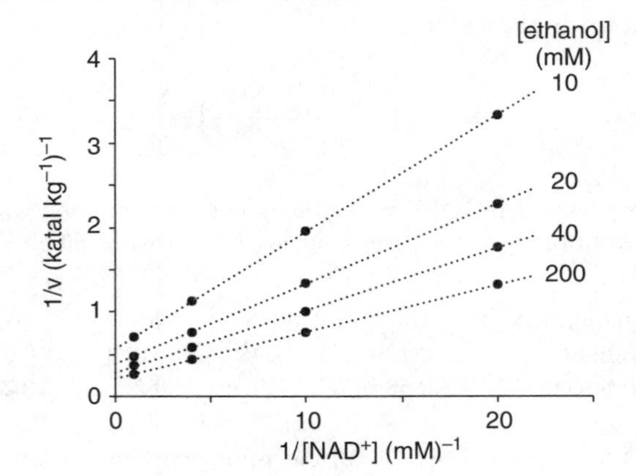

Fig. 12.4 Primary plots of 1/v versus 1/[NAD$^+$] at different ethanol concentrations for the oxidation of ethanol by yeast alcohol dehydrogenase.

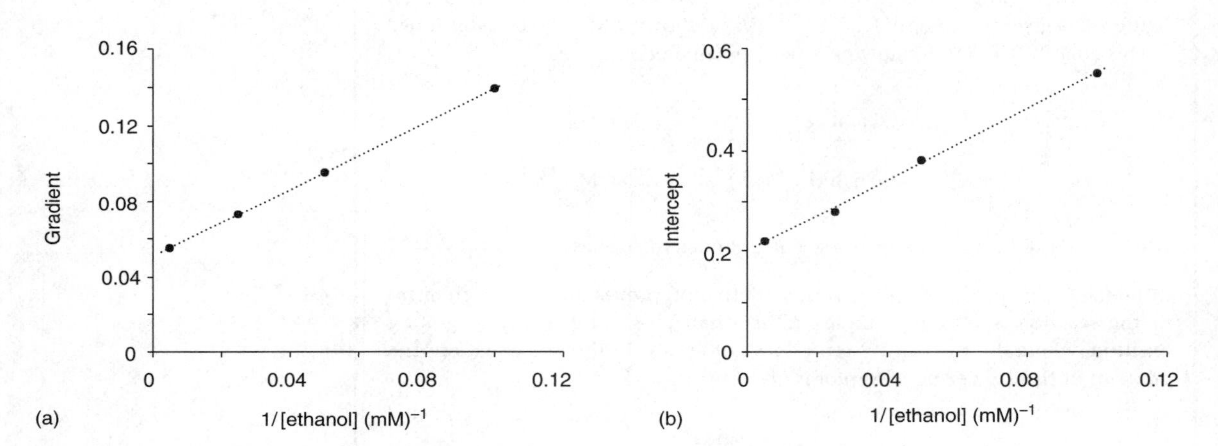

Fig. 12.5 Secondary plots of gradient and intercept versus 1/[ethanol] for the primary plots in figure 12.4. These plots show the reaction to be via a ternary complex mechanism and can be used to extract V_{max} and the Michaelis constants for the reaction.

Enzyme inhibition

The last topic we will cover here also involves more than one species binding to an enzyme but in this case one of the species is an inhibitor that reduces the rate of the reaction. Again this is done by the use of binding energy, the energy available from the inhibitor binding to the enzyme being used to prevent the enzyme from acting as a catalyst either by blocking or altering the active site. The role of enzyme inhibitors *in vivo* is very important because it is one mechanism whereby enzyme activities, and thus cell machinery, can be controlled. Inhibitors are also useful tools for enabling biochemists to alter biochemical processes selectively and to probe the details of enzyme mechanisms. Finally, enzyme inhibitors have enormous medical importance, for example as antibiotics (such as penicillin) that selectively inhibit bacterial enzymes.

Inhibitors can be initially divided into covalent and non-covalent inhibitors, depending on whether or not they form covalent bonds to the enzyme. These are sometimes referred to as irreversible and reversible inhibitors, but this is misleading as covalent modification can be reversible (such as enzyme phosphorylation) whereas non-covalent inhibitors can bind so tightly as to be effectively irreversible. Covalent inhibitors can be treated as reducing the enzyme concentration.

Non-covalent inhibitors can be classified by their effect on the kinetics of the reaction (see Table 12.1 and figure 12.6). This does not imply anything about the mechanism of inhibition.

For multisubstrate enzymes, an inhibitor may act differently with respect to the different substrates. For example, the enzyme UDP-glucosylceramide transferase catalyses the addition of glucose to ceramide during glycolipid biosynthesis. The glucose analogue *N*-butyl deoxynojirimicin is a competitive

TABLE 12.1

Type of inhibitor	Kinetic effect	Comments
Competitive	Increases K_m but does not affect V_{max}	The inhibitor prevents binding of the substrate and vice versa. Inhibition can be overcome by increasing the substrate concentration because when $[S] \gg K_m$ the rate is independent of K_m.
Non-competitive	Reduces V_{max} but does not affect K_m	The inhibitor does not change substrate binding. Inhibition is independent of substrate concentration because the rate always depends on V_{max}.
Uncompetitive	Reduces both K_m and V_{max}	V_{max} and K_m both decrease but the specificity constant k_{cat}/K_m is unaltered.
Mixed	Affects both K_m and V_{max}	V_{max} is reduced and K_m may increase or decrease.

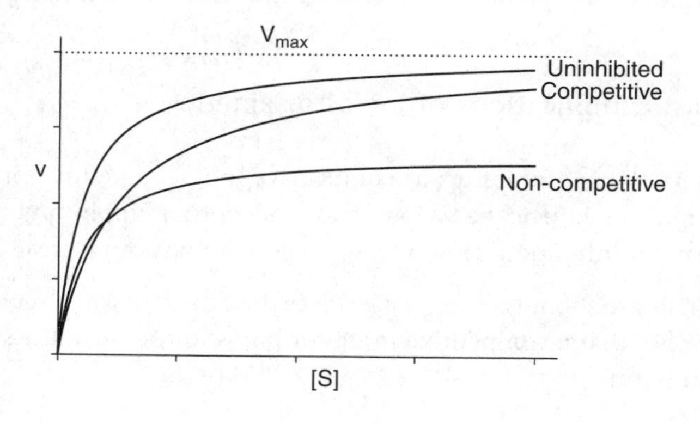

Fig. 12.6 Typical plots of v versus [S] for an enzyme reaction and the same reaction in the presence of a competitive or a non-competitive inhibitor.

inhibitor of the reaction with respect to ceramide and a non-competitive inhibitor with respect to UDP-glucose (see problems).

WORKED EXAMPLE

Discuss the relative advantages of competitive and non-competitive/mixed inhibitors as anti-bacterial agents.

SOLUTION: Inhibiting any step in a metabolic pathway will result in a decrease in the concentration of the products of that step and an increase in the concentration of the reactants because the preceding steps in the pathway will still be working. These changes in concentration are likely to affect the bacteria and may or may not be lethal.

Assuming that these changes are not immediately lethal, the reactant concentration will continue to rise unless the reactants are removed by an alternative pathway. For a competitive inhibitor that increases K_m, the reactant concentration will eventually become large enough for the rate to become independent of K_m and the flux through the pathway will be restored. At this point the reactant concentration will cease to rise and the product concentration will start to increase. For a non-competitive or mixed inhibitor that decreases V_{max}, the inhibition is independent of substrate concentration and so the substrate concentration will continue to rise and the product concentration will remain low. Thus metabolic pathways may be able to adjust reactant concentrations to overcome competitive inhibition but cannot do so for non-competitive or mixed inhibition.

COMMENT: The active component in the herbicide Roundup (or glyphosate) is *N*-phosphonomethylglycine. This is an inhibitor of an enzyme in the shikimate pathway, the pathway in plants for synthesising aromatic amino acids. As aromatic amino acids are necessary for protein synthesis, inhibition of this pathway leads to plant death. As glyphosate is a non-competitive inhibitor, its effects cannot be overcome by increasing substrate concentrations.

We shall only consider inhibition of single-substrate enzyme reactions in detail, but the same approach will work for multisubstrate reactions as well.

Mechanistic implications of inhibitor kinetics

The classification of inhibitors as competitive, non-competitive or mixed is made purely on kinetic behaviour and it does not imply any particular mechanism for inhibition. However, some conclusions can be reached.

- An inhibitor that binds to the same site as the substrate will be competitive. The reverse, that a competitive inhibitor binds to the substrate site, is not necessarily true.

- A non-competitive inhibitor must bind to a site completely distinct from the substrate, the two binding events are independent.
- A mixed inhibitor may have a binding site distinct from the substrate, partially overlapping with the substrate or the substrate may even be part of the binding site. However, the two binding sites will not overlap completely.

Rate equations for inhibition of single-substrate enzyme reactions

For the inhibition of single-substrate enzyme reactions, we can use the following general kinetic scheme (although this is not the only possible one, as ESI may produce products but at a different rate from ES).

$$\begin{array}{ccccc} E + S + I & \rightleftharpoons & ES + I & \xrightarrow{k_2} & E + \text{Products} + I \\ \upharpoonleft\downharpoonright & & \upharpoonleft\downharpoonright & & \\ EI + S & \rightleftharpoons & ESI & & \end{array}$$

The rate equation for this scheme can be derived either by using the steady-state approximation or the equilibrium approximation. We will use the latter in this case as the former gives a very long and complicated derivation. Using the equilibrium approximation, we can define dissociation constants for all the complexes

$$K_d^{S/ES} = \frac{[E][S]}{[ES]} \qquad K_d^{S/ESI} = \frac{[EI][S]}{[ESI]} \qquad K_d^{I/EI} = \frac{[E][I]}{[EI]} \qquad K_d^{I/ESI} = \frac{[ES][I]}{[ESI]}$$

which gives

$$[ES] = \frac{[E][S]}{K_d^{S/ES}} \qquad [EI] = \frac{[E][I]}{K_d^{I/EI}} \qquad [ESI] = \frac{[E][S][I]}{K_d^{S/ES}K_d^{I/ESI}}$$

The fraction, θ, of enzyme in the active form, ES, is given by

$$\theta = \frac{[ES]}{[E]_{total}} = \frac{[ES]}{[E] + [ES] + [EI] + [ESI]} = \frac{\dfrac{[E][S]}{K_d^{S/ES}}}{[E] + \dfrac{[E][S]}{K_d^{S/ES}} + \dfrac{[E][I]}{K_d^{I/EI}} + \dfrac{[E][S][I]}{K_d^{S/ES}K_d^{I/ESI}}}$$

and so the rate, after dividing by [E] and multiplying by $K_d^{S/ES}$, is given by

$$v = \theta V_{max} = \frac{V_{max}[S]}{K_d^{S/ES} + [S] + \dfrac{[I]K_d^{S/ES}}{K_d^{I/EI}} + \dfrac{[S][I]}{K_d^{I/ESI}}}$$

which can be rearranged to give

$$v = \frac{V_{max}[S]}{K_d^{S/ES}\left(1 + \frac{[I]}{K_d^{I/EI}}\right) + [S]\left(1 + \frac{[I]}{K_d^{I/ESI}}\right)}$$

This is a general solution to the above kinetic scheme and is the rate equation for mixed inhibition. This can be shown by rewriting the equation in a form similar to the Michaelis–Menten equation

$$v = \frac{\dfrac{V_{max}}{\left(1 + \frac{[I]}{K_d^{I/ESI}}\right)}[S]}{K_d^{S/ES}\dfrac{\left(1 + \frac{[I]}{K_d^{I/EI}}\right)}{\left(1 + \frac{[I]}{K_d^{I/ESI}}\right)} + [S]} = \frac{V_{max}^{app}[S]}{K_m^{app} + [S]}$$

The maximum velocity measured in the presence of the inhibitor is given by

$$V_{max}^{app} = \frac{V_{max}}{\left(1 + \frac{[I]}{K_d^{I/ESI}}\right)}$$

and so V_{max}^{app} is less than V_{max}. The Michaelis constant measured in the presence of the inhibitor is given by

$$K_m^{app} = K_d^{S/ES}\frac{\left(1 + \frac{[I]}{K_d^{I/EI}}\right)}{\left(1 + \frac{[I]}{K_d^{I/ESI}}\right)}$$

and so K_m^{app} may be smaller or larger than K_m depending on the relative values of $K_d^{I/EI}$ and $K_d^{I/ESI}$.

For other classes of inhibitors, the kinetic scheme is not as complicated as the general scheme and so the rate equation is simpler than the general solution. We shall deal with competitive and non-competitive inhibition here.

Binding of S and I are mutually exclusive (complex ESI cannot be formed)

If the complex ESI cannot be formed then $K_d^{S/ESI}$ and $K_d^{I/ESI}$ must be infinite and the general solution reduces to

$$v = \frac{V_{max}[S]}{K_d^{S/ES}\left(1 + \frac{[I]}{K_d^{I/EI}}\right) + [S]}$$

Alternatively, we can derive this equation from scratch (and will do so just for practice) by using the following kinetic scheme.

$$\begin{array}{lll} E + S + I & \rightleftharpoons ES + I & \xrightarrow{k_2} E + \text{Products} + I \\ \updownarrows & & \\ EI + S & & \end{array}$$

The fraction, θ, of enzyme in the active form, ES, is given by

$$\theta = \frac{[ES]}{[E]_{total}} = \frac{[ES]}{[E] + [ES] + [EI]}$$

Substituting for [ES] and [EI] gives

$$\theta = \frac{\frac{[E][S]}{K_d^{S/ES}}}{[E] + \frac{[E][S]}{K_d^{S/ES}} + \frac{[E][I]}{K_d^{I/EI}}} = \frac{[S]}{K_d^{S/ES} + [S] + \frac{[I]K_d^{S/ES}}{K_d^{I/EI}}}$$

and the rate is given by

$$v = \theta V_{max} = \frac{V_{max}[S]}{K_d^{S/ES} + [S] + \frac{[I]K_d^{S/ES}}{K_d^{I/EI}}} = \frac{V_{max}[S]}{K_d^{S/ES}\left(1 + \frac{[I]}{K_d^{I/EI}}\right) + [S]}$$

This can be rewritten in the same form as the Michaelis–Menten equation

$$v = \frac{V_{max}[S]}{K_m^{app} + [S]}$$

where the apparent Michaelis constant, K_m^{app}, is given by

$$K_m^{app} = K_d^{S/ES}\left(1 + \frac{[I]}{K_d^{I/EI}}\right)$$

This corresponds to competitive inhibition, the value of V_{max} remaining unaltered and the apparent K_m increasing as the concentration of inhibitor

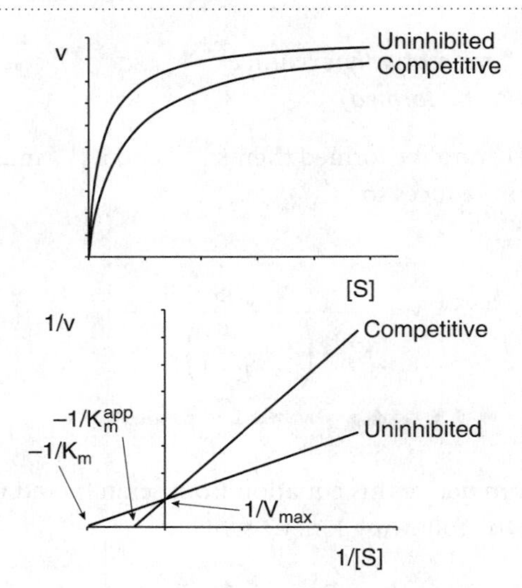

Fig. 12.7 Typical plots of v versus [S] and 1/v versus 1/[S] (Lineweaver–Burk plot) for competitive inhibition. The presence of the inhibitor does not change V_{max} but does change K_m.

increases. At low values of [S], the rate is reduced. At high values of [S], when $[S] \gg K_m^{app}$, the rate approaches V_{max} and so is unaffected (figure 12.7).

The binding of S is unaffected by the presence of I

If S binds with the same affinity to both E and EI then $K_d^{S/ES} = K_d^{S/ESI}$. Because E, ES, EI and ESI are all in equilibrium, this means that $K_d^{I/EI}$ must equal $K_d^{I/ESI}$. The general solution then becomes

$$v = \frac{V_{max}[S]}{K_d^{S/ES}\left(1 + \frac{[I]}{K_d^{I/EI}}\right) + [S]\left(1 + \frac{[I]}{K_d^{I/EI}}\right)}$$

which can be rewritten as

$$v = \frac{V_{max}^{app}[S]}{K_m + [S]}$$

where

$$V_{max}^{app} = \frac{V_{max}}{\left(1 + \frac{[I]}{K_d^{I/EI}}\right)}$$

This corresponds to non-competitive inhibition, the value of K_m remaining unaltered and the apparent V_{max} decreasing as the concentration of inhibitor increases. The rate is reduced at both low and high values of [S] (figure 12.8).

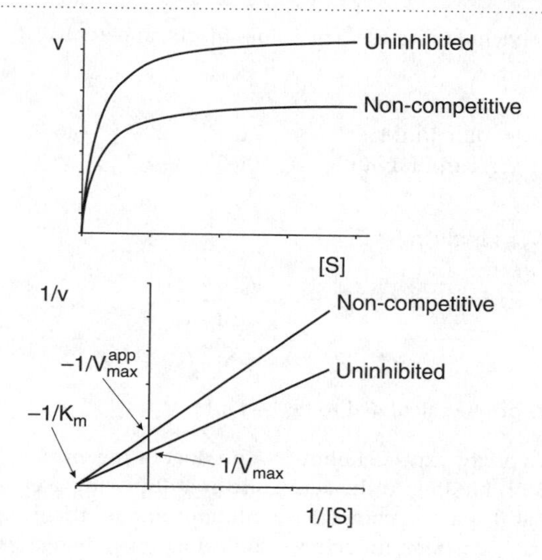

Fig. 12.8 Typical plots of v versus [S] and 1/v versus 1/[S] (Lineweaver–Burk plot) for non-competitive inhibition. The presence of the inhibitor does not change K_m but does change V_{max}.

The analysis of uncompetitive inhibition we shall leave as an exercise for the reader (see Problem 5).

WORKED EXAMPLE

Insect acetylcholinesterase catalyses the reaction

$$\textbf{acetylcholine + water} \rightleftharpoons \textbf{acetate + choline}$$

As we have seen before, this reaction is inhibited by its substrate at high concentrations. The following data were obtained for this reaction in the presence of differing amounts of the product, choline.

[choline]/mM	Relative velocity/arbitrary units				
	[acetylcholine]/mM				
	0·10	0·15	0·25	0·40	0·70
0	28·5	37·5	50·0	61·5	73·7
20	11·9	15·6	20·9	25·7	30·7
40	7·5	9·9	13·2	16·2	19·4

Determine the type of inhibition that is being observed in this case and calculate the value of K_{EI}.

SOLUTION: The plot of 1/v versus 1/[acetylcholine] at the three different choline concentrations is shown in figure 12.9. All three plots give the same value of K_m (0·25 mM) but different values for V_{max}. Thus, choline is a non-competitive inhibitor of acetylcholinesterase. The values for the apparent V_{max} are

[choline]/mM	0	20	40
V_{max}^{app}/arbitrary units	100·0	41·7	26·3

The apparent V_{max} is given by

$$V_{max}^{app} = \frac{V_{max}}{\left(1 + \frac{[I]}{K_d^{I/EI}}\right)}$$

From this, $K_d^{I/EI}$ can be calculated to be 14·3 mM.

COMMENT: We might expect choline as a product of the enzyme to be a competitive inhibitor, binding to the active site and displacing acetylcholine. The observation that it is a non-competitive inhibitor implies that it binds at a different site to the substrate, inducing a conformational change that alters the active site to reduce V_{max}. It is likely that acetylcholine can also bind to this secondary site, which would explain the substrate inhibition observed at high acetylcholine concentrations (see previous chapter).

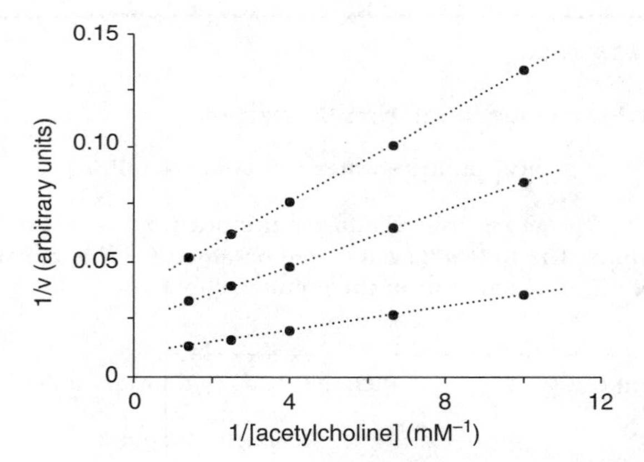

Fig. 12.9 Plots of 1/v versus 1/[acetylcholine] at different choline concentrations for the reaction catalysed by acetylcholinesterase.

FURTHER READING

As for chapter 11:

A. Cornish-Bowden, 1995, 'Fundamentals of enzyme kinetics', Portland Press—*a lucid and rigorous description of enzyme kinetics.*

A. Fersht, 1999, 'Structure and mechanism in protein science', Freeman—*a comprehensive account of the molecular mechanisms of enzyme action, including a good discussion of the use of binding energies in catalysis.*

N.C. Price and L. Stevens, 1999, 'Fundamentals of enzymology' Oxford University Press—*a user-friendly account of enzymes and enzyme kinetics.*

PROBLEMS

1. Derive the following equation for the rate of a random-order ternary complex enzyme-catalysed reaction

$$v = \frac{V_{max}[A][B]}{K_m^{A/EA}K_m^{B/EAB} + K_m^{B/EAB}[A] + K_m^{A/EAB}[B] + [A][B]}$$

[Hint: the rate is given by $v = V_{max} \times \theta$ where θ is the fraction of the total enzyme that has both substrates bound. See Chapter 4 for the calculation of θ.]

2. The enzyme nucleoside diphosphokinase will catalyse the following reaction.

$$\text{GTP} + \text{dGDP} \xrightarrow{Mg^{2+}} \text{GDP} + \text{dGTP}$$

The following data were obtained using the enzyme isolated from erythrocytes.

[dGDP]/μM	Velocity/katal kg^{-1}			
	[GTP]/μM			
	22	30	50	200
20	0·095	0·112	0·141	0·196
25	0·102	0·120	0·155	0·223
40	0·112	0·136	0·180	0·284
100	0·125	0·156	0·218	0·385

What can you deduce from these data regarding the likely mechanism for the enzyme-catalysed reaction?

3. The enzyme creatine kinase catalyses the following reaction.

$$\text{creatine} + \text{ATP} \xrightarrow{Mg^{2+}} \text{phosphocreatine} + \text{ADP}$$

The following data were obtained using enzyme isolated from rabbit muscle.

[Creatine]/mM	Velocity/katal kg^{-1}			
	[ATP]/mM			
	0·46	0·62	1·23	3·68
6	0·377	0·463	0·660	0·968
10	0·555	0·678	0·950	1·308
20	0·845	1·005	1·338	1·803
40	1·180	1·378	1·718	2·295

Evaluate the kinetic parameters for this reaction. From other data it appears that the reaction proceeds via a random-order ternary complex mechanism. What can you deduce about the binding of substrates to the enzyme?

4. What are the effects of competitive and non-competitive inhibitors as shown in Eadie–Hofstee and Hanes plots?

5. Discuss the mechanistic implications of the observation that *N*-butyl deoxynojirimicin is a competitive inhibitor of UDP-glucosylceramide transferase with respect to ceramide but a non-competitive inhibitor with respect to UDP-glucose.

6. Uncompetitive inhibition is defined as a decrease in V_{max} and K_m but the ratio V_{max}/K_m remains the same. Starting from the following kinetic scheme

$$\begin{array}{ccccc} E + S + I & \rightleftharpoons & ES + I & \xrightarrow{k_2} & E + \text{Products} + I \\ \upharpoonleft\downharpoonright & & \upharpoonleft\downharpoonright & & \\ EI + S & \rightleftharpoons & ESI & & \end{array}$$

show that this occurs when the complex EI cannot be formed but ESI can.

What is the effect of an uncompetitive inhibitor on the reaction rate at (a) low and (b) high concentrations of substrate? Sketch the Lineweaver–Burk, Eadie–Hofstee and Hanes plots for this type of inhibition. Discuss what the mechanism for this type of inhibition could be.

7. The enzyme fructose bisphosphatase catalyses the hydrolysis of fructose bisphosphate to fructose 6-phosphate and phosphate and is inhibited by AMP. The following data were obtained using the enzyme isolated from rat liver.

[AMP]/μM	Velocity/katal kg^{-1}				
	[FBP]/μM				
	4	6	10	20	40
0	0·059	0·076	0·101	0·125	0·150
8	0·034	0·043	0·056	0·071	0·083

Comment on these data.

8. The enzyme lactate dehydrogenase catalyses the following reaction.

$$NAD^+ + \text{lactate} \rightarrow NADH + \text{pyruvate}$$

The following data were obtained for the effect of pyruvate, one of the products, on the reaction catalysed by the enzyme isolated from rabbit muscle.

At a fixed NAD^+ concentration (1·5 mM)

[Pyruvate]/μM	Velocity/katal kg^{-1}			
	[Lactate]/mM			
	1·5	2·0	3·0	10·0
0	1·88	2·36	3·10	5·81
40	1·05	1·34	1·88	4·19
80	0·73	0·94	1·34	3·27

At a fixed lactate concentration (15 mM)

[Pyruvate]/μM	Velocity/katal kg^{-1}			
	$[NAD^{+}]$/mM			
	0·5	0·7	1·0	2·0
0	3·33	3·91	4·50	5·42
30	2·65	3·13	3·60	4·33
60	1·97	2·30	2·66	3·21

What types of inhibition are being observed in these cases?

9. The enzyme pyruvate kinase catalyses the following reaction.

$$\text{phosphoenolpyruvate} + \text{ADP} \xrightarrow{Mg^{2+}} \text{pyruvate} + \text{ATP}$$

The velocity of the reaction was studied as a function of phosphoenolpyruvate (PEP) concentration with the results shown below (assume that ADP is present in saturating concentrations). Also given are data on the effects of two inhibitors of this reaction, 10 mM 2-phosphoglycerate and 10 mM phenylalanine.

Inhibitor	Velocity/katal kg^{-1}				
	[PEP]/mM				
	0·25	0·3	0·5	1·0	2·0
None	1·10	1·25	1·63	2·05	2·43
10 mM 2-phosphoglycerate	0·73	0·83	1·15	1·68	2·08
10 mM phenylalanine	0·60	0·68	0·90	1·25	1·50

Comment on these data.

10. Data of the initial rate of an enzyme reaction versus inhibitor concentration, repeated at different initial substrate concentrations, can be used to determine both the type of inhibition and the values of the dissociation constants for the inhibitor.

By rearranging the equation for mixed inhibition of a single-substrate enzyme reaction

$$v = \frac{V_{max}[S]}{K_d^{S/ES}\left(1 + \frac{[I]}{K_d^{I/EI}}\right) + [S]\left(1 + \frac{[I]}{K_d^{I/ESI}}\right)}$$

show that (a) the straight-line plots of 1/v versus [I] at different substrate concentrations intersect when $-[I] = K_d^{I/EI}$ and (b) the straight-line plots of [S]/v versus [I] at different substrate concentrations intersect when $-[I] = K_d^{I/ESI}$. [Hint: In (a), called a Dixon plot, the lines will intersect when 1/v is independent of [S], i.e. at the value of [I] when the coefficient of [S] in the equation of the straight line is zero.]

What would the plots (a) and (b) look like for a competitive inhibitor and an uncompetitive inhibitor?

The enzyme xanthine oxidase is inhibited by an inhibitor I. The following kinetic data were obtained at three different inhibitor concentrations.

Initial substrate concentration [S]/M	**Initial rate v/μM min^{-1}**		
	Inhibitor concentration [I]/M		
	4×10^{-9}	8×10^{-9}	16×10^{-9}
50×10^{-6}	13·7	9·6	6·0
25×10^{-6}	11·0	7·4	4·4
$16{\cdot}7 \times 10^{-6}$	9·2	5·9	3·5
$12{\cdot}5 \times 10^{-6}$	7·8	5·0	2·9

Show that I is a mixed inhibitor of xanthine oxidase and determine $K_d^{I/EI}$ and $K_d^{I/ESI}$.

13

Coupled reactions and biochemical pathways

KEY POINTS

- Chemical reactions can be coupled *sequentially* or *in parallel* to alter positions of equilibrium.
- Metabolic pathways consist of series of enzyme-catalysed reactions that are energetically favourable as the result of sequential and parallel coupling.
- *Metabolic control analysis* is a powerful tool for analysing the flux through a metabolic pathway. The analysis shows that control of flux is normally shared by many enzymes in the pathway.
- The *flux control coefficient* for an enzyme measures the extent to which it determines the flux through a pathway.
- The *elasticity coefficients* for an enzyme measure the extent to which the activity depends on the concentrations of effectors (i.e. ions and metabolites that bind to the enzyme).
- For a metabolic pathway, the values of the flux control coefficients add up to 1. The values of the flux control and elasticity coefficients for each enzyme are linked by a *connectivity theorem*.

So far we have been largely concerned with developing a thermodynamic and kinetic framework for the analysis of single reactions. However, in biological systems there are many reactions occurring simultaneously and the interactions between them have an important bearing on the behaviour of the whole system. The thermodynamic analysis of simple reactions indicates that some are energetically favourable while others are not. In cells every flux through a reaction pathway and every net movement of material across a membrane has to be energetically favourable. So if a process is required that would not be expected to occur spontaneously, then some way has to be found to drive the reaction forward using the work energy that is available from the processes that do occur spontaneously.

Sequential coupling of chemical reactions

An important feature of metabolic pathways is the way in which the product of one reaction is often the substrate of the next. Consider a simple pathway

$$A \rightarrow B \rightarrow C$$

In principle, the concentrations of A and B in the cell could be such as to favour a net flux from A to B, or a net flux from B to A, or no net flux at all (when the concentrations would be the equilibrium concentrations of A and B). However, the effect of the second step, $B \rightarrow C$, is to reduce the concentration of B if the concentrations of B and C are such as to favour a net flux from B to C. This reduction in B will favour a net flux from A to B once the concentration of B falls below the equilibrium value for the first step in the pathway. Thus the second reaction effectively drives the first reaction by reducing the concentration of B to the point where the first reaction becomes thermodynamically favourable. Note that the sequential coupling of successive steps in a metabolic pathway depends only on the existence of the shared metabolite B, and so there is no requirement for the two steps to occur at the same time or in the same place.

WORKED EXAMPLE

The phosphorolysis of sucrose with inorganic phosphate (P_i) gives fructose and glucose 1-phosphate.

$$\text{sucrose} + P_i \rightleftharpoons \text{fructose} + \text{G-1-P} \qquad K_1 = 20$$

Calculate the concentration of fructose at equilibrium if the initial concentrations of sucrose and phosphate are 5 and 10 mM respectively. Phosphoglucomutase catalyses the conversion of glucose 1-phosphate to glucose 6-phosphate.

$$\text{G-1-P} \rightleftharpoons \text{G-6-P} \qquad K_2 = 19$$

What happens to the equilibrium concentration of fructose when phosphoglucomutase is added to the solution?

SOLUTION: The equilibrium constant for the phosphorolysis of sucrose is

$$K_1 = \frac{[\text{fructose}][\text{G-1-P}]}{[\text{sucrose}][P_i]} = 20$$

If equilibrium is reached after the phosphorolysis of x mM of sucrose, then the equilibrium concentrations will be: [sucrose] = $(5 - x)$ mM; $[P_i] = (10 - x)$ mM; and [fructose] = [G-1-P] = x mM.

Substituting these values, expressed in M, into the equation for K_1 gives

$$20 = \frac{(x \times 10^{-3})^2}{(5 - x) \times 10^{-3} \times (10 - x) \times 10^{-3}}$$

and this can be rearranged to give

$$19x^2 - 300x + 1000 = 0$$

This equation has two solutions for x, 4·78 and 11·01, but since x must be between 0 and 5 mM the equilibrium concentration of fructose is

$$[\text{fructose}] = 4{\cdot}78\,\text{mM}$$

Adding phosphoglucomutase to the solution generates a pathway with two sequentially coupled reactions. There are now two equilibria to consider and thus two equilibrium equations that need to be satisfied simultaneously

$$K_1 = \frac{[\text{fructose}][\text{G-1-P}]}{[\text{sucrose}][P_i]} = 20$$

and

$$K_2 = \frac{[\text{G-6-P}]}{[\text{G-1-P}]} = 19$$

If equilibrium is reached after the phosphorolysis of x mM of sucrose, and if y mM of the resulting glucose 1-phosphate is converted to glucose 6-phosphate, then the equilibrium concentrations will be: [sucrose] = (5 – x) mM; $[P_i]$ = (10 – x) mM; and [fructose] = x mM; [G-1-P] = (x – y) mM; and [G-6-P] = y mM.

Substituting into the equations for K_1 and K_2 gives

$$20 = \frac{x \times 10^{-3} \times (x - y) \times 10^{-3}}{(5 - x) \times 10^{-3} \times (10 - x) \times 10^{-3}} \quad \text{and} \quad 19 = \frac{y \times 10^{-3}}{(x - y) \times 10^{-3}}$$

From the second equation y = 19x/20 and substituting this into the first equation gives

$$19{\cdot}95x^2 - 300x + 1000 = 0$$

Again the solution must lie between 0 and 5 and the only solution that meets this criterion is x = 4·99. Thus at equilibrium in the coupled system

$$[\text{fructose}] = 4{\cdot}99\,\text{mM}$$

COMMENT: The phosphorolysis of sucrose is a thermodynamically favourable reaction that goes to about 95% completion. However, this value is increased still further, to 99·8%, on coupling to a second favourable reaction.

Parallel coupling of chemical reactions

Many reactions are so thermodynamically unfavourable that it would be impracticable to make them occur by reducing the concentration of the products to a very low level through the action of sequential coupling. In this situation the strategy adopted by cells is to arrange for the unfavourable reaction to occur simultaneously with a favourable reaction in a process known as parallel coupling. For example, the synthesis of glucose 6-phosphate by the reaction

$$\text{glucose} + P_i \rightleftharpoons \text{glucose 6-phosphate} + H_2O$$

is energetically very unfavourable and it would be impracticable to drive it forward *in vivo* by manipulating the concentrations of the reactants and products. On the other hand, the reaction

$$\text{ATP} + H_2O \rightleftharpoons \text{ADP} + P_i$$

is usually energetically favourable, and if it could be made to occur simultaneously with the first reaction then the net effect would be to allow the phosphorylation of glucose to take place

$$\text{glucose} + \text{ATP} \rightleftharpoons \text{glucose 6-phosphate} + \text{ADP}$$

since this reaction would have a favourable ΔG value.

This is precisely how the phosphorylation of glucose occurs in cells, and it requires the use of an enzyme, for example hexokinase or glucokinase, to provide the physical coupling between the two reactions. Thus the energetically favourable hydrolysis of ATP drives the energetically unfavourable phosphorylation of glucose, but it only works if the two events occur at the same point in time and space, i.e. in the active site of a suitable enzyme.

WORKED EXAMPLE

Glutamine is used both as an amino acid in protein synthesis and as a non-toxic source of ammonia for other biochemical pathways. Glutamine can be produced from glutamate, but the energetics of the reaction

$$\textbf{glutamate(aq)} + \textbf{NH}_3\textbf{(aq)} \rightleftharpoons \textbf{glutamine(aq)} + \textbf{H}_2\textbf{O}$$

are unfavourable since $\Delta G^{\circ\prime} = +14{\cdot}3\,\text{kJ mol}^{-1}$. Suggest how this reaction could be made to occur in cells.

SOLUTION: Sequential coupling is unlikely to be useful here, since it may be necessary to allow cells to build up a significant pool of glutamine for metabolic use. However, the energy required to drive the reaction is considerably less than the energy available from the hydrolysis of ATP ($\Delta G^{\circ} = -30{\cdot}5\,\text{kJ mol}^{-1}$). This suggests that the two reactions should be coupled in parallel to give

$$ATP(aq) + glutamate(aq) + NH_3(aq) \rightleftharpoons ADP(aq) + P_i(aq) + glutamine(aq)$$

with $\Delta G^{o\prime} = 14{\cdot}3 - 30{\cdot}5 = -16{\cdot}2\,kJ\,mol^{-1}$. The expectation is that this coupled reaction would occur spontaneously, and indeed this reaction is catalysed by glutamine synthetase *in vivo*.

COMMENT: The actual driving force of the coupled reaction is likely to differ from the value calculated above, because reactions in cells do not occur under standard conditions, and so it is the value of ΔG and not ΔG^o that provides the driving force.

Coupled reactions and biochemical pathways

The principles of sequential and parallel coupling of chemical reactions can be applied to the complex networks of coupled reactions that make up biochemical systems. The complexity of the system, be it a relatively simple metabolic pathway such as glycolysis, or a physiological response such as muscle contraction, is not really an issue. Thermodynamic arguments are based on the values of state functions and changes in state functions only depend on the initial and final states of the system (Chapter 2). So when metabolic pathways are scrutinised from this perspective it is possible to draw the following conclusions:

- Biochemical and biosynthetic pathways can be constructed from sequentially coupled reactions. Thus, in principle, there is no need for any restriction on when and where each reaction occurs, and thus no need for specific coupling mechanisms. In practice, the real pathways that occur in cells are subject to multiple controls in space and time to allow regulation of the biological activity of the organism.
- Reactions that are only slightly unfavourable thermodynamically are driven by the sequential coupling in the pathway, i.e. the consumption of the product of the reaction by a subsequent favourable step drives the unfavourable reaction forward. This has the effect of converting the energetically unfavourable step into a favourable one.
- Reactions that are very unfavourable thermodynamically are coupled in parallel to energy-producing reactions, such as the hydrolysis of ATP. Biological systems use very few of these energy-producing reactions and so the same reactions are used in many different pathways. Parallel coupling requires the energetically favourable and unfavourable processes to be tightly coupled, and this is achieved through the enzyme that catalyses the step in the pathway.

- The energy from energy-producing reactions is only available in certain amounts and so it is not possible to drive steps in a pathway that require an input of more energy than is available. Similarly, if ATP is used to drive a reaction that requires rather little work energy, then the remainder is lost as heat, because it is not possible to hydrolyse a fraction of an ATP molecule.
- While it can be assumed that selection pressures over an evolutionary time scale will have influenced the energetic efficiency of pathways, consumption of energy is only one factor in the overall utility of a pathway. Thus the energy-consuming processes are not perfectly matched to the energy-producing processes and some waste of work energy as heat is inevitable.

Thus in thermodynamic terms, a biochemical pathway is a series of steps that are energetically favourable as a result of sequential and parallel coupling between reactions, and it necessarily supports a net flux of material in the thermodynamically favoured direction.

Kinetic control of biochemical pathways

Metabolic pathways have to be regulated if a cell is to function properly, but a thermodynamic analysis gives little information on how this can be achieved. For example, a thermodynamic analysis of the glycolytic pathway in a cell or organ requires an assessment of the ΔG value for each step in the pathway. These values can be calculated from the corresponding ΔG^o values and the intracellular concentrations of the substrates and products. If the concentrations are measured in a metabolic steady state in which there is a net flux through glycolysis, then the calculated ΔG values should all be negative, confirming that a forward flux is possible. However, what is missing from this analysis is any indication as to what controls the rate of the overall pathway.

Initial attempts to solve this problem revolved around the idea that there ought to be a rate-determining step in the pathway, allowing the pathway to be regulated by controlling the activity of a particular enzyme. It was further argued that potential candidates for these regulatory steps could be identified by using a thermodynamic argument based on measurements of the mass action ratio (Γ, Chapter 3).

If a step in a pathway is at equilibrium, then $\Delta G = 0$, and $\Gamma = K$. In this situation the kinetic behaviour of the step is determined by the concentrations of the reactants and products: if the reactant concentrations increase, the reaction goes forwards, whereas if the product concentrations increase the reaction goes backwards. Under these conditions the flux through the reaction is said to be under thermodynamic control.

Alternatively, if a step is not at equilibrium, then $\Delta G < 0$ and $\Gamma < K$. In this situation, the reaction can only go forwards and the flux is said to be under kinetic control because the net conversion of reactants to products depends on the rate of the reaction.

WORKED EXAMPLE

The concentrations of glucose 6-phosphate, fructose 6-phosphate and unbound fructose 1,6-bisphosphate were found to be 0·17 mM, 0·04 mM and 0·68 μM respectively in a working rat heart. Calculate the values of Γ for the reactions catalysed by phosphoglucoisomerase (K′ = 0·30) and phosphofructokinase (K′ = 1·32 × 10^3) given that the intracellular pH was 7·0 and the ATP/ADP ratio was 45·4.

SOLUTION: The reaction catalysed by phosphoglucoisomerase is

$$\text{glucose 6-phosphate} \rightleftharpoons \text{fructose 6-phosphate}$$

and so

$$\Gamma = \frac{[\text{fructose 6-phosphate}]}{[\text{glucose 6-phosphate}]} = \frac{0{\cdot}04}{0{\cdot}17} = 0{\cdot}235$$

Similarly, the reaction catalysed by phosphofructokinase is

$$\text{fructose 6-phosphate} + \text{ATP} \rightleftharpoons \text{fructose 1,6-bisphosphate} + \text{ADP}$$

and so

$$\Gamma = \frac{[\text{fructose 1,6-bisphosphate}][\text{ADP}]}{[\text{fructose 6-phosphate}][\text{ATP}]} = \frac{0{\cdot}68 \times 10^{-3}}{0{\cdot}04 \times 45{\cdot}4} = 3{\cdot}74 \times 10^{-4}$$

COMMENT: The reaction catalysed by phosphoglucoisomerase is close to equilibrium because Γ/K′ = 0·78, which is close to 1, whereas the reaction catalysed by phosphofructokinase is far from equilibrium because Γ/K′ = 2·8 × 10^{-7}. Thus the flux through the glycolytic pathway can be said to be under thermodynamic control in the first case and kinetic control in the second.

It is the non-equilibrium steps in a pathway that were of interest in the hunt for the rate-determining step. In particular, it was argued that if the enzyme catalysing a non-equilibrium step was saturated with substrate, then the enzyme could have a regulatory or controlling function if its activity was sensitive to other effectors. Thus the conversion of fructose 6-phosphate to fructose 1,6-bisphosphate was identified as a rate-determining step in glycolysis because the activity of phosphofructokinase, which was found to be saturated *in vivo* and to catalyse a non-equilibrium reaction, was shown to be influenced by a range of endogenous inhibitors and activators.

The need for a systems-based analysis of kinetic control

Although attractively simple, the description of the kinetic control of metabolic pathways developed in the preceding section has several shortcomings.

- How far from equilibrium does a reaction have to be, for it to be a candidate for the rate-determining step? The model does not lead to a quantitative criterion for deciding whether $\Gamma \ll 1$.
- How can one step, the rate-determining step, be slower than any of the others in a linear pathway when by definition the flux through each step must be the same in a metabolic steady state? This point has already been explored in the analysis of steady-state kinetics in Chapter 10.
- Overexpression of phosphofructokinase, the classical rate-determining enzyme of glycolysis (see Problem 4, Chapter 3), has no effect on the glycolytic flux in genetically modified yeast.
- The rates of many of the steps in a pathway will depend on the concentrations of the substrates and products, since the enzymes in a pathway are not necessarily saturated. It follows that the rates of the individual steps in the pathway can influence each other, because the product of one enzyme will be the substrate of the next.

It is the last point that indicates the way forward. Contrary to the idea of a rate-determining step, it is highly likely that the flux through a pathway will in fact be determined by the activities of all the enzymes in the system. Some may have a greater influence than others, and some may have no influence at all, but there is no *a priori* reason why one enzyme should control the flux. Thus the flux through a pathway is a systemic property, i.e. it depends on the behaviour of all the components in the system, and not just on the properties of a particular component.

In retrospect, the need for a systems-based analysis of kinetic control should not be too surprising and there are clearly limits to the extent to which a full understanding of a system can be obtained by simply focusing on the properties of the individual components of a system. Elsewhere in this book, it has been shown that the notion of the high-energy bond in ATP arises from a failure to consider the energetics of the hydrolysis of ATP in its proper context (Chapter 2); that the thermodynamics of ions in solution can only be analysed rigorously in terms of activities, where the activity coefficient for an ion depends on the composition of the whole solution (Chapter 8); and that the pH of a solution is determined by the composition of the solution and not simply by the number of protons added to the solution (Chapter 5). Each of these examples emphasises the role of the system in determining the properties of particular components of the system, and it is this approach that is needed to understand the control of metabolic pathways.

Metabolic control analysis

One approach to systems-based analysis is to construct a kinetic model of the entire pathway. A model of this kind has to take into account every chemical reaction in which the metabolites are involved, as well as the transport steps between different pools of the same metabolite. Kinetic equations are set up for every step, and these equations are found to be coupled together by the concentration terms. In principle, the equations can be solved numerically, provided that sufficient information is available about the system. This information includes: the Michaelis constants and catalytic constants for all the enzymes; a knowledge of the *in vivo* effects of all the enzyme activators and inhibitors that are present; the activities of the enzymes; and the concentrations of all the effectors and substrates in the cell. The output of the model is a value for the flux through the pathway, and by changing the input parameters it is possible to investigate the dependence of the flux on the activities of the individual enzymes and the concentrations of the metabolites.

In practice, the approach outlined in the preceding paragraph may be too difficult to implement because of a shortage of data. However, it does lead to two new concepts—control coefficients and elasticities—and to a method for analysing the flux through pathways that is known as metabolic control analysis. The control coefficient measures the degree to which the activity of a particular enzyme controls the flux through a pathway, while the elasticities measure the extent to which the activity of an enzyme is affected by the ions and metabolites in the system. Further detail is given in the following sections, but the general picture that emerges from this type of analysis is that the control of flux through a metabolic pathway is usually distributed over many enzymes, rather than being vested in one particular step. Moreover, it is not obvious which enzymes will have the highest control coefficients, emphasising the point that it is necessary to look at the integrated system if the operation of the pathway is to be understood.

Control coefficients

The systems-based analysis of kinetic control assumes that the observed flux through a pathway is likely to be the result of many interacting components rather than a single rate-determining step. Metabolic control analysis is the most commonly encountered development of this idea, and this approach provides a framework for quantifying the contribution that individual components of a metabolic network make to the overall operation of the system. The first step is to define the flux control coefficient, $C^J_{E_i}$, for the effect of a change in the activity of an enzyme (E_i) on the flux (J) through a pathway

$$C^J_{E_i} = \frac{dJ}{J} \Big/ \frac{dE_i}{E_i} = \frac{d(\log_e J)}{d(\log_e E_i)}$$

Thus the flux control coefficient can be measured from the gradient of a plot of J against E_i (figure 13.1a) or $\log_e J$ against $\log_e E_i$ (figure 13.1b). The flux control coefficient is a measure of the extent to which E_i determines the flux through the pathway. It varies between 0, where the activity of the enzyme has no influence on the flux, and 1, where the flux is directly proportional to the activity of the enzyme. Although not apparent from the graphs in figure 13.1, it is also possible for the flux control coefficient to take negative values. For example, if the pathway is branched, then increasing the activity of an enzyme may increase the flux down one branch at the expense of the other, leading to a positive flux control coefficient for one branch and a negative flux control coefficient for the other.

A flux control coefficient can be defined for every enzyme that influences the flux through the pathway, but the values of these coefficients are constrained by the summation theorem

$$\sum_{i=1}^{n} C^J_{E_i} = 1$$

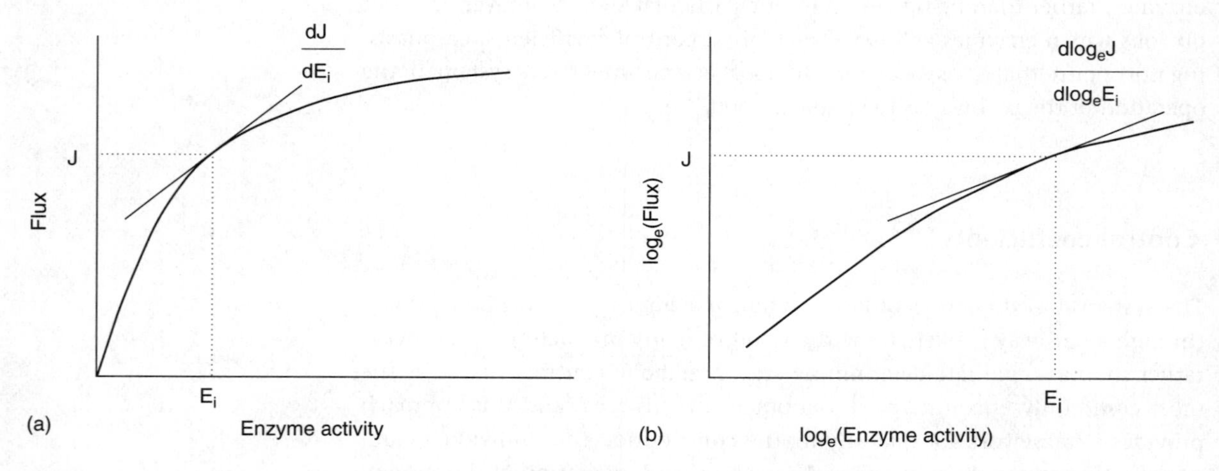

Fig. 13.1 Plots of (a) flux versus activity of enzyme i and (b) $\log_e$(flux) versus $\log_e$(enzyme activity) for a metabolic pathway. The control coefficient for the enzyme is given by either $(dJ/dE_i)(E_i/J)$ from plot (a) or by $d \log_e J/d \log_e E_i$ from plot (b).

The important point about this relationship is that it shows that the control of the flux through the pathway can be shared by more than one enzyme. Of course if the flux control coefficient for a particular enzyme is 1, then the flux control coefficients for all the other enzymes must be zero (assuming an unbranched pathway), and it would be appropriate to identify the enzyme with the flux control coefficient of 1 as the rate-determining enzyme. However, there is no fundamental reason why this should be the case, and in fact experimental measurements suggest that it is much more likely for the control to be shared between several enzymes, with the exact balance being strongly influenced by the physiological conditions prevailing in the cell or organ.

WORKED EXAMPLE

Prove the summation theorem for flux control coefficients.

SOLUTION: From the definition of the flux control coefficient, the change in flux when the activity of an enzyme increases by a small fraction $x = \delta E_i/E_i$ is

$$\frac{\delta J}{J} = C^J_{E_i}\frac{\delta E_i}{E_i} = C^J_{E_i}x$$

If n enzymes have non-zero control coefficients for the flux through a pathway, and if the activity of each enzyme is simultaneously increased by the same fraction x, then the net effect will be to increase the flux through the pathway by the same fraction x. Thus,

$$\sum\frac{\delta J}{J} = \sum_{i=1}^{n} C^J_{E_i}\frac{\delta E_i}{E_i} = \sum_{i=1}^{n} C^J_{E_i}x = x$$

and eliminating x gives the summation theorem

$$\sum_{i=1}^{n} C^J_{E_i} = 1$$

COMMENT: The summation theorem emphasises the fact that the flux control coefficient of an enzyme is determined by the whole system and not just by the properties of the enzyme itself. Accordingly, flux control coefficients cannot be determined by studying enzymes in isolation; they can only be obtained from experiments on intact systems.

Elasticities

It is also necessary to consider the way in which the composition of the system can modify the properties of the enzymes in a metabolic network. Metabolic control analysis does this by introducing quantities known as elasticities.

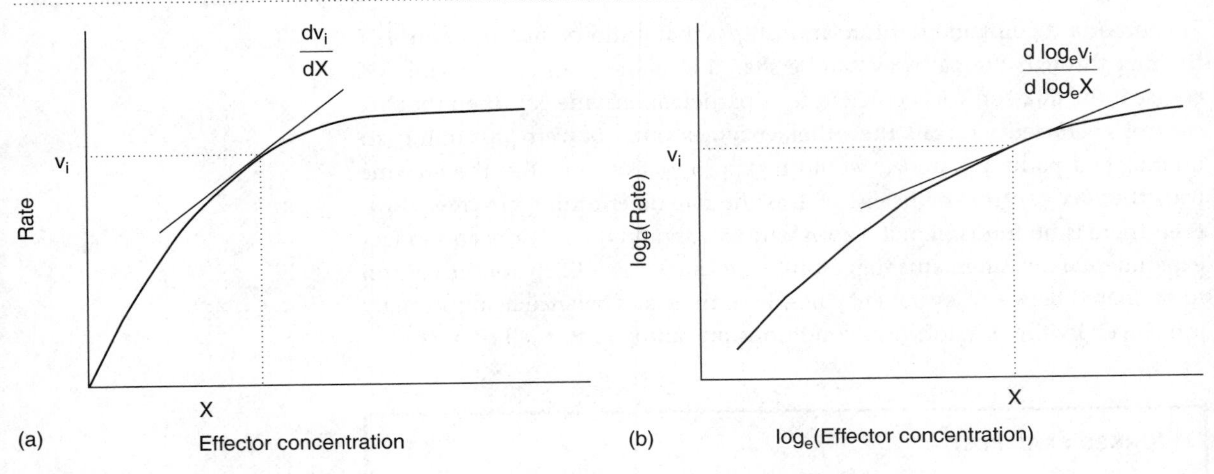

Fig. 13.2 Plots of (a) rate versus effector concentration and (b) $\log_e$(rate) versus $\log_e$(effector concentration) for one step in a metabolic pathway. An effector molecule can be a reactant or product of the reaction or an activator or inhibitor of the enzyme. The elasticity coefficient of the effector for the enzyme is given by either $(dv_i/dX)(X/v_i)$ from plot (a) or by $d \log_e v_i/d \log_e X$ from plot (b).

In contrast to control coefficients, which are properties of the whole system, the elasticities are properties of individual enzymes and they provide the link between the kinetic properties of an enzyme and its potential for flux control. So if a change in the concentration of an effector molecule X alters the rate (v_i) of a step catalysed by the enzyme E_i, then the elasticity coefficient, $\varepsilon_X^{v_i}$, for the effect is defined by

$$\varepsilon_X^{v_i} = \frac{dv_i}{v_i} \bigg/ \frac{d[X]}{[X]} = \frac{d(\log_e v_i)}{d(\log_e X)}$$

The elasticity coefficients for an enzyme indicate the extent to which different metabolites influence its activity, and non-zero elasticities are expected with respect to the substrates and products of the reaction catalysed by the enzyme, as well as for activators and inhibitors. Elasticities have the same form as control coefficients, and in this case the elasticity coefficient can be determined from the gradient of a plot of v_i against [X] (figure 13.2a) or $\log_e v_i$ against $\log_e$[X] (figure 13.2b).

WORKED EXAMPLE

Obtain an expression for the elasticity coefficient of an enzyme with respect to its substrate on the assumption that the enzyme follows Michaelis–Menten kinetics.

SOLUTION: From the definition of the elasticity coefficient

$$\varepsilon_S^v = \frac{[S]}{v}\frac{dv}{d[S]}$$

and since

$$v = \frac{V_{max}[S]}{K_m + [S]}$$

the elasticity coefficient can be rewritten as

$$\varepsilon_S^v = (K_m + [S])\frac{d}{d[S]}\left\{\frac{[S]}{K_m + [S]}\right\}$$

Recalling that

$$\frac{d}{dx}\left\{\frac{f}{g}\right\} = \frac{1}{g^2}\left\{g\frac{df}{dx} - f\frac{dg}{dx}\right\}$$

leads to the following expression for the elasticity coefficient

$$\varepsilon_S^v = \frac{K_m}{K_m + [S]}$$

This expression shows that the elasticity coefficient depends both on the kinetic properties of the enzyme, represented by K_m, and by the prevailing conditions in the system in which it is measured, represented by [S]. Note also that while it is possible, as here, to derive theoretical expressions for elasticity coefficients on the basis of the equations of enzyme kinetics, it is not necessary to do so and elasticity coefficients are usually determined empirically.

Several enzymes in a pathway may respond to the concentration of a particular metabolite X giving rise to several non-zero elasticity coefficients, one for each of the enzymes. It turns out that the values of these elasticity coefficients are constrained by the connectivity theorem

$$\sum_{i=1}^{n} C_{E_i}^J \varepsilon_X^{v_i} = 0$$

This very important equation demonstrates that there is a link between the kinetic properties of individual enzymes in a pathway, represented by their elasticity coefficients, and their contribution to the flux through the pathway, represented by their control coefficients.

In the simplest case, where a metabolite acts as a product for one enzyme (E_1) and as a substrate for the following enzyme (E_2), and where no other enzyme with a non-zero control coefficient for the flux through the pathway has a non-zero elasticity coefficient for the metabolite, applying the connectivity theorem leads to the expression

$$\frac{C^J_{E_1}}{C^J_{E_2}} = -\frac{\varepsilon^{v_2}_S}{\varepsilon^{v_1}_S}$$

Here $\varepsilon^{v_1}_S$ is the (negative) elasticity coefficient for product inhibition of E_1 and $\varepsilon^{v_2}_S$ is the (positive) elasticity coefficient for substrate activation of E_2. Note that when this equation is combined with the summation equation for the two-step pathway

$$C^J_{E_1} + C^J_{E_2} = 1$$

it becomes clear that the flux control coefficients can be deduced from the elasticity coefficients. Thus the properties of the metabolic network can be understood in terms of the properties of the individual enzymes.

FURTHER READING

D. Fell, 1997, 'Understanding the control of metabolism', Portland Press—*an excellent textbook written from a biochemical perspective.*

http://gepasi.dbs.aber.ac.uk—*a convenient source for user-friendly software for metabolic control analysis.*

PROBLEMS

1. The conversion of glucose 6-phosphate to fructose 6-phosphate is catalysed by the enzyme phosphoglucoisomerase. $\Delta G^{o\prime}$ for this reaction is 2·1 kJ mol^{-1}. If we start with 0·1 M glucose 6-phosphate and add phosphoglucoisomerase at 298 K, the final composition is [glucose 6-phosphate] = 0·07 M and [fructose 6-phosphate] = 0·03 M (see worked example in Chapter 3). What would be the composition of the solution if phosphoglucomutase, which catalyses the reaction

$$\text{glucose 1-phosphate} \rightleftharpoons \text{glucose 6-phosphate} \qquad \Delta G^{o\prime} = -7{\cdot}27\ \text{kJ mol}^{-1}$$

were added to the mixture?

2. The control of gluconeogenesis by isocitrate lyase (ICL) in the endosperm of germinating castor bean seedlings was investigated using 3-nitropropionate as a selective, tight-binding inhibitor of the enzyme activity. The following measurements of gluconeogenic flux were obtained by determining the incorporation of [2-^{14}C]acetate into sucrose

Relative ICL activity	1	0·88	0·61	0·44	0·38	0·27
Relative gluconeogenic flux	1	0·94	0·68	0·53	0·49	0·47

Determine the flux control coefficient for ICL on gluconeogenesis and comment on the result.

3. Consider an enzyme E catalysing a flux J_1 at level E_1 and J_2 at level E_2. If the pathway is linear, and if the enzyme obeys Michaelis—Menten kinetics, then it can be shown that

$$C^J_E = \frac{(J_2 - J_1)E_2}{(E_2 - E_1)J_2}$$

where C^J_E is the flux control coefficient corresponding to level E_1.

By writing $J_2 = f \times J_1$ and $E_2 = r \times E_1$, rearrange this equation to obtain a relationship between f and r. Use this relationship to determine the increase in enzyme activity that would be required to double the flux for: (a) $C^J_E = 0{\cdot}5$; (b) $C^J_E = 0{\cdot}75$; and (c) $C^J_E = 1{\cdot}0$. Comment on the significance of your results for the field of metabolic engineering.

ATOMIC AND MOLECULAR STRUCTURE

14 Quantum mechanics: particles, waves and the quantisation of energy

KEY POINTS

- The classical Newtonian ideas of matter and energy do not work at the atomic scale.
- Both matter and energy can be described as collections of waves or *wave-packets*.
- The behaviour of wave-packets leads directly to (i) the uncertainty principle, (ii) the concept of zero-point energy and (iii) the quantisation of energy.
- Whenever matter or energy is localised spatially, only specific waves are allowed, giving rise to the quantisation of energy.
- *Quantum numbers* are labels given to allowed energy levels and they also give information about the shape of the allowed wave.
- The energy separations between quantised energy levels decrease as the mass of the particle and the size of the region in which it is localised increase.
- The *Boltzmann equation* allows the occupancies of energy levels to be calculated.

The thermodynamics and kinetics that we have looked at so far only apply to bulk systems, that is those containing many molecules. Quantities such as equilibrium constants or rate constants are *average* properties of a system and so only apply when the system is large enough to be able to define an average. A chemical reaction between two molecules will either have occurred or not after a given time. The reaction cannot have gone part-way and thus it cannot have a half-time or an equilibrium constant. At the molecular level these ideas no longer apply. In fact, at the molecular level a lot of other ideas that are based on observation of the bulk systems around us are also incorrect. Possibly the most important of these is what we think of as matter and energy. As we shall see, our *classical* picture of matter and energy is incorrect. Although it is often a good first approximation for bulk systems there are many important biochemical systems, such as molecular bonding, structures of macromolecules, absorption of light during vision and photosynthesis, etc., that we cannot understand unless we have a more accurate picture of matter and energy.

The classical picture of matter and energy

The classical picture splits everything into particles (matter) and waves (energy). These two forms are completely distinct.

Particles

A particle (think of a billiard ball) is anything that has mass. The behaviour of particles is governed by classical mechanics, which is based on the ideas of Isaac Newton. The basic properties of a particle are

- It has a well-defined mass
- It exists at a single location, i.e. its position in space at any given time can be defined exactly
- Its total energy is given by the sum of its kinetic and potential energies
- Its total energy can have any value

Since the mass, position and velocity of the particle are completely defined, the kinetic and potential energies of the particle must also be completely defined. Thus, we know both its position and energy exactly.

Waves

All forms of energy not associated with particles are waves (think of a sound wave in air). The basic properties of a wave are

- It has no mass
- It has no location, i.e. it will go on to infinity unless it is stopped
- Its energy E is simply related to its frequency ν. For light, $E = h\nu$ where h is the Planck constant
- It can have any frequency and so can have any energy

Two waves that occupy the same space at the same time will add together to give a resultant wave. This process is called interference (figure 14.1). If the resultant wave is larger than the original waves it is called constructive interference, if the resultant is smaller it is called destructive interference. Whether we get constructive or destructive interference depends on the relative wavelengths and phases of the two waves, i.e. where the maxima of the two waves come in relation to each other.

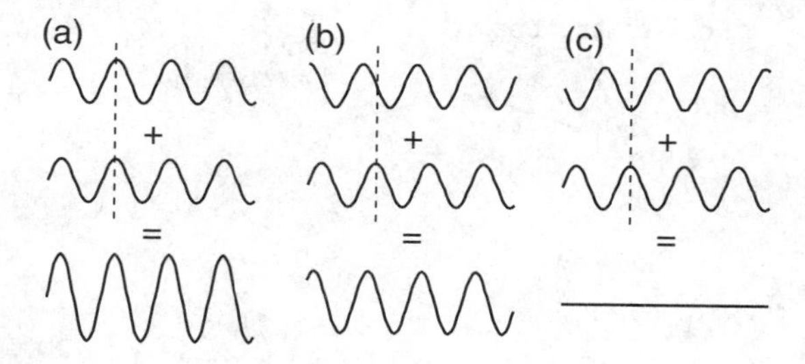

Fig. 14.1 Constructive and destructive interference of two waves. (a) The two waves are exactly in phase and so add at all points (pure constructive interference). (b) The two waves are 90° out of phase (partial destructive interference). (c) The two waves are 180° out of phase and so cancel at all points (pure destructive interference).

Breakdown of the classical picture

The above picture is a good first approximation, but it is wrong. There are many observations that cannot be explained by the classical picture. We shall not go through them all here, but just give a few general examples.

Waves can behave as particles

A very strong gravitational field affects the direction of light. For instance, light from distant stars is deflected by the Sun (figure 14.2). Thus, light must have mass and momentum.

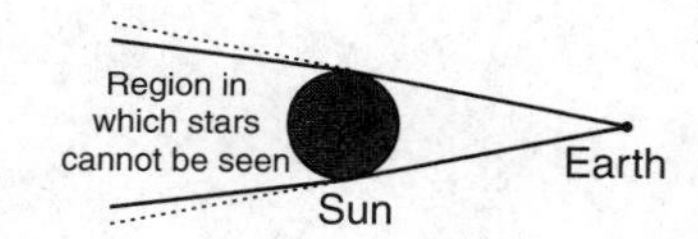

Fig. 14.2 The gravitational field of the Sun is strong enough to deflect photons (particles of light) and so stars passing behind the Sun are visible when the direct path from the Earth to the star is blocked by the Sun. This can only occur if photons are affected by gravity, i.e. they have gravitational mass.

The momentum of light changes on reflection or refraction by an object. This change of momentum requires the object to exert a force on the light and so the light must be exerting an equal and opposite force on the object. Light from an intense source, such as a laser, has enough momentum to be able to move objects the size of single cells. One such technique is 'optical tweezers' where a focused laser beam can be used to hold a cell or bead at the focus point of the beam. An example of the application of this technique to measure the force generated by a bacterial molecular motor is shown in figure 14.3.

Particles can behave as waves

A diffraction pattern can be obtained by shining light through a regular grating due to the constructive and destructive interference of the scattered light (figure 14.4). Similarly, a diffraction pattern can be obtained by shining X-rays on a regular crystal. A very similar diffraction pattern can be obtained from a crystal by using a beam of electrons or neutrons or protons or even helium nuclei. Thus, these species must be being scattered and can interfere constructively and destructively with each other.

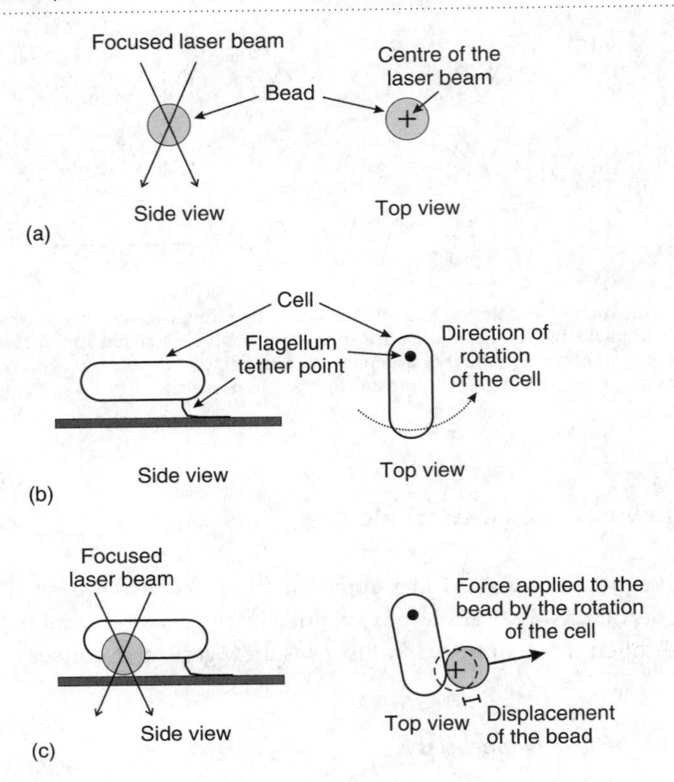

Fig. 14.3 (a) The technique of optical tweezers involves using a laser beam that is focused to a point and defocuses again to hold small objects, such as a bead or cell. The changes in momentum of the light as it is reflected and refracted produce a force on the object pushing it towards the focus point. The force increases as the distance of the object from the focus point increases until the object is out of the beam. If the laser is moved, the object moves to remain at the focus point. (b) Bacteria with flagella move by rotating the long thin 'tails' that are connected to the molecular motor embedded in the cell envelope. If one of the flagella is tethered to a plate, then the molecular motor will cause the cell to rotate around the tether point. (c) If a tethered bacterial cell is allowed to push against a bead held in optical tweezers, it will push the bead away from the focus point of the laser. The broken line gives the position of the bead in the absence of the cell. The displacement of the bead is proportional to the force exerted by the cell. This allows the force generated by a single molecular motor to be measured directly. A flagella motor generates a torque of about 4800 pN nm (equivalent to $\sim 10^{-18}$ brake horsepower). [R. M. Berry and H. C. Berg, 1997, *Proceedings of the National Academy of Sciences USA*, **94**, 14433.]

Particle energies can only have certain values

When light is shone on almost any material only selected frequencies, or energies, are absorbed, i.e. the absorption spectrum consists of discrete lines (figure 14.5). If the particles making up the material could have any energy then they could absorb any amount of energy. The fact that they only absorb selected energies implies that their possible energies are not continuous but that there is a set of distinct energy levels.

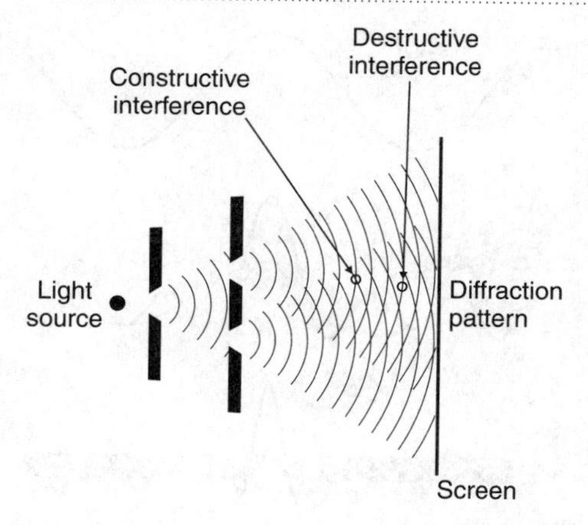

Fig. 14.4 A diffraction pattern is obtained when a wave is scattered from two or more points and the resulting waves interfere. At specific points where the maxima of waves, represented by the curved lines, coincide, the two waves interfere constructively to give increased intensity. At other points, where the maximum of one wave coincides with the minimum of another, the two waves interfere destructively to give complete cancellation. The pattern of spots produced by this is called a diffraction pattern. The production of a diffraction pattern is a property of waves and so anything that can be scattered to give a diffraction pattern must have wave-like properties.

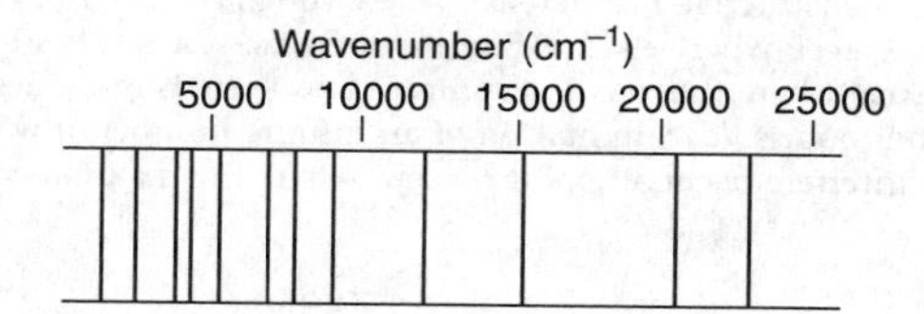

Fig. 14.5 The absorption and emission spectra of atomic hydrogen show discrete lines. If a hydrogen atom could have any energy then continuous absorption or emission would be seen. This is not observed and so a hydrogen atom must have specific energy levels.

The wave–particle duality and wave-packets

As we have seen, what we usually think of as particles can also behave as waves and what we usually think of as waves can also behave as particles. This is called the *wave–particle duality* of matter and energy. What we need is a single description for both matter and energy that has both wave-like properties and particle-like properties.

Since it is easier to start with a wave and show how it can be made to behave as a particle, that is what we shall do here. Figure 14.6a shows the classical picture of a wave. This has a perfectly defined frequency but no position, it extends from $-\infty$ to $+\infty$. If we mix this wave with waves of other frequencies, the resultant wave will be given by the summation of all the separate components. Figure 14.6b shows the summation of three waves with the same

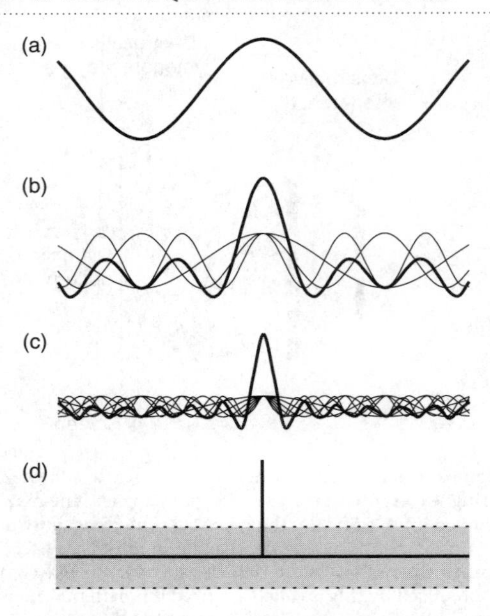

Fig. 14.6 Adding together continuous waves of different frequencies can give particle-like properties. The waves are chosen so that at one point their maxima coincide. The bold line shows the summation of the individual waves. (a) A single pure wave. (b) Summation of three waves gives pure constructive interference where the maxima coincide and partial destructive interference everywhere else. (c) Summation of seven waves still gives pure constructive interference where the maxima coincide and there is more destructive interference at all other points. (d) Summation of an infinite number of waves leads to complete destructive interference at all points except where the maxima coincide.

intensity but different frequencies. The resultant wave is shown by the bold line. Where all the maxima of the individual waves coincide we get perfect constructive interference and a resultant intensity much higher than the individual waves. Where the maxima do not coincide we get partial or complete destructive interference. The resultant still looks like a wave but it has a single, broad, more intense peak. Figure 14.6c shows the superposition of seven waves. At the point where the maxima coincide we still get perfect constructive interference. Elsewhere the destructive interference is more efficient. The dominant peak in the resultant is more intense and narrower than before but the resultant looks less like a wave. At the limit of adding an infinite number of waves (figure 14.6d), the peak has become infinitely narrow and the destructive interference is perfect at all other points. The resultant wave now only exists at a single point and has no wavelength. This is the classical picture of a particle. So by adding waves together we can turn something that looks like a classical wave (no location, perfect frequency) into something that looks like a classical particle (absolute location, no frequency).

The resultant waves in figures 14.6b and 14.6c have both wave-like properties (frequency and particle-like properties (location). It is these resultant waves that have the properties we want in order to explain the wave–particle duality of matter and energy. This description can be summarised as follows

All matter and energy can be described as the resultant of the superposition of a finite number of waves.
This collection of waves is called a *wave-packet.*
The resultant wave is called the *wave-function.*
A wave-function has properties of both a classical wave and a classical particle.

The last step we need is to give waves mass (or give mass a wave-length). This is done by the *de Broglie equation,* which gives a relationship between the momentum (p) of a classical particle and its associated wavelength (λ).

$$\lambda = \frac{h}{\mathrm{p}}$$

WORKED EXAMPLE

Calculate the wavelength of electrons travelling at a velocity of (a) $0{\cdot}2 \times 10^6\,\mathrm{m\,s^{-1}}$ and (b) $20 \times 10^6\,\mathrm{m\,s^{-1}}$. Comment on these values with respect to the resolution that can be achieved by an electron microscope. The mass of an electron is $9{\cdot}109 \times 10^{-31}$ kg.

SOLUTION: The wavelength of the electron is given by the de Broglie equation.

$$\lambda = \frac{h}{\mathrm{p}} = \frac{h}{\mathrm{mv}}$$

where m is mass and v is velocity.

Thus, for (a)

$$\lambda = \frac{6{\cdot}626 \times 10^{-34}}{(9{\cdot}109 \times 10^{-31}) \times (0{\cdot}2 \times 10^{6})} = 3{\cdot}6 \times 10^{-9}\,\mathrm{m}$$

and for (b)

$$\lambda = \frac{6{\cdot}626 \times 10^{-34}}{(9{\cdot}109 \times 10^{-31}) \times (20 \times 10^{6})} = 0{\cdot}036 \times 10^{-9}\,\mathrm{m}$$

To a first approximation, a microscope cannot resolve two objects that are closer together than the wavelength of the radiation used. Thus, electrons travelling at $0{\cdot}2 \times 10^6\,\mathrm{m\,s^{-1}}$ can give a maximum resolution of 3·6 nm, whilst electrons travelling at $20 \times 10^6\,\mathrm{m\,s^{-1}}$ can give a resolution of 0·036 nm. A typical atomic bond is of the order of 0·1 nm.

Consequences of the wave-packet nature of matter

The behaviour of matter and energy as wave-packets leads directly to important properties that could not be predicted by the classical picture. We shall deal with three of these consequences, two here and one in the next section.

Heisenberg's uncertainty principle

A single pure wave (as in figure 14.6a) has a perfectly defined wavelength, and thus an exact energy, but has no position. A wave-packet (figure 14.6b) has a position but it is not perfectly defined, the wave-packet is spread over a finite distance. Similarly, a wave-packet has a wavelength but again it is not perfectly defined, the distance from one maximum to the next is not quite constant. As more waves are added to make the wave-packet (figure 14.6c), its position becomes better defined (narrower main maximum) but its wavelength becomes less well defined (more variation between the minor maxima). A classical particle (figure 14.6d) could be made by adding together an infinite number of waves and would have a perfectly defined position but no definable wavelength and thus an undefined energy.

This leads to one statement of Heisenberg's uncertainty principle.

'It is impossible to specify energy and position simultaneously, with arbitrary precision.'

This can be expressed quantitatively as follows

$$\Delta p \, \Delta q \geq \frac{\hbar}{2}$$

where Δp is the uncertainty in the momentum, Δq is the uncertainty in the position and $\hbar$ is Planck's constant (h) divided by 2π.

This principle applies to certain other pairs of quantities as well, such as energy and time (see Chapter 17).

WORKED EXAMPLE

Calculate the uncertainty in the position of (a) a 65 kg person and (b) a glucose molecule moving at a velocity measured at 10 ms^{-1} to an accuracy of 0·001 ms^{-1}. Comment on the magnitude of the results.

SOLUTION: The uncertainty in momentum is given by $\Delta p = m\Delta v$. The uncertainty in position is then given by Heisenberg's principle as

$$\Delta q = \frac{h}{4\pi\Delta p} = \frac{h}{4\pi m\Delta v}$$

Thus, for (a)

$$\Delta q = \frac{6{\cdot}626 \times 10^{-34}}{4\pi \times 65 \times 0{\cdot}001} = 8{\cdot}1 \times 10^{-34}\ \text{m}$$

The mass of the glucose molecule is given by the molecular weight (180 Da) divided by the Avogadro number (L). Thus, for (b), remembering that we need mass in kg

$$\Delta q = \frac{6{\cdot}626 \times 10^{-34}}{4\pi \times \left(\frac{180 \times 10^{-3}}{6{\cdot}022 \times 10^{23}}\right) \times 0{\cdot}001} = 1{\cdot}8 \times 10^{-7}\ \text{m}$$

COMMENT: For a large object, such as a person, the uncertainties inherent in the wave-packet nature of matter are insignificant and can be ignored. However, for very small masses this is not true. In the example given here, the uncertainty in position of the molecule is several orders of magnitude greater than the diameter of the molecule (5×10^{-10} m).

Zero-point energy

One consequence of the Heisenberg uncertainty principle is that matter cannot have zero energy. If a particle has no energy then it must be stationary, since it has no kinetic energy. We must know both its energy (zero) and position, which breaks the uncertainty principle. To put it another way, if the energy (momentum) is zero, then the Δp must be zero and so Δq must be infinitely large. Thus, a molecule must always have some energy, even at absolute zero (0 K). This residual energy is called the zero-point energy. As for the uncertainty principle, the zero-point energy is inversely proportional to mass and so is only significant at the atomic level. We have already seen one example of this in the effect of isotope changes on reaction rates (see Chapter 9).

Localising waves in space gives quantisation of energy

Whenever we try to localise, or confine, a wave to a small region of space we find that only certain waves occur. Although this sounds complex, we are used to seeing it all around us. A sound wave travelling through the air can have any frequency, i.e. pitch. If we localise a sound wave, for example in a trumpet, then we can only get certain frequencies, i.e. well-defined notes. If we change the size of the region we localise the wave in, for our trumpet change the length of the tube by using valves, then we change the frequency of the waves we get, i.e. change the note. This is a general property of waves.

> Spatial localisation of a wave to a specific region leads to only certain allowed frequencies. The size and shape of the region determine which frequencies are allowed.

To try to explain this behaviour, consider an arbitrary wave that can travel between the walls of a box but cannot escape the box. We shall start the wave at one side of the box and let it travel to the other side (figure 14.7a). On reaching the far wall it will be reflected with the same amplitude, but a reversal of phase, and travel back again (figure 14.7b). It will now interfere with itself causing partial cancellation, the resultant being shown by the heavy line in figure 14.7b. Each time the wave is reflected there will be partial cancellation, until eventually the wave has cancelled itself out. Even if we only get a very small amount of cancellation on each reflection, eventually the wave will cancel itself once it has been reflected enough times (figure 14.7c). The only way we can get a wave that will be stable in the box is if there is perfect constructive interference on every reflection. This only happens when the amplitude of the wave is zero at the edges of the box. Some of these allowed waves are shown in figure 14.8. Thus, only certain waves can exist in the box.

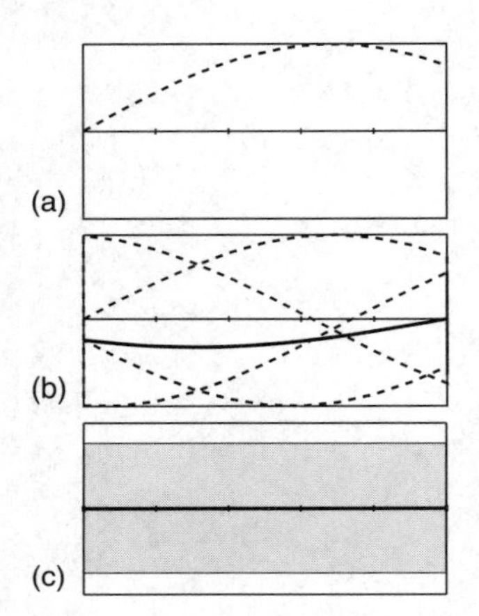

Fig. 14.7 A wave of arbitrary wavelength that is reflected backwards and forwards (or is localised) in a one-dimensional box interferes with itself. The broken line shows the wave and the bold line the resultant. (a) Before the first reflection. (b) After four reflections there is partial destructive interference. (c) After an infinite number of reflections there is complete destructive interference and the wave has cancelled itself out. This wave cannot exist for a significant length of time in this box. The same will be true for any wave that leads to even a small amount of destructive interference on its first reflection.

As the energy of a wave depends on its frequency (or wavelength), if we only have certain allowed waves then we only have certain allowed energies. The energy of a spatially localised wave is quantised. The same is true of wave-packets, the energy of a spatially localised wave-packet is quantised. Thus, quantisation of energy is a simple and natural consequence of the wave-like properties of matter.

WORKED EXAMPLE

Derive a general equation for calculating the allowed energy levels for a particle of mass m confined in a one-dimensional box of length a.

SOLUTION: The energy of a particle is related to its wavelength by the de Broglie relationship. Thus, the problem comes down to deriving a formula for the allowed wavelengths in a one-dimensional box.

As we have seen, the allowed waves in a one-dimensional box must all have zero intensity at the edges of the box. The lowest frequency, and thus lowest energy, wave that meets this criterion is where the wavelength is twice the length of the box (see figure 14.8). The second wave that meets this criterion is where the wavelength is equal to the length of the box, and so on.

Lowest frequency wave/first energy level	$\frac{1}{2}\lambda = a$
Next lowest frequency wave/second energy level	$\lambda = a$
Next lowest frequency wave/third energy level	$1\frac{1}{2}\lambda = a$
Next lowest frequency wave/fourth energy level	$2\lambda = a$

In general, for the nth allowed wave we have n half-wavelengths in the box. Thus, the general criterion for allowed waves is

$$n\frac{\lambda}{2} = a$$

The wavelength of a particle is related to its momentum by the de Broglie relationship

$$\lambda = \frac{h}{p} = \frac{h}{mv}$$

The kinetic energy of a particle is given by

$$E = \frac{1}{2}mv^2 = \frac{1}{2m}(mv)^2$$

Combining these two equations gives

$$E = \frac{h^2}{2m\lambda^2}$$

We can now substitute in for λ to give the energy of the nth level.

$$E_n = \frac{n^2h^2}{8ma^2}$$

The separation between two energy levels, n and n′, is given by

$$\Delta E = E_n - E_{n'} = \frac{(n^2 - n'^2)h^2}{8ma^2}$$

COMMENT: As the mass of the particle and the size of the box in which it is localised increase, the separation between energy levels decreases. Eventually when masses are large enough and distances long enough ΔE is so small as to be insignificant. So at a macroscopic level quantisation of energy is not important.

In the above worked example, the energy of a particle in a one-dimensional box is given by

$$E_n = \frac{n^2h^2}{8ma^2}$$

The integer n is called a *quantum number*. As well as being a label for each energy state (first, second, third, etc.), it also tells us something about the

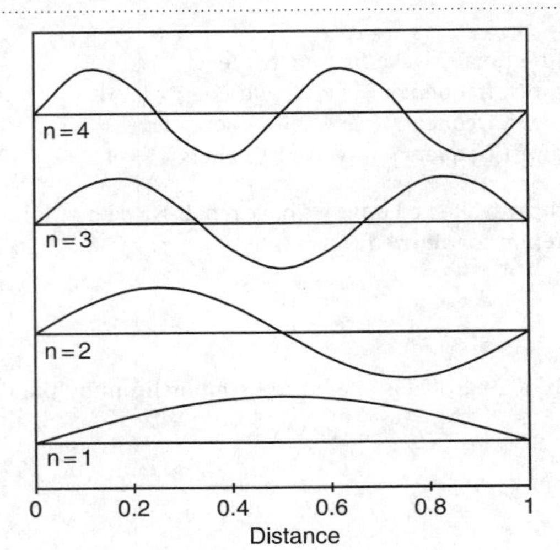

Fig. 14.8 Allowed waves in a one-dimensional box. These waves give perfect constructive interference on each reflection. The wave labelled n = 1 is the lowest frequency wave which can exist in the box, n = 2 the next lowest, etc.

shape of the allowed wave for each state (in this case, the number of half-wavelengths present in the box). For a one-dimensional wave, we only need one quantum number to tell us the shape of the wave. As we shall see in the next chapter, for a three-dimensional wave we need three quantum numbers.

Quantisation of energy in molecules

Molecules contain many different forms of energy, for instance kinetic energy associated with their motion (translational energy), electrostatic interactions between the nuclei and electrons (electron energy), etc. All of these forms of energy are quantised, but as each form of energy is localised to a different degree, we would expect very different energy level separations. Typical values for the separations of the different energy levels within a molecule are given in Table 14.1. As energy is localised to a smaller region, the energy separation goes

Table 14.1

Type of energy	Region of localisation	Energy level separation
Nuclear	Protons and neutrons within a nucleus	Very, very large: $\sim 10^{-16}$ J
Electronic	Few atoms in a molecule (involves electrons)	Very large: $\sim 10^{-18}$ J
Vibrational	Few atoms in a molecule (involves moving atoms)	Medium $\sim 10^{-20}$ J
Rotational	Whole molecule	Small: $\sim 10^{-22}$ J
Translational	Whole molecule moving over large distances	Effectively a continuum (not quantised)

up. The only apparent anomaly in Table 14.1 is that electronic and vibrational energy are localised over roughly the same sized region, i.e. a few atoms in the molecule. However, energy level separations are also inversely proportional to mass and as electrons are much lighter than atoms electronic energy levels have a much greater separation than vibrational energy levels.

Occupancy of energy levels

A single molecule must exist in a particular energy level since, for example, one molecule cannot have two different amounts of rotational energy. For a very large number of molecules, the thermal collisions between molecules enable energy to pass from one molecule to another, for example a collision could increase or decrease the rotational energy of a molecule. Thus, each molecule will not necessarily have the same energy. At low temperatures there will be very little thermal energy available and so all molecules will be in low energy levels. At high temperatures some molecules could be in high energy states but others will be in low energy states (figure 14.9). The thermal energy at a temperature T is given by kT, where k is the Boltzmann constant, and is approximately 5×10^{-21} J at room temperature.

At any given temperature the molecules will be distributed over the available energy levels, the distribution being dependent on the temperature and the spacing between the energy levels (figure 14.10). The mathematical form of this distribution is given by the Boltzmann equation. The Boltzmann equation states that at equilibrium at a temperature T, the ratio of the number of molecules with energies E_i and E_j is given by

$$\frac{n_i}{n_j} = \exp\left(\frac{-\Delta E}{kT}\right)$$

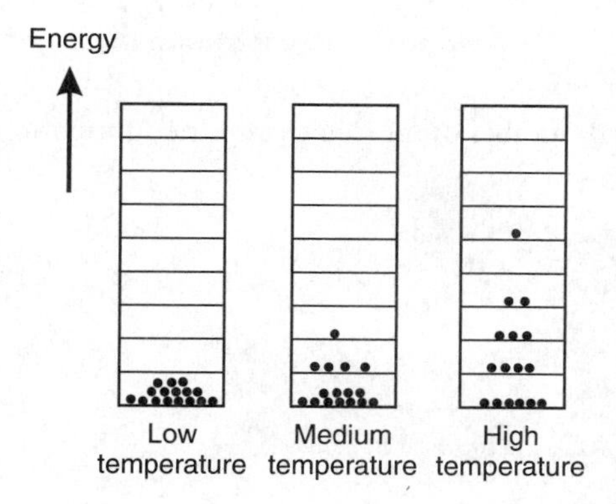

Fig. 14.9 For a given set of energy levels, the distribution of molecules between the different levels depends on the temperature. At very low temperatures all molecules will be in the lowest energy level, the ground state. As the temperature is increased, molecules will start to occupy excited states.

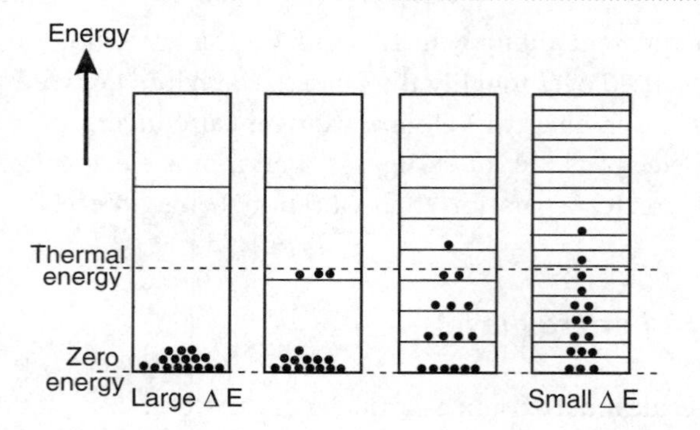

Fig. 14.10 At any given temperature the occupancy of a set of energy levels depends on their spacing. If the separation between the energy levels is greater than the thermal energy, for example electronic energy levels, then all the molecules will be in the ground state. If the separation between the energy levels is less than the thermal energy, for example rotational and translational energy levels, then all the levels up to the thermal energy, and beyond, will be populated to some degree. The number of molecules in a given energy level decreases as the energy of the level increases and this is given by the Boltzmann distribution.

where $\Delta E = E_i - E_j$. When ΔE is much greater than kT this ratio approaches zero since there are then no molecules in the higher energy state i. When ΔE is much smaller than kT this ratio approaches one, indicating that there are equal numbers in the higher and lower energy states. Thus at room temperature, on the basis of the values given in Table 14.1, molecules can be in a number of different translational, rotational and vibrational states and can only be in the lowest electronic and nuclear states, because the ΔE values for the latter are much greater than kT. At equilibrium, there can never be more molecules in a higher energy state than a lower energy state.

An alternative version of this equation is that the probability, $P(E_i)$, of a molecule having a specific energy, E_i, is

$$P(E_i) = \frac{n_i}{N} = \frac{\exp\left(\frac{-E_i}{kT}\right)}{\sum_j \exp\left(\frac{-E_j}{kT}\right)}$$

where N is the total number of molecules and the summation is taken over all energy levels (j).

WORKED EXAMPLE

The separation between adjacent energy levels relevant to three different forms of spectroscopy is given below.

NMR spectroscopy	$\Delta E = 1{\cdot}19 \times 10^{-2}\,J\,mol^{-1}$
Rotational spectroscopy	$\Delta E = 11{\cdot}9\,J\,mol^{-1}$
Electronic spectroscopy	$\Delta E = 119 \times 10^{3}\,J\,mol^{-1}$

Calculate the difference in population between the ground state and first excited state at room temperature in each case and comment on the implication of the results for the sensitivity of the spectroscopic techniques.

SOLUTION: The relative population of two states (in numbers of molecules) is given by the Boltzmann equation. As ΔE is in $J\,mol^{-1}$, we have to divide by the Avogadro number (L) to give ΔE in J per molecule. Thus

$$\frac{n_{upper}}{n_{lower}} = \exp\left(\frac{-\Delta E}{LkT}\right)$$

At room temperature (298 K), this gives the ratio for n_{upper} to n_{lower} as follows:

NMR spectroscopy	0·9999952
Rotational spectroscopy	0·9952
Electronic spectroscopy	$1{\cdot}86 \times 10^{-21}$

The first excited states relevant for NMR and rotational spectroscopy are populated at room temperature, whilst very nearly all molecules are in the electronic ground state.

COMMENT: The sensitivity of a spectroscopic technique depends on the population difference between the ground and excited states. If this difference is small then only a few molecules can be excited before the populations of the two levels are equalised and the system becomes saturated. For NMR spectroscopy only one molecule in every one hundred thousand will give a signal, whereas in electronic spectroscopy every molecule will give a signal. Thus, electronic spectroscopy is potentially one hundred thousand times more sensitive than NMR.

FURTHER READING

P.W. Atkins, 1998, 'Physical chemistry', Oxford University Press—*an excellent introduction to waves and particles, but continuing to a much higher level than is necessary for most biochemists.*

P.A. Cox, 1996, 'Introduction to quantum theory and atomic structure', Oxford University Press—*an accessible introduction to quantum theory.*

PROBLEMS

1. Explain, briefly, the following ideas or concepts.

(a) The de Broglie relationship
(b) The uncertainty principle
(c) Zero-point energy

2. Why does the quantisation of energy occur and when is this important?

3. Derive the energies of the first three energy levels for an electron localised in a one-dimensional box of length 5 nm (the mass of an electron is $9{\cdot}1 \times 10^{-31}$ kg).

Calculate the relative populations of these levels at 300 K.

4. To a first approximation, the behaviour of the π-electrons in a system of conjugated double bonds, such as found in biological carotenoids,

can be treated as electrons in a one-dimensional box.

Assuming that the total length of the conjugated system is 0·7 nm, corresponding to four –CH=CH– units, calculate the energies of the first two π-electron transitions.

Experimentally, it is observed that the π-electron absorption peaks of conjugated alkenes shift to lower frequencies as the number of conjugated double bonds increase. Explain this observation.

5. Derive an equation that relates the energies of the allowed waves on a circle to the radius of the circle. [Hint: Consider a wave travelling around a circle. This wave will interfere with itself on the second and subsequent times round. The only allowed waves will be those that give complete constructive interference (cf. a wave in a 1D box).]

15

Electrons in atoms

KEY POINTS

- Localising an electron around a nucleus gives rise to electron orbitals with well-defined shapes.
- Each electron is characterised by four quantum numbers, three giving the shape of the orbital and the fourth giving the spin.
- The *Pauli exclusion principle* means that only two electrons can occupy each orbital.
- The energy of any given electron depends on the attraction with the nucleus and the repulsion with all the other electrons. This leads to electrons in different orbitals having different energies.
- The electronic configuration of a multi-electron atom or ion is obtained by putting the electrons in orbitals to give the minimum energy.
- The electronic structures of atoms and ions determine almost all their chemical properties.

Classical picture of an atom

Atoms are made up of positively charged protons, negatively charged electrons and neutrons with no charge. The protons and neutrons form the nucleus of the atom, which is thus positively charged. The electrons are held close to the nucleus by the electrostatic attraction between their negative charges and the positive charge of the nucleus. In the classical model, electrons are small particles that orbit about the nucleus in much the same way that the planets orbit about the sun. They cannot escape from the nucleus because of the electrostatic attraction and they do not fall into the nucleus because of their kinetic energy. As long as the electrostatic attraction and kinetic energy balance then the electron will remain in orbit around the nucleus.

Wave-theory model of an atom

The classical model is attractive because it is simple, easy to understand and easy to visualise. Unfortunately, it is also incorrect. The model cannot account for the fact that electrons in atoms only have certain energies. It also cannot account for the fact that electrons occupy orbitals with very specific shapes and the shape of an orbital plays a key role in bond formation (see next chapter).

Electrons should not be treated as particles but as waves or wave-packets. When a wave is localised to a small region of space, such as around an atom, then only particular waves are allowed and so we get quantisation of energy. In the last chapter, we saw what happened when we localised a wave in a one-dimensional box. The treatment of electrons in atoms is rather more complex, but only because we have to deal with three-dimensional waves rather than one-dimensional waves. We won't attempt to do this here, but we can get a good qualitative idea of what is going on by looking at what happens in one dimension at a time. Because an atom is spherical, it is easier to do this using spherical polar (r, θ, ϕ) coordinates rather than Cartesian (x, y, z) coordinates (figure 15.1).

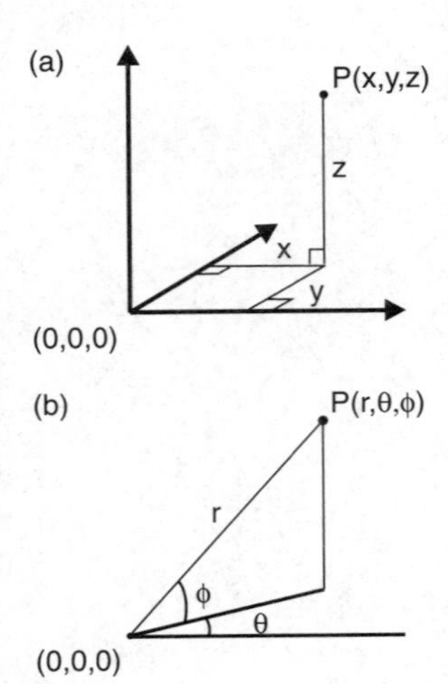

Fig. 15.1 A point in three-dimensional space can be identified uniquely by coordinates along three axes. (a) Cartesian coordinates are useful to describe linear motion: x, y and z are the distances along each axis to a perpendicular dropped from the point. (b) Polar coordinates are useful to describe circular motion: r (radial coordinate) is the distance from the origin to the point, θ and ϕ (angular coordinates) are angles of rotation relative to an axis and a plane respectively.

In the radial (r) dimension, we have something very similar to a one-dimensional box. The electron is kept in the box by attraction to the nucleus, which is in the middle of the box, rather than by repulsion from the walls and this changes the shapes of the waves slightly, but to a first approximation the allowed radial waves (figure 15.2) look similar to the allowed waves in a one-dimensional box (figure 14.8).

In the angular dimensions (θ and ϕ), we have waves that are localised on circles centred on the nucleus. We can treat this in exactly the same way as the one-dimensional box by considering what happens if a wave is started at one point on the circle and allowed to travel round and round (figure 15.3a). Once the wave has gone round once it will meet itself and cause either destructive or constructive interference, and similarly when it has gone round twice, three times, etc. For a wave that causes partial destructive interference the first time

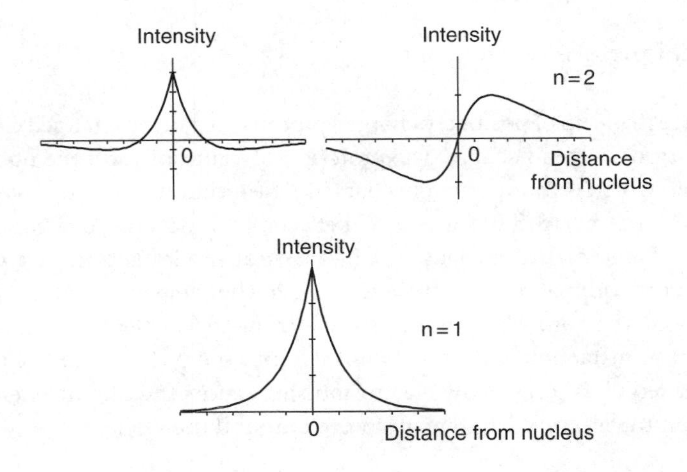

Fig. 15.2 Radial components of allowed waves for an electron in an atom. These are labelled by the quantum number n. For n = 2 there are two solutions at the same energy.

round, then by the time it has gone round the circle many times there will be complete destructive interference (figure 15.3b) and the wave will cease to exist. Thus, the only allowed waves will be those that always interfere constructively. The criterion for this to happen is that there must be an integral number of wavelengths per circumference (figure 15.4).

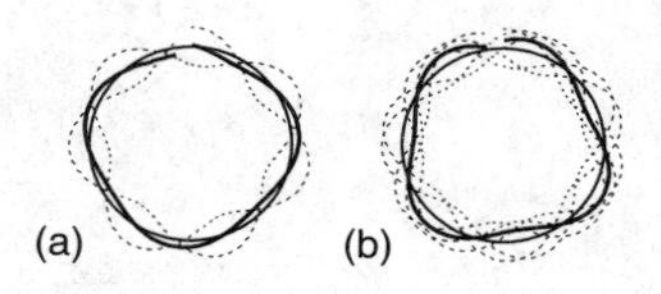

Fig. 15.3 A wave (dotted line) of arbitrary wavelength travelling around a circle will interfere with itself on its second and subsequent circuit to give a resultant wave (solid line). (a) The wave has gone round twice. (b) The wave has gone round five times. Eventually, after enough circuits, this wave will cancel itself out completely.

Allowed electron orbitals and quantum numbers

In order for a wave to exist in an atom, the wave must interfere constructively in all three dimensions at once. There are only certain combinations of the allowed waves in each separate dimension that are allowed in all three dimensions at once. These allowed three-dimensional waves are called *atomic orbitals* (figure 15.5).

As with the waves in a one-dimensional box, we can use quantum numbers to label an orbital and also to tell us what shape an orbital has. Since the wave is three dimensional we need three spatial quantum numbers, one radial and two angular for the three polar coordinates (see Table 15.1).

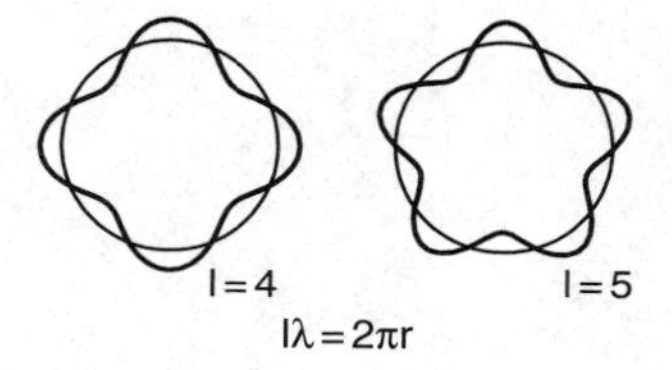

Fig. 15.4 Two of the allowed waves on a circle. The criterion for constructive interference is that the circumference of the circle must be an integral number of wavelengths. This number is called the quantum number l.

TABLE 15.1

Quantum number	Value		Description
Spatial quantum numbers for an electron in an atom			
n	1, 2, 3, ...	radial	distance from nucleus
l	0, 1, 2, ..., (n − 1)	angular	number of wavelengths around circumference
m	−l, −(l − 1), ... (l − 1), l	angular	angular direction
Spin quantum number for an electron in an atom			
s	$+\frac{1}{2}, -\frac{1}{2}$	spin	time evolution of the wave

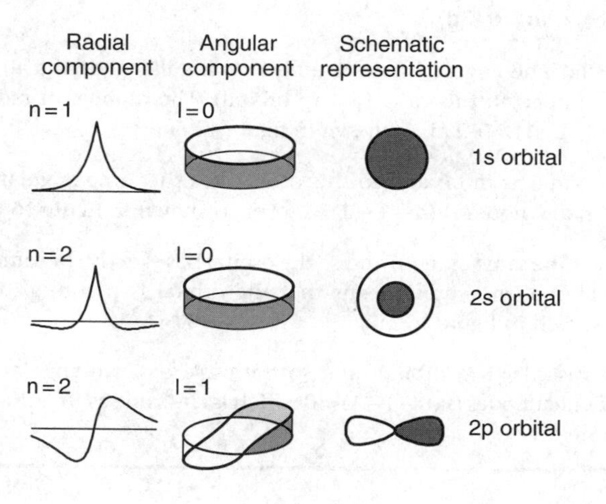

Fig. 15.5 Atomic orbitals are obtained by combining radial and angular components. In the schematic representations, shaded areas represent regions where the wave is positive and unshaded areas where the wave is negative.

These three quantum numbers, n, l and m, give us the shape of any atomic orbital. The two most useful classifications of orbitals are on the basis of their distance from the nucleus (n, principal quantum number) and their angular shape (l, figure 15.4).

$n=1$	1st shell	$l=0$	s orbital
$n=2$	2nd shell	$l=1$	p orbital
$n=3$	3rd shell	$l=2$	d orbital
$n=4$	4th shell	$l=3$	f orbital

These two can be combined to give a description of a given orbital (e.g. 2s, 2p, 4d, etc.). There is only one 2s orbital ($n=2$, $l=0$, $m=0$) in an atom, but there are three 2p orbitals ($n=2$, $l=1$, $m=-1, 0, 1$) with different m quantum numbers. Similarly, there are five 4d orbitals.

Each allowed orbital corresponds to a specific wavelength and so to a specific energy. As a general rule, the more maxima and minima that the orbital has (both radial and angular), the shorter the wavelength and so the higher the energy. This can be quantified by counting the number of *nodes* or points at which the wave goes through zero (ignoring long distances from the nucleus where the wave becomes and then stays zero). For a wave in a one-dimensional box, each allowed wave has $(n-1)$ nodes (figure 14.8). For a wave on a circle, each allowed wave has l nodes (figure 15.4). For a wave localised in an atom, each allowed orbital has a total of $(n-1)$ nodes (l angular nodes and $(n-l-1)$ radial nodes).

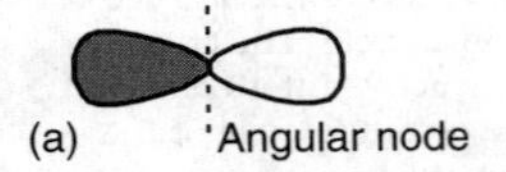

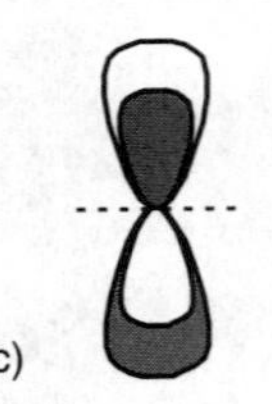

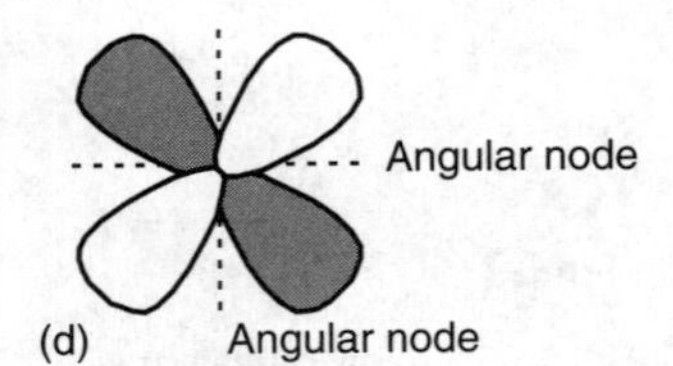

Fig. 15.6 Schematic representations of a (a) $2p_z$ (b) $3p_z$, (c) $3p_x$ and (d) 3d orbital.

WORKED EXAMPLE

Describe the shapes of the orbitals with the following sets of quantum numbers

(a) $n=2$, $l=1$, $m=0$ (2p)
(b) $n=3$, $l=1$, $m=0$ (3p)
(c) $n=3$, $l=1$, $m=1$ (3p)
(d) $n=3$, $l=2$, $m=0$ (3d)

SOLUTION: (a) The angular shape (number of angular nodes) is given by the l quantum number. In this case, $l=1$ (p orbital). The number of radial nodes is given by $(n-l-1)=0$. This is shown in figure 15.6(a).

(b) 1 is the same as in (a) and so this orbital has the same angular shape. The number of radial nodes is $(n-1-1)=1$. This is shown in figure 15.6(b).

(c) n and l are the same as in (b) and so the orbital has exactly the same shape. The different value of m simply means that the orbital is pointing in a different direction, shown in figure 15.6(c)

(d) In this case, $l=2$ (d orbital) and so the wave has two angular nodes. The number of radial nodes is $(n-l-1)=0$ which is the same as in (a). This is shown in figure 15.6(d).

As the wavelength of an orbital depends on the number of nodes, the energy of a single electron in an atom will only depend on the value of n for the orbital it occupies. Thus, an electron placed in a 3p orbital or a 3d orbital will have the same energy but this will be higher than the energy if it were placed in a 2p orbital. Note that in multi-electron atoms things are rather more complicated and this is discussed below.

Electron spin

The n, l and m quantum numbers give the shape of an orbital or, alternatively, the spatial properties of an electron in that orbital. However, electrons have an extra property, a *magnetic moment*. This arises from the 'internal motion' of the electron (not the 'motion' round the nucleus which is given by l), called the *electron spin*. The nearest classical idea is of an electron as a particle spinning on its axis. This is not true but it can be a useful picture. In terms of waves, spin is a description of the time evolution of the wave. As an analogy, we only know if a classical particle is spinning by its time evolution, i.e. by looking at its orientation at two different time points.

The four quantum numbers, n, l, m and s, are sufficient to describe the wavefunction of an electron in an atom completely.

In order to be able to keep track of the spin of an electron we need a fourth quantum number s (see Table 15.1). It is also useful to define the net electron spin, S, of an atom or molecule that has many electrons. This is the sum of all the s values for the individual electrons.

$$S = \sum_i s_i$$

The effects of electron spin can be detected directly. One example of this is electron spin resonance (ESR) spectroscopy, a technique based on monitoring the way that the spin of an unpaired electron behaves in a magnetic field. There are many biochemical applications of ESR, frequently involving labelling biomolecules with probes that have an unpaired electron and using these to give information on the environment of the biomolecule.

Pauli exclusion principle

Each electron in an atom is characterised by four quantum numbers that describe the shape of the space that it occupies and its time evolution. The Pauli exclusion principle states that

> 'No two electrons in an atom can have the same set of quantum numbers'

No two electrons can continuously occupy the same space. If they do occupy the same space (n, l and m the same), then their spin s must be different. As s can have only two values ($\pm\frac{1}{2}$), only two electrons can occupy any one orbital. For two electrons in the same orbital, the net electron spin s is zero.

Electron energies

As we saw in the last chapter, quantum numbers give the shape of allowed waves but not their energy. In order to determine the energy of a wave, we need to consider all the energetic interactions present. For electrons in atoms, there are two types of interaction that we need to consider

- Attraction of the nucleus for the electron. This will increase with increasing nuclear charge but will decrease with the distance of the electron from the nucleus, given by the n quantum number.
- Repulsion between electrons in the atom. This will increase with the number of electrons present. This is a complex function that depends on precisely which orbital each electron is in. However, to a first approximation the larger the value of the l quantum number the more repulsion an electron will experience, because it is localised to a smaller region and so cannot move as far to minimise repulsions.

The energy of an electron in a specific orbital will thus depend on the size and shape of the orbital, the nuclear charge and the number of other electrons present.

WORKED EXAMPLE

The following lists give the order of orbital energies for selected atoms or ions.

(a) H $\mathbf{1s < 2s = 2p < 3s = 3p = 3d}$
(b) $\mathbf{Li^{2+}}$ $\mathbf{1s < 2s = 2p < 3s = 3p = 3d}$
(c) Li $\mathbf{1s < 2s < 2p < 3s < 3p < 3d}$
(d) Li(2s) $\mathbf{\sim Na(3s)}$
(e) Ca $\mathbf{3s < 3p < 4s < 3d < 4p}$

Provide an explanation for these orders and comment on their importance.

SOLUTION: (a) H only has one electron and so there is no electron–electron repulsion to consider, just the attraction with the nucleus. The energy of the single electron in any given orbital will simply depend on the distance of the electron from the nucleus. As the distance increases the attraction by the nucleus decreases and so the energy of the electron becomes higher (less stable). Thus, the order of the orbitals is

$$1s < 2s = 2p < 3s = 3p = 3d$$

(b) Li^{2+} also only has one electron so the order of orbitals is identical to (a). Note that the energies are not the same as in (a) because the nuclear charge is different.

(c) Li has three electrons and so we need to consider electron–electron repulsion as well. The repulsion will be bigger in the orbitals with larger l, but this will be a smaller effect than that of the change in distance of the electron from the nucleus as n increases. Thus, the order of the orbitals becomes

$$1s < 2s < 2p < 3s < 3p < 3d$$

(d) The 2s orbital in Li is closer to the nucleus than the 3s orbital in Na, leading to reduced nuclear attraction in Na. The nuclear charge for Li is only $+3$ compared to $+11$ for Na, leading to increased nuclear attraction in Na. These two factors approximately cancel, thus

$$\mathrm{Li(2s)} \sim \mathrm{Na(3s)}$$

(e) Ca has 20 electrons and so the electron–electron repulsion terms will be large. In this case, the increase in electron–electron repulsion in a d orbital relative to an s orbital is larger than the decrease in electrostatic attraction of a fourth shell relative to a third shell. Thus, the 4s orbital is lower in energy (more stable) than the 3d orbital.

$$3s < 3p < 4s < 3d < 4p$$

COMMENT: It is these sorts of variations in relative orbital energies that give rise to the shape of the periodic table. Each successive element has one more electron than the previous one. Electrons are put into the orbitals with the lowest energy, starting with 1s (see next section). After filling the 1s (elements H and He), 2s (Li and Be), 2p (B to Ne), 3s (Na and Mg) and 3p (Al to Ar) orbitals the next electrons go into the 4s orbitals (K and Ca) and then the 3d orbitals to give the transition metals (Sc–Zn) followed by the 4p orbitals (Ga to Kr).

There is also a specific spin effect involving electrons having the same n and l quantum numbers. If their spins are different, called antiparallel, then their m quantum numbers can be the same and so they can occupy the same orbital, i.e. get very close together. If their spins are the same, called parallel, then their m quantum numbers must be different due to the Pauli exclusion principle and so they cannot occupy the same orbital. Thus, the pair with parallel spins will have slightly less repulsion because they cannot be as close together.

Electronic configurations of atoms and ions

The electronic configuration of an atom (or ion) is a list of the orbitals occupied by electrons. As there are many orbitals available, any atom can have many different electronic configurations. For example, a lithium atom has three electrons and the possible electronic configurations include $1s^1\ 2s^1\ 2p^1$, $2s^1\ 3p^1\ 4s^1$ and $1s^1\ 3d^2$.

The electronic configuration that gives the lowest overall energy is called the *ground state*. All other electronic configurations will give a higher energy and are called *excited states*. We can find the ground state electronic configuration by using the *Aufbau (building up) principle*, which states that we put each electron into the lowest energy orbital available.

WORKED EXAMPLE

Determine the ground state electronic configurations of Li and Ca.

SOLUTION: This is done in three stages: (i) determine the order of the orbital energies, (ii) determine the number of electrons present and (iii) feed the electrons into the lowest energy orbitals available.

As we have seen above, the order of orbital energies in Li is $1s < 2s < 2p$. Li has three electrons. The first electron goes into the lowest energy orbital, the 1s orbital. The second electron also goes into the 1s orbital. This orbital is now full and so the third electron goes into the 2s orbital. Thus the electronic configuration of Li is $1s^2\ 2s^1$.

As we have seen above, the order of orbital energies in Ca is $1s < 2s < 2p < 3s < 3p < 4s < 3d$. Ca has 20 electrons. The first two go into the 1s, the next two into the 2s, the next six into the 2p, and so on. Thus the electronic configuration of Ca is $1s^2\ 2s^2\ 2p^6\ 3s^2\ 3p^6\ 4s^2$.

WORKED EXAMPLE

The ground state configurations of Fe and Fe^{2+} are $1s^2\ 2s^2\ 2p^6\ 3s^2\ 3p^6\ 4s^2\ 3d^6$ and $1s^2\ 2s^2\ 2p^6\ 3s^2\ 3p^6\ 3d^6$ respectively. Comment on these.

SOLUTION: The order of electron energies in Fe is the same as in Ca above. Fe has six more electrons that Ca. Thus the electron configuration of Fe is $1s^2\ 2s^2\ 2p^6\ 3s^2\ 3p^6\ 4s^2\ 3d^6$.

The order of electron energies in Fe^{2+} is slightly different. Because of the overall positive charge, the attraction for the nucleus is now more significant than the electron-electron repulsion in determining the orders of the 4s and 3d electron energies. Thus, the order of electron energies is $1s < 2s < 2p < 3s < 3p < 3d < 4s$ and the electronic configuration is $1s^2\ 2s^2\ 2p^6\ 3s^2\ 3p^6\ 3d^6$.

COMMENT: The same behaviour is seen for all first-row transition metals. The atoms have electronic configurations $4s^2 3d^n$, but the 4s electrons are lost first on ionisation.

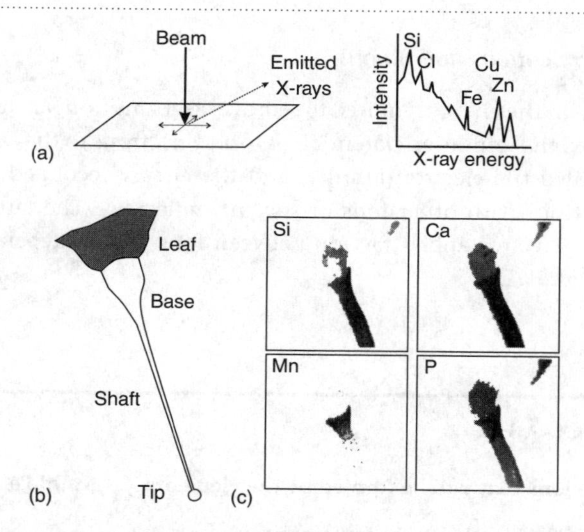

Fig. 15.7 Use of a scanning proton microprobe to map the distribution of elements during the biomineralisation of a growing nettle hair. (a) A very narrow beam of high energy protons is scanned across the sample. This beam causes X-rays to be emitted by the core electrons of the atoms present. Each element has a unique set of core electron energies and so gives a unique peak in the resulting spectrum. This can be used to identify the elements that are present at any given point and their concentrations. (b) A nettle hair consists of a hollow tube, or shaft, that contains a solution of toxins. The top of the shaft and the tip are made of silica, whilst the bottom of the shaft and the base are made of calcium oxalate plus smaller amounts of silica. On touching the hair, the tip breaks off to leave a sharp point that penetrates the skin and injects the toxins. (c) Results from a microprobe analysis of a developing nettle hair, aligned as in (b), show both silicification (Si) and calcification (Ca) to have started by this point of growth. Phosphorus (P) and manganese (Mn) are also present at this stage. [N. P. Hughes, C. C. Perry, R. J. P. Williams, F. Watt and G. W. Grime, 1988, *Nuclear Instruments and Methods in Physics Research*, **B30**, 383.]

Each element has a unique electronic configuration and a unique set of orbital energies because each element has a different nuclear charge and a different number of electrons. This pattern of orbital energies can be used as a 'fingerprint' of an element, which is the basis of spectroscopic techniques for elemental analysis.

As we shall see in the next chapter, the energies of the outermost orbitals may change when atoms are involved in bonding but the energies of the lowest orbitals remain unchanged and so bonded atoms still retain part of their 'fingerprint'. This can be exploited to perform elemental analysis on intact samples. One example of this type of technique applied to the spatial mapping of elemental distributions is shown in figure 15.7.

Atomic and ionic properties

All the chemical properties of atoms and ions derive from the energies of the electrons and the size and shape of the orbitals that they occupy. We shall mention just two here.

Ability to form cations and anions

This depends on the energy required to remove electrons from an atom to form a cation, called the ionisation potential, or to add electrons to the atom to form an anion, called the electron affinity, and the energy recouped in forming ionic interactions with other ions or solvent molecules. The latter is determined by the electrostatic attraction between ions, which depends on their charge and radius.

WORKED EXAMPLE

Predict the ions that will be formed by the elements K, Ca and Fe.

SOLUTION: Consider the ionisation process

$$M^{n+} \rightarrow M^{(n+1)+} + e^-$$

If this process requires little energy, then we can easily regain it by electrostatic interactions with other ions. If this process requires a lot of energy, then we shall never be able to regain the energy by electrostatic interactions and the electron will not be lost.

For K, the electronic configuration is $1s^2\ 2s^2\ 2p^6\ 3s^2\ 3p^6\ 4s^1$. The first electron (4s) is very easy to remove. The second electron (3p) is very difficult to remove because it is much closer to the nucleus. Thus, K will not form a $+2$ ion and will only form the $+1$ ion.

For Ca, the electronic configuration is $1s^2\ 2s^2\ 2p^6\ 3s^2\ 3p^6\ 4s^2$. The first and second electrons (4s) are very easy to remove, but the third (3p) is very difficult. Thus, Ca will not form a $+3$ ion. In addition, Ca will not form a $+1$ ion because removal of the second electron is so easy that the energy required can always be recouped through electrostatic interactions. Thus, Ca will only form the $+2$ ion.

For Fe, the electronic configuration is $1s^2\ 2s^2\ 2p^6\ 3s^2\ 3p^6\ 3d^6\ 4s^2$. The first and second electrons (4s) are very easy to remove and so Fe will form a $+2$ ion but not a $+1$ ion, like Ca. Removal of the next electron (3d) requires about the same amount of energy as we can regain from electrostatic interactions, so in some cases this will be favourable and in some not. Removal of the fourth electron (3d) will be more difficult (because of the increased charge) but in some compounds we can still get enough energy back to make it possible. Removal of the fifth electron (3d) requires too much energy to ever recoup. Thus, Fe will form the $+2$ and $+3$ ions and, in very particular circumstances, the $+4$ ion.

COMMENT: K and Ca only show a single oxidation state and so cannot be used for oxidation or reduction reactions. In contrast, Fe can adopt more than one oxidation state in solution and as a result it has an extensive biological role in a very large number of oxidation/reduction reactions. As well as being very useful for electron transfer reactions, the ability of Fe to undergo oxidation/reduction reactions also makes it much more reactive than K^+ or Ca^{2+}. Thus whilst the latter two can be found in free solution in biological systems, Fe is always present bound to macromolecules.

Atomic/ionic radius

This depends on the size of the largest occupied orbital, which in turn depends on the value of the n quantum number and the nuclear charge. In general, orbitals with a larger value of n are larger, and ions with a larger positive charge are smaller.

WORKED EXAMPLE

Put the following in order of increasing ionic radius: Na^+, K^+, Ca^{2+} and Zn^{2+}.

SOLUTION: The occupied orbitals with the highest energy in these ions are Na^+ (2p), K^+ (3p), Ca^{2+} (3p) and Zn^{2+} (3d). For the singly charged ions, the Na^+ (2p) orbital will be smaller than the K^+ (3p) orbital, because of the smaller value of the n quantum number. For the doubly charged ions, the Zn^{2+} (3d) orbital will be smaller than the Ca^{2+} (3p) orbital, because the nuclear charge of Zn^+ is 10 times greater than Ca^{2+}. When comparing Na^+ (2p) with Ca^{2+} (3p), the increase in n quantum number (leading to an increase in size) is offset by an increase in formal charge (leading to greater attraction for the electrons and so a decrease in size). Thus, the Na^+ (2p) orbital is about the same size as the Ca^{2+} (3p). The order of ionic radii is

$$Zn^{2+} < Ca^{2+} \sim Na^+ < K^+$$

COMMENT: The small size and high charge of Zn^{2+} means that it has stronger electrostatic interactions with other molecules than any other ion available in biochemical systems and so it is frequently used in the active site of enzymes, such as carboxypeptidase.

WORKED EXAMPLE

Cytochrome c is an electron transport protein. The protein contains an Fe in a porphyrin ring buried in the middle of the protein, and the Fe is also coordinated to His 18 and Met 80 above and below the ring respectively. The reduced form of cytochrome c contains Fe^{2+} whilst the oxidised form contains Fe^{3+}. Explain how the change from the reduced to the oxidised form can lead to a conformational change on the surface of the protein.

SOLUTION: Fe^{3+} has one less electron that Fe^{2+}. The remaining electrons on Fe^{3+} experience slightly less electron-electron repulsion and so are held slightly more tightly. This makes Fe^{3+} smaller than Fe^{2+}. As a result, the histidine and methionine residues are pulled slightly towards the porphyrin ring. This effect can then be relayed along the peptide backbone around these residues to other parts of the structure, such as the protein surface.

COMMENT: In fact, the protein structure is rather rigid around His 18 and does not change significantly on oxidation of cytochrome c. In contrast, the protein structure around Met 80 is less stable and there are significant changes to the orientations of two helices on this side of the protein.

The size and shape of the individual d orbitals of the Fe are also important to the function of the protein. During reduction and oxidation of cytochrome c, an electron has to 'hop' to and from the Fe. One factor that determines the rate at which this hopping occurs is how far the d orbital containing the electron extends towards the source of, or target for, the electron.

FURTHER READING

P.W. Atkins, 1998, 'Physical chemistry', Oxford University Press.—*a more advanced account of the electronic structure of atoms.*

P.A. Cox, 1996, 'Introduction to quantum theory and atomic structure', Oxford University Press.—*a clear description of the principles underlying atomic structure.*

PROBLEMS

1. Explain why electrons in atoms occupy discrete orbitals with well-defined shapes and energies.

2. Explain, briefly, the following ideas or concepts:

(a) Principal quantum number
(b) Pauli exclusion principle
(c) Aufbau principle

3. Write down the ground-state electronic configurations for the following atoms:

H C P Ni Mo

4. The following graph shows plots of the first ionisation potential versus atomic number for the elements of the first short period, Li to Ne, and the second short period, Na to Ar. Explain the major features of the plots.

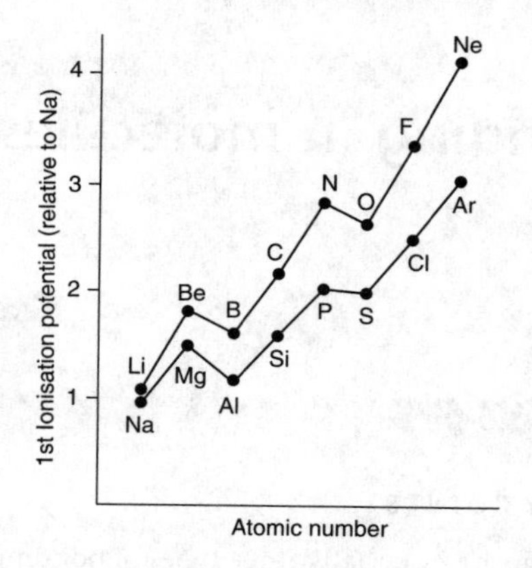

5. Put the following atoms in order of increasing first ionisation potential and increasing first electron affinity, giving your reasoning.

Na H Cu K S Ca Cl Br Ne

Which of these atoms are likely to form positive ions in compounds and which are likely to form negative ions?

16

Bonding in molecules

KEY POINTS

- There are essentially three types of bonding between atoms, *ionic*, *covalent* and *metallic*. These are not mutually exclusive.
- Covalent bonding occurs when an electron in an atomic orbital is affected by the presence of other atoms close by, leading to a reduction in energy.
- The innermost, or core, electrons are almost unaffected whereas the outermost, or valence, electrons are strongly perturbed by other atoms.
- *Crystal-field theory* can be used to describe the behaviour of the d electrons in transition metal complexes.
- *Molecular orbital theory* is used to describe the behaviour of valence electrons in covalent molecules.
- The properties of molecules depend on the nature of the bonding between the atoms.

Definition of bonding

Most atoms, at least in this world, do not exist as isolated species but in well-defined clusters called molecules. The atoms in molecules are held together by strong forces called *bonds*. The criterion for whether a bond exists between two atoms is a simple energetic one.

> If two or more atoms are more stable when close together than when far apart, then there is a bond between the atoms.

In order to determine if specific atoms will form a molecule, all we have to do is to determine the energy of the molecule and compare it with the energies of the separate atoms. There are two specific points worth making at this stage.

- This definition makes no assumption about the type of forces involved, and in particular, bonds do not have to involve electron pairs.
- Bonds do not have to form specifically between two atoms, but can involve three or more.

The concept of each bond being a pair of electrons shared between two nuclei and other electrons forming lone-pairs on the individual atoms, which is the basis of valence bond theory and valence shell electron pair repulsion, or VSEPR, theory, will not be dealt with here. Whilst this concept provides a very 'easy to use' pictorial model of molecules, it is wrong and in some cases directly misleading.

Types of molecular bonding

There are essentially three limiting types of bonding.

- Ionic bonding. This is a direct electrostatic attraction between two ions with opposite charges. The strength of the interaction is proportional to the charges on the ions and inversely proportional to their separation.
- Metallic bonding. This consists of positively charged ions in a 'sea' of 'free' electrons. The ions are held in place by their mutual electrostatic attraction for the electrons. The strength of this interaction depends on the number of 'free' electrons per ion and the charge on the ion.
- Covalent bonding. In this case, electron orbital energies are changed by the presence of other atoms. This change can either decrease their energy, in which case bonding occurs, or increase their energy, in which case no bonding occurs.

Bonding in most molecules is a mix of these effects, some being more important than others (see Table 16.1).

TABLE 16.1

Compound	Description of bonding
$NaCl$	Pure ionic bonding between Na^+ and Cl^-
$FeCl_6^{3-}$	Chiefly ionic interactions between Fe^{3+} and Cl^- but there is also a smaller covalent contribution from the 3d orbitals of the Fe
$SiCl_4$	Chiefly covalent bonding but there is also a smaller ionic contribution
Cl_2	Pure covalent bonding between chlorine atoms

Electrons in molecules

The easiest way to discuss the behaviour of electrons in molecules is to consider what happens to the electrons in an atom when a second atom is brought close to the first. In a single atom

- The electron is localised to a spherical region around the nucleus by the electrostatic attraction of the nucleus.
- The shape of an electron orbital is determined by the allowed waves in the sphere.
- The energy of an electron is determined by the electrostatic attraction of the nucleus and the electrostatic repulsion of the other electrons.

When a second atom is brought up to the first atom, the electrons will experience an extra electrostatic attraction from the second nucleus and extra electrostatic repulsions from the electrons around the second atom. The effect of this extra atom on the electron will depend on how large these new interactions are compared to the interactions already present. In general, we can classify electrons into three groups (figure 16.1). These are

- Core electrons. These are electrons in the (n – 2), (n – 3), etc., shells, where n is the principal quantum number of the highest occupied orbital. They are

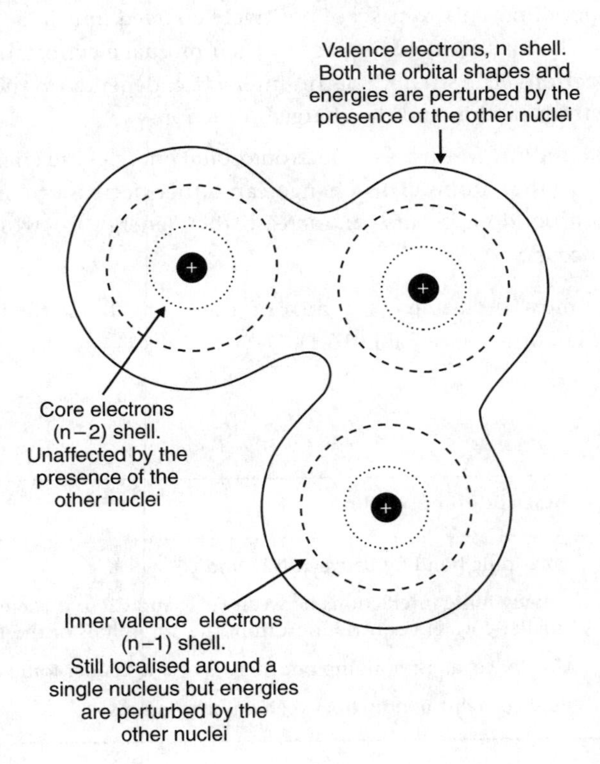

Fig. 16.1 Schematic representation of the behaviour of different electrons in a molecule. The nuclei are the small black circles. The lines represent the sizes of the atomic orbitals.

very close to their own nucleus and far away from the second nucleus. The orbital shapes and electron energies are unaffected by the presence of the second atom and so cannot contribute to bonding.

- Tightly bound valence electrons. These are electrons in the (n – 1) shell. They are further away from their own nucleus and closer to the second nucleus than the core electrons and so will be affected by the presence of the second atom. They will still interact most strongly with their own nucleus. The orbital shapes are not significantly altered because the electron is still localised in a spherical region around one nucleus but the electron energies will be perturbed by the extra electrostatic interactions and so could make a small contribution to bonding.
- Valence electrons. These are electrons in the n shell. They are equally affected by both nuclei and so cannot be treated as localised around one atom in a spherical region of space. Both the orbital shapes and electron energies will be altered and so could make a large contribution to bonding.

d orbitals in transition metal complexes

Transition metals have incomplete (n – 1)d shells. For instance, the configuration of the outer electrons of Fe in the ground state is $3d^6\,4s^2$, and for Fe^{3+} it is $3d^5$. Transition metal cations will form stable complexes with a wide variety of negatively charged ions and molecules with regions of negative charge (figure 16.2). The major interaction between the cation and ligands will be ionic, but there is also a significant covalent contribution from the (n – 1)d-orbital electrons. From the discussion above, the energy of the (n–1)d-orbital electrons will be perturbed by the presence of the ligand atoms but, to a first approximation, the orbital shapes will remain the same. If this perturbation leads to a reduction in the energy of the complex, it will contribute to the bonding.

Consider what happens to the energy of the electrons in a metal as a ligand approaches. As the ligand will be negatively charged in the region which is close to the cation, the electrostatic repulsion between the electrons on the metal and the ligand electrons will be greater than the electrostatic attraction between the electrons on the metal and the ligand nucleus. Thus, all the electrons in the metal cation will be slightly destabilised. This effect will be greatest if the metal electron is in an orbital that points towards the ligands and least if it is in an orbital that points away from the ligands.

Orbital points toward ligands	⇒	Electron in the orbital will be destabilised relative to other orbitals
Orbital points away from the ligands	⇒	Electron in the orbital will be stabilised relative to other orbitals

There are five (n – 1)d orbitals in a transition metal atom. They all have the same energy in free atoms or ions, i.e. they are energetically degenerate.

Cys S Cu N His
Met S N His
Plastocyanin

Cys S S Cys
Fe Fe
Cys S S S Cys
Ferredoxin

N His
N N
Fe
N N porphyrin
S Met
Cytochrome c

Fig. 16.2 Coordination geometries and ligands of metals in the active sites of three electron-transfer proteins. In plastocyanin and ferredoxin the metal environment is approximately tetrahedral (Td), in cytochrome c it is approximately octahedral (Oh).

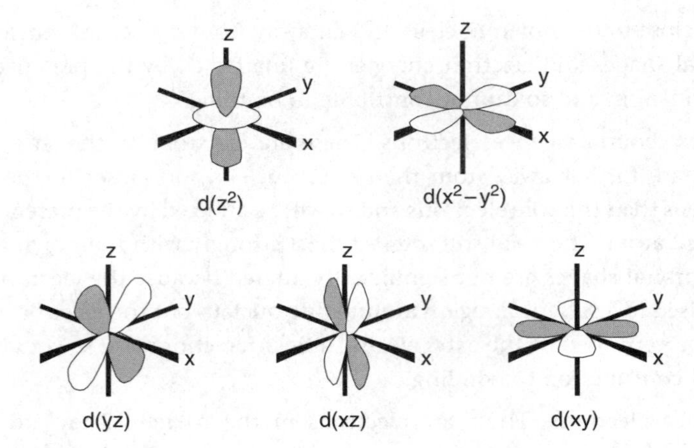

Fig. 16.3 Shapes of the five 3d orbitals. The $d(x^2 - y^2)$ and $d(z^2)$ orbitals point along the x,y and z axes. The d(yz), d(xz) and d(zy) orbitals point between the axes.

Since the d orbitals are highly directional and point in different directions (figure 16.3), they can be affected to different degrees by ligands. Thus, in complexes the five d orbitals can have different energies.

This qualitative approach to the behaviour of transition metal d-electrons is called *crystal-field theory* (CFT).

WORKED EXAMPLE

Predict the effects on the 3d orbital electrons of a first-row transition metal if the metal is (a) octahedrally (Oh) coordinated or (b) tetrahedrally (Td) coordinated.

SOLUTION: The shapes of the 3d orbitals are given in figure (16.3). The xz, yz and xy orbitals point between the x-, y- and z-axes. The $x^2 - y^2$ and z^2 orbitals point along the axes.

If a 3d orbital points towards a ligand, an electron in that orbital will be destabilised relative to the other electrons. If a 3d orbital points between the ligands, an electron in that orbital will be stabilised relative to the other electrons.

(a) In an octahedral (six-coordinate) complex, the ligands are positioned along the x-, y- and z-axes (figure 16.4a). The xz, yz and xy orbitals point between the ligands and so will be stabilised. The $x^2 - y^2$ and z^2 orbitals point towards the ligands and so will be destabilised. The five d orbitals, which are degenerate in the free ion, are split into two groups of three and two (called the t_{2g} and e_g respectively, figure 16.4b). The t_{2g} orbitals are slightly bonding (electrons are lower in energy than in the free ion) and the e_g orbitals are slightly antibonding (electrons are higher in energy than in the free ion).

(b) In a tetrahedral (four-coordinate) complex, the ligands are positioned between the x-, y- and z-axes (figure 16.4c). The $x^2 - y^2$ and z^2 orbitals point between the ligands and so will be stabilised. The xz, yz and xy orbitals point towards the ligands and so will be destabilised. The five d orbitals are again split into two groups of two and three (called the e and t_2 orbitals respectively, figure 16.4d) but the order of the orbitals is reversed compared to (a). The e orbitals are slightly bonding and the t_2 orbitals are slightly antibonding.

In both octahedral and tetrahedral complexes, the d orbitals are split into two groups, one slightly bonding and the other slightly antibonding. The energy gap between these two groups is called the *crystal-field splitting*, Δ. The size of this splitting (and hence the amount of bonding) depends on the amount of electrostatic repulsion experienced by the d-orbital electrons. Δ is always much smaller for tetrahedral complexes than for octahedral complexes because the presence of four ligands rather than six leads to less electrostatic repulsion in total. Δ will increase if the ligands get nearer to the metal. Thus, transition metals with higher charges and increased electrostatic attraction for the ligands will have higher values of Δ. The value of Δ also depends on the ligands. Ligands can be placed in order (called the spectrochemical series) according to the amount of splitting that they cause (figure 16.5).

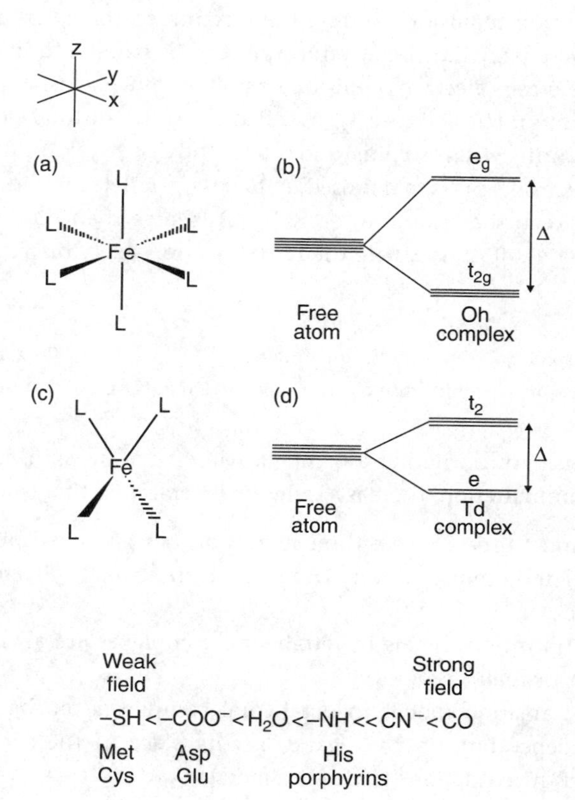

Fig. 16.4 In an octahedral complex (a) the orbitals which point along the axes are raised in energy and the orbitals which point between the axes are lowered in energy (b). In a tetrahedral complex (c) the orbitals which point between the axes are raised in energy and the orbitals which point along the axes are lowered in energy (d).

Fig. 16.5 The spectrochemical series for ligands found in biological systems. Weak-field ligands cause a small splitting of the d orbitals, strong-field ligands cause a large splitting.

d-orbital ground-state electronic configurations

As we have seen in Chapter 15, the ground-state electronic configuration is determined by placing each electron sequentially in the lowest energy available orbital. All that needs to be considered for atoms and isolated ions is the electron orbital energy because of the very large differences in energy between different orbitals. Precisely the same principle of minimising total energy applies for transition metal ion complexes. However, as the differences in orbital energy (Δ) are small we also need to consider other forms of energy, such as electron–electron repulsion.

$$\text{Total energy} = \text{Orbital energy} + \text{Electron–electron repulsion energy}$$

Consider an octahedral complex of a transition metal with the d orbitals split into three lower energy (t_{2g}) and two higher energy (e_g) orbitals (figure 16.4b). The first d-electron will occupy a t_{2g} orbital, as will the second and third, all with parallel spins. For the fourth electron we have a choice, either we can put it in a t_{2g} orbital or an e_g orbital (figure 16.6). Adding the electron to the t_{2g} orbital gives it a lower orbital energy than if it were in the e_g but the electron spin must be antiparallel to the others which gives us a higher electron–electron repulsion. Adding the electron to the e_g orbital gives it a higher orbital energy but the electron spin can be parallel to the other three giving less electron–electron repulsion (from the Pauli exclusion principle, see previous chapter). The first case, t_{2g}^4, is called low spin (the total electron spin is reduced by pairing electron spins) and the second case, $t_{2g}^3e_g^1$, is called high spin (the total electron spin is maximised by having parallel electron spins). If the splitting between the t_{2g} and e_g orbitals (Δ) is large then the orbital energy term dominates, if Δ is small then the electron–electron repulsion term dominates.

t_{2g}^4	low spin	minimise orbital energy	favoured by large Δ
$t_{2g}^3e_g^1$	high spin	minimise electron–electron repulsion	favoured by small Δ

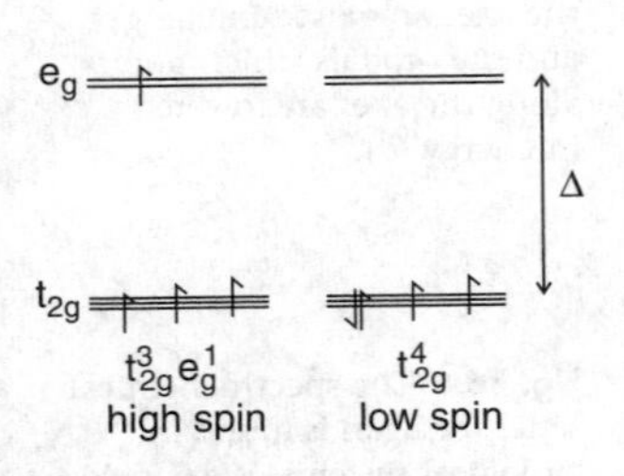

Fig. 16.6 The two possible d-electron configurations of a d^4 ion in an octahedral complex.

When considering whether a complex will give a low-spin or a high-spin d-electron configuration, there is a series of general rules that can be followed.

- Second- and third-row transition metals always give low-spin complexes because their d orbitals are very large and so are strongly affected by ligands (very large Δ).
- First-row transition metals in tetrahedral complexes are always high-spin because Δ is small (see above).
- First-row transition metals in octahedral complexes can be high-spin or low-spin depending on the ligand, i.e. its place in the spectrochemical series, and the oxidation state of the metal.

High-spin and low-spin first-row transition metal complexes can have different physical and chemical properties, for instance their spectroscopic and magnetic behaviour, size and reactivity to ligand exchange.

WORKED EXAMPLE

Haemoglobin is a protein involved in oxygen transport. It consists of four identical subunits, each of which binds one oxygen molecule reversibly. The oxygen binds to an Fe^{2+} ion coordinated in the centre of the protein. Oxygen binding is cooperative (see Chapter 4). This cooperativity is due to the binding of oxygen to one subunit causing a conformational change in that subunit which then induces a conformational change in the other subunits. Explain how oxygen binding to Fe^{2+} can trigger this conformational change.

SOLUTION: The oxygen-binding site of haemoglobin contains an Fe(II) ion in a porphyrin ring. There are five ligands (four from the porphyrin and one from the protein). The sixth site is empty (to be filled by O_2). The Fe(II) is in the high-spin state ($t_{2g}^4e_g^2$) because there are only five ligands (figure 16.7a). On binding O_2 the oxidation state of the Fe does not change.

$$[\mathrm{Fe(II)}] + O_2 \rightleftharpoons [\mathrm{Fe(II)}{\cdot}O_2]$$

However, the number of ligands increases to six. This leads to more electrostatic repulsion between the ligands and the d electrons, a higher value of Δ and thus a low-spin (t_{2g}^6) complex (figure 16.7b).

Deoxy-form	5 ligands	Low Δ	High-spin	Radius = 0·078 nm
Oxy-form	6 ligands	High Δ	Low-spin	Radius = 0·061 nm

The t_{2g} orbitals are slightly smaller than the e_g orbitals because they are lower in energy. Thus, low-spin Fe(II) is slightly smaller than high-spin Fe(II).

The high-spin Fe(II) ion is slightly too large to fit in the centre of the porphyrin ring (figure 16.7c). The low-spin Fe(II) ion can fit into the middle of the porphyrin ring. This displacement of the Fe(II) ion causes a displacement of the fifth ligand that is attached to the peptide backbone (figure 16.7d). The peptide backbone can then relay this conformational change to remote regions of the structure such as the protein surface involved in the interfaces between the subunits.

COMMENT: It is interesting to compare the conformational changes caused by O_2 binding to haemoglobin with the conformational changes caused by oxidation of cytochrome c (see previous chapter). Both of these are transport proteins, haemoglobin uses Fe to carry O_2 while cytochrome c uses Fe to carry an electron, and in both cases the change at the Fe ion is relayed to the rest of the protein structure by a ligand attached to the peptide backbone, resulting in subtle but important conformational changes remote from the Fe ion. In both cases, there is a coupling between the atomic and bonding properties of the Fe and the three-dimensional structure of the protein.

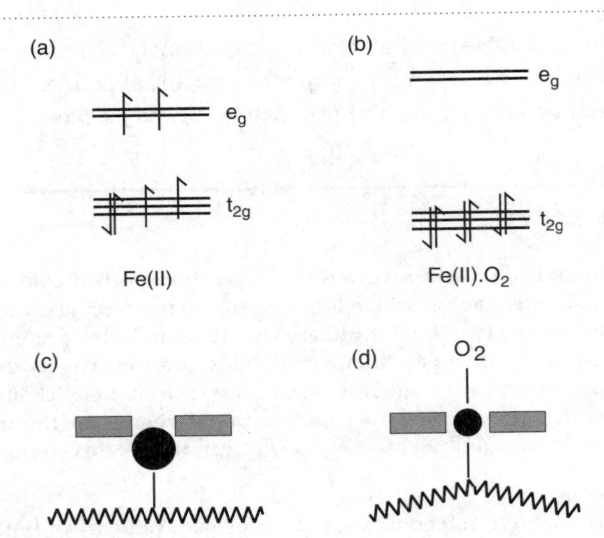

Fig. 16.7 The change in d-electron configuration of the Fe(II) ion between deoxy-haemoglobin (a) and oxy-haemoglobin (b) is responsible for the conformational change in the protein that leads to allosteric triggering ((c) and (d)). [C. Cooper and G. Moore, 1999, *The Biochemist*, **21** (4), 7.]

Valence electron bonding in molecules

The situation becomes more complex when considering the effect of introducing another atom on the valence electrons (n-shell). These electrons will experience a similar amount of electrostatic attraction from both nuclei. Rather than the electron being localised around one atom, it is now localised around two or more atoms. The allowed waves for electrons localised around two or more atoms are called *molecular orbitals* (the allowed waves for electrons localised around one atom are atomic orbitals). The shapes and energies of these molecular orbitals will be very different from atomic orbitals.

We can derive the shapes and energies of molecular orbitals from the atomic orbitals of the individual atoms using the ideas of constructive and destructive interference of waves. We shall start by considering two 1s electrons on two hydrogen atoms. If we overlap the two atomic orbitals, by bringing the two atoms very close together, then the electron wave in one orbital will interfere with the electron wave in the other orbital. This could result in either constructive or destructive interference, depending on the relative phases of the two waves (figure 16.8). In either case, the resultant wave is spread over both atoms. These resultant waves are molecular orbitals for H_2.

The next step is to determine the energies of these molecular orbitals relative to the initial atomic orbitals. The molecular orbital formed by constructive interference has increased electron density between the two nuclei where the electron will experience the maximum electrostatic attraction. An electron in this orbital will be of lower energy compared to (or more stable than) an electron in one of the atomic orbitals. This is called a *bonding*

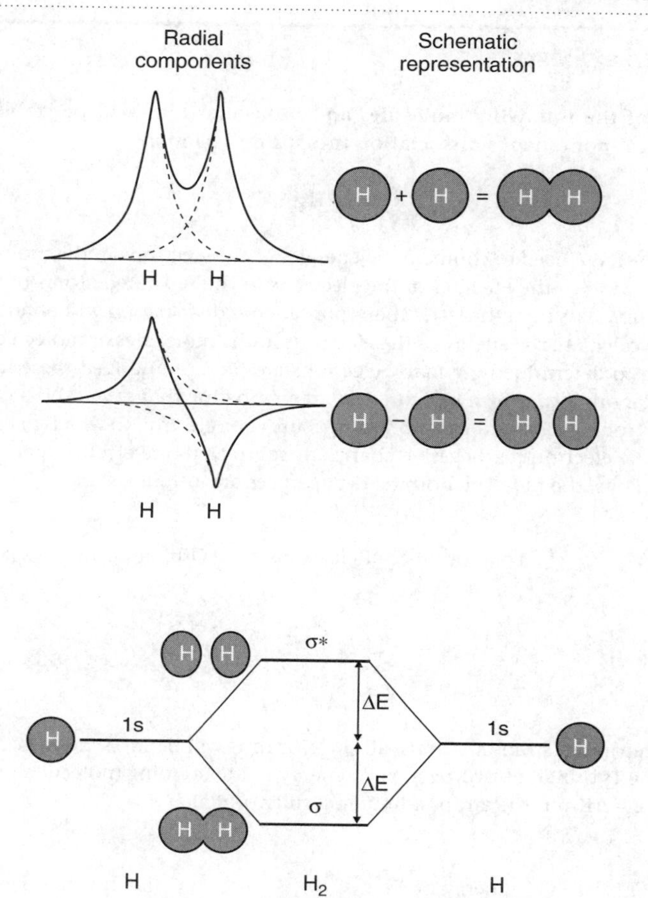

Fig. 16.8 Two hydrogen atom 1s orbitals (dotted lines) can interfere either constructively (top) or destructively (bottom). In both cases the resultant molecular orbital (solid lines) is spread over both atoms.

Fig. 16.9 Molecular orbital diagram for H_2.

molecular orbital. The molecular orbital formed by destructive interference has reduced electron density between the atoms and increased electron density further away from the nuclei. An electron in this orbital will be of higher energy compared to (or less stable than) an electron in one of the atomic orbitals. This is called an *antibonding* molecular orbital, usually denoted by an asterisk. The energies of these atomic and molecular orbitals are shown schematically in the molecular orbital diagram in figure 16.9. Each molecular orbital can accommodate two electrons, just like an atomic orbital. To a first approximation, the bonding orbital is stabilised by an amount ΔE relative to the atomic orbitals and the antibonding orbital is destabilised by the same amount (figure 16.9).

To decide whether bonding occurs or not, we simply have to 'feed' electrons into the available molecular orbitals and determine whether this gives a lower energy than 'feeding' them into atomic orbitals. If the total energy is lower for the molecule than for the atoms then the molecule is stable.

WORKED EXAMPLE

Which of the following molecules and molecular ions will be stable with respect to spontaneous dissociation into atoms or ions?

$$H_2^+ \quad H_2 \quad H_2^- \quad H_2^{2-}$$

SOLUTION: We need to compare the energies of the electrons in the molecule or molecular ion to the energies of the electrons in the separated atoms or ions. If the former is less than the latter then spontaneous dissociation will not occur. In order to calculate the energy of the electrons in the molecules or molecular ions, we need to determine how many electrons are present and feed these into the molecular orbitals, shown in figure 16.9, remembering that each orbital can take two electrons. Each hydrogen atom contributes one electron to which needs to be added one electron per negative charge, or subtracted one electron per positive charge. Thus, the molecular orbital occupancies are as follows.

	Total number of electrons	Orbital occupancy
H_2^+	1	σ^1
H_2	2	σ^2
H_2^-	3	$\sigma^2\ \sigma^{*1}$
H_2^{2-}	4	$\sigma^2\ \sigma^{*2}$

In the separated atoms and ions, all the electrons will be in 1s atomic orbitals.

Let the 1s orbital of hydrogen be at energy E, the bonding molecular orbital at energy $E - \Delta E$ and the antibonding molecular orbital at $E + \Delta E$.

	Orbital occupancy	Energy of electrons			Energy of electrons	Difference (atoms – molecule)
H_2^+	σ^1	$(E - \Delta E)$	vs.	$H + H^+$	E	$-\Delta E$
H_2	σ^2	$2 \times (E - \Delta E)$	vs.	$H + H$	$2 \times E$	$-2\Delta E$
H_2^-	$\sigma^2\ \sigma^{*1}$	$2 \times (E - \Delta E)$ $+(E + \Delta E)$	vs.	$H + H^-$	$3 \times E$	$-\Delta E$
H_2^{2-}	$\sigma^2\ \sigma^{*2}$	$2 \times (E - \Delta E)$ $+2 \times (E + \Delta E)$	vs.	$H^- + H^-$	$4 \times E$	0

The total energies of the electrons in H_2^+, H_2 and H_2^- are less than the total energies of the electrons in the separated atoms/ions and so they will not dissociate spontaneously. In H_2^{2-} the total energy of the electrons is the same in the molecular ion as in the separated ions and so there is no energy holding these ions together. Thus, H_2^{2-} will dissociate spontaneously.

The most stable molecule will be H_2 because this has the largest difference in the total energy of the electrons between the molecule and the separate atoms/ions.

The approach we have used above is called the 'Linear Combination of Atomic Orbitals' or LCAO method and can be summarised as follows.

- Molecular orbitals are made by adding and subtracting atomic orbitals.
- If we combine n atomic orbitals we must make n molecular orbitals.
- Constructive interference between atomic orbitals gives rise to bonding molecular orbitals whilst destructive interference between atomic orbitals gives rise to antibonding molecular orbitals.

This procedure can be extended to atoms with several occupied atomic orbitals. To do this a few further rules for combining atomic orbitals need to be used. In order to interfere constructively or destructively

- The atomic orbitals must have significant spatial overlap. If the orbitals are too small or too far apart the electron waves will not interfere.
- The atomic orbitals must be of the correct symmetry. Only orbitals with complementary shapes can overlap. Figure 16.10 shows a variety of possible overlaps of s and p orbitals. In the first two cases, the positive regions of the orbitals overlap giving constructive interference. In the third case, the positive regions of the two orbitals overlap as do the two negative regions giving constructive overlap. In the fourth case, the positive region of one orbital overlaps with both positive and negative regions of the other orbital giving both constructive and destructive interference that cancel each other out. The orbitals have been given the symmetry labels σ and π. A σ orbital is symmetric about the internuclear axis (it has the same phase on both sides), a π orbital is antisymmetric about the internuclear axis (it has opposite phases on opposite sides).
- The atomic orbitals must be of similar energies. Two waves of very different wavelength cannot interfere constructively or destructively with each other because if they are in phase at one point they will be out of phase at other points (figure 16.11). The net result is no constructive or destructive interference.

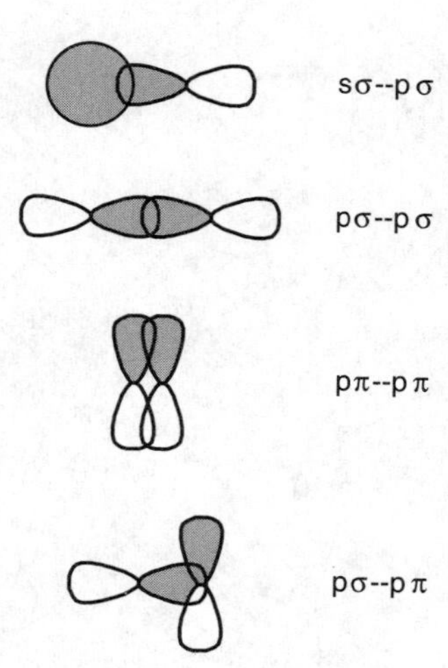

Fig. 16.10 Constructive interference between two atomic orbitals requires that they have the correct symmetry, i.e. overlap must occur only where the waves are the same sign. The first three cases give rise to constructive interference. The last ($p\sigma$–$p\pi$) gives equal constructive and destructive interference and thus has no overall effect.

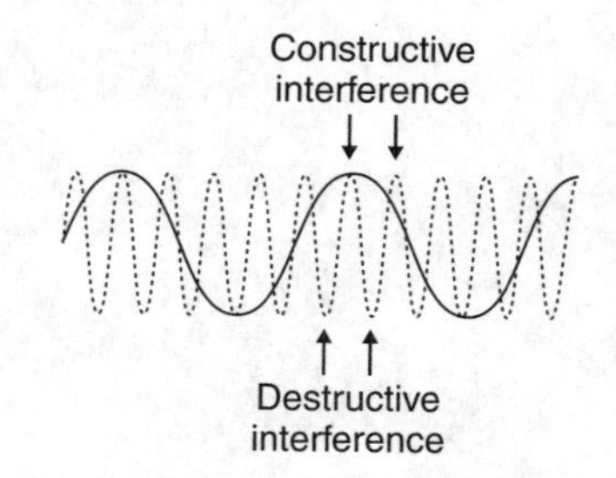

Fig. 16.11 Two waves with different energies (different wavelengths) cannot interfere constructively over significant distance because if they are in phase at one point they will be out of phase at other points. This effect will become more pronounced if the difference between the two wavelengths is greater.

WORKED EXAMPLE

Derive the molecular orbital diagram for O_2.

SOLUTION: Each O atom has occupied 1s, 2s and 2p atomic orbitals and we shall consider each of these in turn.

1s These are very small and core-like and so will not overlap with anything.

2s These orbitals are small but are just large enough to overlap with orbitals on the other atom. They will not interfere with 2p orbitals because the difference in energy is too great but they can interfere with the 2s orbital on the other atom to give bonding and antibonding σ molecular orbitals. These will not vary much in energy relative to the atomic orbitals because of the small spatial overlap.

2p These orbitals are relatively large and so will overlap well with orbitals on the other atom. Because of symmetry considerations (figure 16.10), the pσ orbital which points along the internuclear axis on one atom can only interfere with the pσ orbital on the other atom to give bonding and antibonding σ molecular orbitals. The two pπ orbitals which point at 90° to the internuclear axis can only interfere with the pπ orbitals on the other atom to give bonding and antibonding π molecular orbitals. The constructive or destructive interference between the pσ atomic orbitals will be much stronger than between the pπ atomic orbitals because the former point towards each other.

The final molecular orbital diagram is shown in figure 16.12.

COMMENT: This diagram will work for N_2 and F_2 as well because the atomic orbitals involved are the same in each case, only the electron occupancies will change.

Properties of molecular bonds

All the properties of a molecule depend on the behaviour of its bonding electrons. We shall only consider some of the more simple properties here.

Bond strength

This is given quantitatively by the difference between the energy of the electrons in the free atoms and the energy of the electrons in the molecule. It depends on how many net bonding electrons there are, i.e. on the number of bonding electrons minus the number of antibonding electrons, and on the difference in energy between the atomic and molecular orbitals (ΔE in

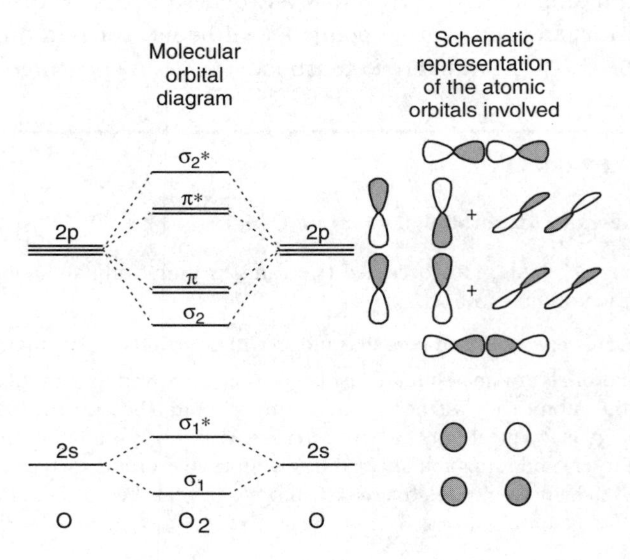

Fig. 16.12 Molecular orbital diagram for O_2, N_2 and F_2.

figure 16.9). The number of net bonding electrons is often indicated by a quantity called the bond order where

$$\text{Bond order} = \frac{\text{Number of net bonding electrons}}{2}$$

In many cases, a larger bond order indicates a larger bond strength. This is not always true as bond order takes no account of the difference in energy between the atomic and molecular orbitals, for instance F_2 with a bond order of one (single bond) has a higher bond energy than Se_2 with a bond order of two (double bond).

WORKED EXAMPLE

Put the following species in order of increasing bond strength

$$O_2^+ \quad O_2 \quad O_2^- \quad O_2^{2-}$$

Explain your reasoning.

SOLUTION: All these species have identical molecular orbitals (as in figure 16.12). The bond strength is determined by the occupancy of the bonding and antibonding orbitals, the more net bonding electrons the greater the bond strength.

	Electron configuration	Net number of bonding electrons	Bond order
O_2^+	$\sigma_1^2\ \sigma_1^{*2}\ \sigma_2^2\ \pi^4\ \pi^{*1}$	5	$2\frac{1}{2}$
O_2	$\sigma_1^2\ \sigma_1^{*2}\ \sigma_2^2\ \pi^4\ \pi^{*2}$	4	2
O_2^-	$\sigma_1^2\ \sigma_1^{*2}\ \sigma_2^2\ \pi^4\ \pi^{*3}$	3	$1\frac{1}{2}$
O_2^{2-}	$\sigma_1^2\ \sigma_1^{*2}\ \sigma_2^2\ \pi^4\ \pi^{*4}$	2	1

Thus, the order of increasing bond strength is

$$O_2^{2-} < O_2^- < O_2 < O_2^+$$

Bond length

The bond length is determined by the position of the nuclei when the molecule is in its lowest energy form. When two atoms are a long way apart, the degree of orbital overlap between their valence orbitals will be very small leading to a very weak bond. As the two atoms get closer together, the orbital overlap increases and the bond gets stronger leading to a lower overall energy. However, when the atoms get very close together the core electrons start repelling each other and so the energy of the molecule goes up again. This gives a typical energy versus distance curve shown in figure 16.13. The equilibrium bond length corresponds to the internuclear distance at lowest energy. In general, the more bonding electrons that there are the shorter the bond and the larger the atoms the longer the bond.

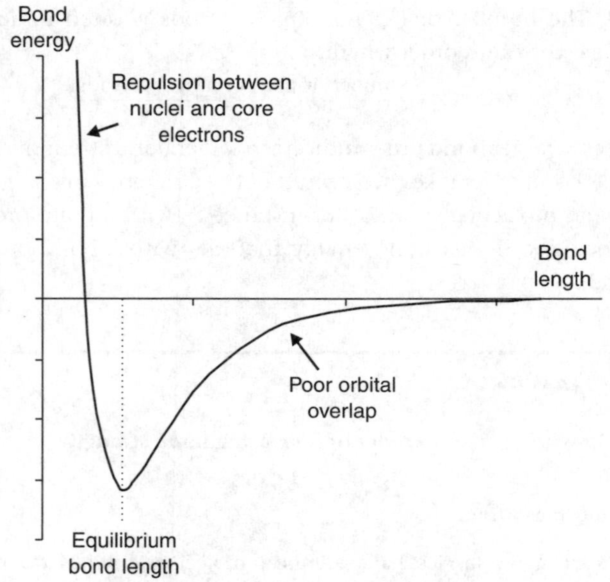

Fig. 16.13 Typical bond length versus energy curve for two bonded atoms. At short distances, repulsion between the two nuclei and core electrons dominate. At very long distances, there is very poor orbital overlap. The lowest point of the curve (minimum energy) gives the equilibrium bond length.

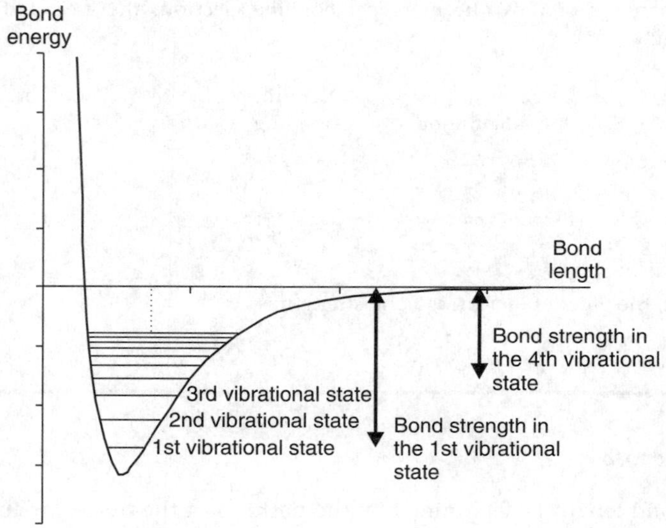

Fig. 16.14 There are many vibrational states for a bond within a single electronic state. During a vibration, the atoms oscillate between two points of equal energy on the curve (represented by the horizontal lines).

Bond vibrations

Nuclei in molecules are not still but are always moving relative to each other, vibrating about the equilibrium bond distance. We can use the energy versus distance curve in figure 16.13 to describe this. There are two values of the bond length at a given total energy. These are the minimum and maximum atomic separations and the two nuclei oscillate, or vibrate, between these values. If we increase the vibrational energy (which is quantised, so we can only increase it by set amounts), the molecule vibrates over a larger distance and the bond becomes weaker (figure 16.14).

Bond angles

These are determined in exactly the same way as bond lengths, it is simply the arrangement of atoms that gives the lowest energy, maximising orbital overlap whilst minimising steric repulsions.

Paramagnetism

Paramagnetism (an atom or a molecule having a permanent magnetic moment) is caused by the presence of unpaired electrons. The size of the magnetic moment depends on the number of unpaired electrons, the more unpaired electrons that are present the greater the magnetic moment.

WORKED EXAMPLE

Put the following species in order of increasing magnetic moment

$$O_2^+ \quad O_2 \quad O_2^- \quad O_2^{2-}$$

Explain your reasoning.

SOLUTION: Once we know the orbital occupancy for each species, all we have to do is count the number of unpaired electrons, remembering that the first two electrons to go into each π orbital will have parallel spins and so will be unpaired.

	Electron configuration	Number of unpaired electrons
O_2^+	$\sigma_1^2\,\sigma_1^{*2}\,\sigma_2^2\,\pi^4\,\pi^{*1}$	1
O_2	$\sigma_1^2\,\sigma_1^{*2}\,\sigma_2^2\,\pi^4\,\pi^{*2}$	2
O_2^-	$\sigma_1^2\,\sigma_1^{*2}\,\sigma_2^2\,\pi^4\,\pi^{*3}$	1
O_2^{2-}	$\sigma_1^2\,\sigma_1^{*2}\,\sigma_2^2\,\pi^4\,\pi^{*4}$	0

Thus, the order of increasing magnetic moment is

$$O_2^{2-} < O_2^- \sim O_2^+ < O_2$$

COMMENT: The valence bond picture of bonding in O_2 as involving two pairs of electrons forming the double bond and two lone pairs on each atom predicts that O_2 will have no unpaired electrons and so will be diamagnetic. As O_2 is paramagnetic, this is a simple demonstration of the invalidity of the valence bond approach.

Bonding in heteronuclear molecules

Bonds between two different atoms, called heteronuclear bonds, are formed in the same way as bonds between two identical atoms, called homonuclear bonds, discussed above. The only difference is that the atomic orbitals on the

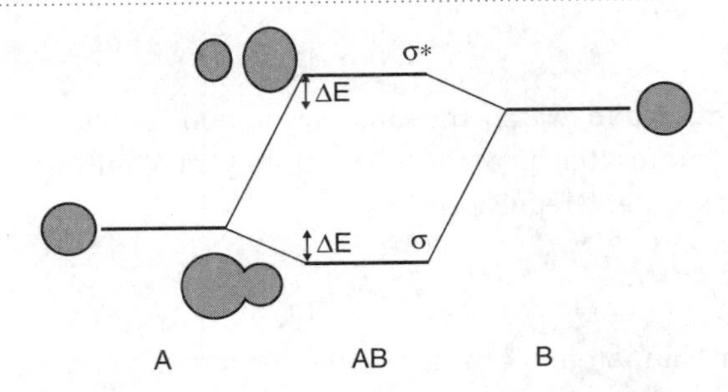

Fig. 16.15 The overlap between two atomic orbitals at different energies leads to asymmetric molecular orbitals. The molecular orbitals are closer in character to the atomic orbital which they are nearer to in energy. The electrons in the σ orbital have a greater density on atom A and the electrons in the σ^* orbital have a greater density on atom B. As the difference in energy between the two atomic orbitals increase, the value of ΔE decreases, because the interference between the two waves becomes less (as in figure 16.11).

two atoms that are used to form the molecular orbitals are not at the same energy. The bonding molecular orbital formed will be closer in energy to one of the atomic orbitals and the antibonding molecular orbital will be closer in energy to the other atomic orbital (figure 16.15). The two atomic orbitals do not make identical contributions to the molecular orbitals. The bonding orbital is not a 50:50 mix of the two atomic orbitals but contains a larger contribution from the lower energy atomic orbital and a smaller contribution from the higher energy atomic orbital.

For the AB diatomic molecule in figure 16.15, the bonding molecular orbital contains a higher proportion of atomic orbital A and a smaller proportion of atomic orbital B. This is equivalent to saying that an electron in the bonding molecular orbital behaves more like an A electron than a B electron or that the electron wavefunction is localised more towards A than B. The result of this is that the bonding electron density is not spread uniformly over the molecule but is higher near nucleus A and lower near nucleus B. This non-uniform distribution of electron charge can have very considerable effects on the chemical properties of a molecule.

WORKED EXAMPLE

Nitric oxide, NO, has now been recognised as a very important second messenger in a wide variety of biochemical systems. By considering the electronic structure of the molecule, predict

(a) whether the molecule will have a dipole moment and, if so, which end of the molecule will be negatively charged; and

(b) which end of the molecule will be more reactive.

Compare these results with the behaviour of N_2 and O_2.

SOLUTION: The molecular orbital diagram for NO differs from that for N_2 and O_2 (figure 16.12) because the atomic orbitals on the N and the O are at different energy levels. The equivalent atomic orbitals for N are higher in energy than those for O because the latter has a higher nuclear charge. These atomic orbitals overlap in a similar pattern to homonuclear diatomics with two significant differences.

(i) The O 2p orbital is at an appropriate energy to overlap with both the N 2s and 2p orbitals. This leads to mixing of the σ_1^* and σ_2 orbitals, with the σ_1^* dropping in energy (it has an extra bonding contribution from the σ_2) and the σ_2 orbital rising in energy above the π orbital (it has an extra antibonding contribution from the σ_1^*).

(ii) The bonding orbitals are localised more on the O and the antibonding orbitals localised more on the N.

The resulting molecular orbital diagram is shown in figure 16.16. Putting in the appropriate number of electrons give a valence electron configuration of

$$\sigma_1^2\ \sigma_1^{*2}\ \pi^4\ \sigma_2^2\ \pi^{*1}$$

(a) All the molecular orbitals have an asymmetric distribution of charge over the two atoms, thus the molecule will have a permanent dipole moment. The bonding electrons are localised between the nuclei and more on the O, tending to make the O more negative. The antibonding electrons are localised away from the two atoms and more on the N, tending to make the N more negative. Balancing these two factors leads to a very small dipole moment. However, the antibonding electrons are further from the centre of the molecule and hence have a larger effect, leading to the N being slightly more negative than the O. This is in direct contradiction to simple predictions made on the basis of electronegativities.

(b) The partially occupied π^* is both the highest occupied molecular orbital (HOMO) and the lowest unoccupied molecular orbital (LUMO). If NO is going to react with another species by either gaining, donating or sharing electrons, this must involve the π^* orbital and its single electron. The π^* orbital is localised more on the N than the O and thus the N end of the molecule will be the more reactive. This effect is much more important than the overall dipole moment.

COMMENT: Neither N_2 nor O_2 has a permanent dipole moment, the electron distributions in both are symmetrical. This leads to these molecules being less reactive and interacting more weakly with other molecules compared to NO.

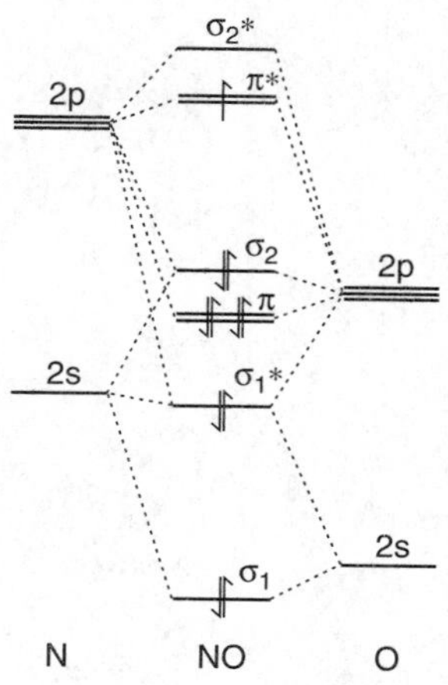

Fig. 16.16 Molecular orbital diagram and electron occupancies for NO.

Bonding in multi-atom molecules

The principles of bonding in multi-atom molecules are the same as for diatomic molecules. Molecular orbitals, each of which can contain two electrons, are formed by overlapping atomic orbitals of appropriate symmetry and energy. The only differences are that more nuclei and more orbitals are involved in making the molecular orbitals and that it may also be necessary to consider steric repulsion between atoms that are not directly bonded. The molecular orbitals that are formed are spread over all the atoms in the molecule, not just localised between a single pair of nuclei. For example, CH_4 contains eight valence electrons. These eight valence electrons are contained in four bonding orbitals, giving an overall bond order of four. The total energy of the valence electrons is lower in the molecule than in the free atoms and so the molecule is stable (there is overall bonding). The average C–H bond energy is simply a quarter of this stabilisation energy. The bonding molecular orbitals are spread over all five atoms and thus each bonding electron contributes to all four C–H bonds.

For small molecules, it is quite possible to consider the overlap of atomic orbitals from all the nuclei at once and obtain the appropriate molecular orbitals. However, as the number of nuclei, and atomic orbitals, increases this rapidly becomes impractical (but not incorrect). In these cases, we have to deal with the molecule in small chunks. The size of the chunk we need to consider varies according to the nature of the bonding; in some cases we can limit it to a bond between two atoms (e.g. a C–H single bond), in other cases we have to consider a larger group of atoms (e.g. an aromatic ring in tyrosine). There are no hard and fast rules for this (it comes with experience) except that we need to take a large enough chunk to explain the molecular behaviour that we observe.

WORKED EXAMPLE

Explain why peptide bonds are always planar (the four atoms of the –NH–CO– group are coplanar) and usually *trans* (the H and O atoms are on opposite sides of the N–C bond).

SOLUTION: If a peptide bond is treated as a double bond between the C and O atoms of the carbonyl group of the first residue and a single bond between the C of the carbonyl group and the N of the next residue (figure 16.17a), then we would expect free rotation about the C–N bond and the C, O, N and H atoms would not lie in the same plane.

Molecular orbital theory requires that we consider overlap of all the orbitals of the correct symmetry at the same time. The O, C, N and H atoms have σ-type atomic orbitals that are used to make the single bonds between all the atoms. In addition, the O, C and N atoms each have a single 2p π-type atomic orbital. The overlap between these can be considered independently of the σ orbitals because they have different symmetry. The three 2p π-type atomic orbitals can overlap to give three molecular orbitals, bonding, non-bonding and antibonding (figure 16.17b), as long as all three orbitals are parallel. Feeding electrons into the π molecular orbitals gives a total bond order of one spread over all three atoms.

Rotation around the C–N bond will break the overlap of the N 2p orbital with the other two. This will not affect the energy of the non-bonding molecular orbital. It will, however, reduce the amount of constructive interference in the bonding molecular orbital and so raise it in energy (figure 16.17b). The bond order will still be one, but now spread over two atoms (C and O) and the bond energy will be less. Thus, the most stable configuration is with all three 2p π-type orbitals parallel, which results in the O, C, N and H atoms all being in the same plane.

The O, C, N and H atoms can all be in the same plane with the O and H atoms *trans* (as shown in figure 16.17) or *cis* (rotation around the C–N bond by 180°). In molecular orbital terms, these two configurations will be equally stable. However, the *cis* form will lead to unfavourable steric interactions between the peptide chains preceding and following the peptide bond (figure 16.17c). Thus, the *trans* configuration is favoured on steric grounds.

COMMENT: *Cis* peptide bonds are occasionally found in proteins and more frequently in synthetic circular peptides; however, non-planar peptide bonds are almost never found. Thus, the molecular orbital interactions are much more important than the steric interactions in determining the conformational stability of peptide bonds.

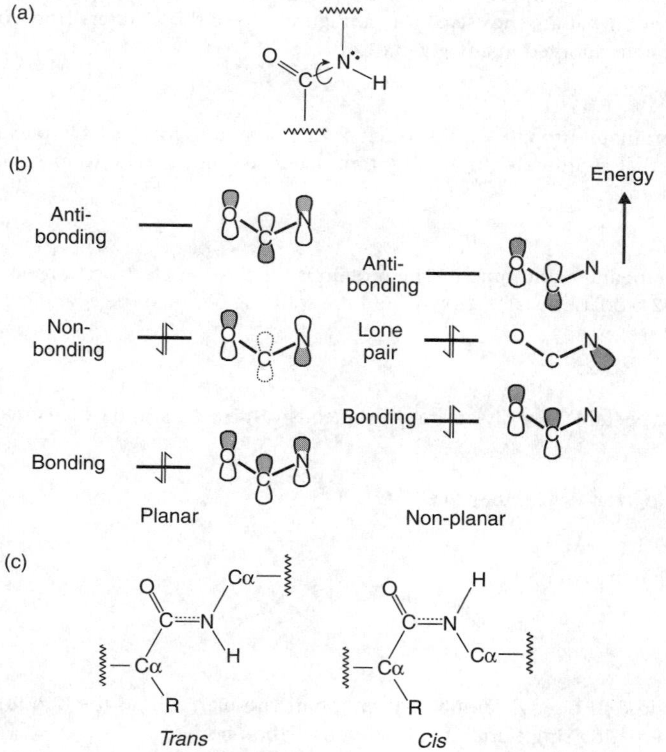

(a)

(b)

Energy

Anti-bonding

Non-bonding

Bonding

Planar

Anti-bonding

Lone pair

Bonding

Non-planar

(c)

Trans

Cis

Fig. 16.17 (a) The schematic representation of a peptide bond (shown in the *trans* configuration) implies free rotation about the C–N bond. (b) If the 2p orbitals on the O,C and N are coplanar, they can overlap to give three π molecular orbitals, bonding, non-bonding and antibonding. In the non-bonding moleular orbital, the relative phase of the C 2p orbital (whether its positive lobe is up or down, indicated by the dotted line) is irrelevant as there are equal amounts of constructive and destructive interference in both cases. Rotation about the C–N bond takes the N 2p orbital out of alignment with the C and O 2p orbitals and so breaks the overlap. The remaining bonding orbital is less stable because there is less constructive overlap between two orbitals than between three. (c) The planar peptide bond can be *cis* or *trans*. The *cis* form leads to greater steric repulsion between the side chain of one residue and the Cα of the next residue.

FURTHER READING

P.W. Atkins, 1998, 'Physical chemistry', Oxford University Press—*a more advanced, and more mathematical introduction to the electronic structure of molecules.*

M.J. Winter, 1994, 'd-block chemistry', Oxford University Press—*a simple, non-mathematical, introducation to crystal field theory.*

M.J. Winter, 1994, 'Chemical bonding', Oxford University Press—*a simple, non-mathematical, introduction to molecular orbital theory.*

PROBLEMS

1. For a first-row transition metal in an octahedral complex with n 3d-electrons, which values of n give rise to potential low-spin and high-spin states and what are the d-electron configurations of those states? How could you determine experimentally which state is adopted in any given case?

2. Cu(II) complexed with ligands that coordinate through S atoms (such as thiols) tend to form Td complexes whereas Cu(II) complexed with ligands that coordinate through N atoms (such as amines) tend to form Oh complexes. Explain the above observation.

3. Plastocyanin contains a Cu ion in a tetrahedral environment. The protein is involved in electron transport with the Cu cycling between the +1 and +2 oxidation states. How would the splitting between the t_2 and e orbitals vary between the two oxidation states?

4. Which of the following molecules would be stable (not dissociate spontaneously into atoms) and which one will be the most stable?

$$He_2^+ \quad He_2 \quad Li_2^{2+} \quad Li_2 \quad Be_2 \quad Be_2^- \quad C_2^+ \quad O_2^{2-} \quad F_2^{2-}$$

5. Put the following in order of decreasing bond strength:

$$N_2^+ \quad N_2 \quad N_2^- \quad N_2^{2-}$$

6. Explain why aromatic rings, such as those in phenylalanine or tryptophan, are planar. Can the bonding in these systems be adequately explained by alternating single and double carbon–carbon bonds?

7. Why are cis-peptide bonds in proteins observed almost exclusively for proline residues?

17

Interaction of molecules with electromagnetic radiation

KEY POINTS

- Electromagnetic radiation consists of oscillating electric and magnetic fields.
- This radiation interacts with molecules by inducing oscillations in electric or magnetic dipoles in the molecule. If this oscillation matches the gap between two energy levels then the radiation can be absorbed.
- Once a molecule has absorbed radiation it can lose the energy in a number of different ways, one of which is emission of electromagnetic radiation.
- Spectroscopy is the study of the absorption and emission of radiation by atoms and molecules.
- The peaks in a spectrum can be characterised by their frequency, intensity, linewidth and fine structure. These give information about the nature, chemical environment and concentration of the atom or molecule.

The interaction between electromagnetic radiation and molecules is of fundamental importance to all living systems, including topics as diverse as photosynthesis, sight and DNA damage caused by the UV component of sunlight. Many of the most powerful experimental techniques for studying biological systems also make use of these interactions, such as measuring protein concentrations, histological staining using fluorescent markers and macromolecular structure determination. The purpose of this chapter is not to cover such specific systems or techniques but to provide the general principles that govern the interaction of electromagnetic radiation with molecules.

Nature of electromagnetic radiation

Electromagnetic radiation consists of an electric wave and a magnetic wave at right angles to each other (figure 17.1). Electromagnetic waves are generated by oscillating electric or magnetic dipoles. A static dipole creates a static field,

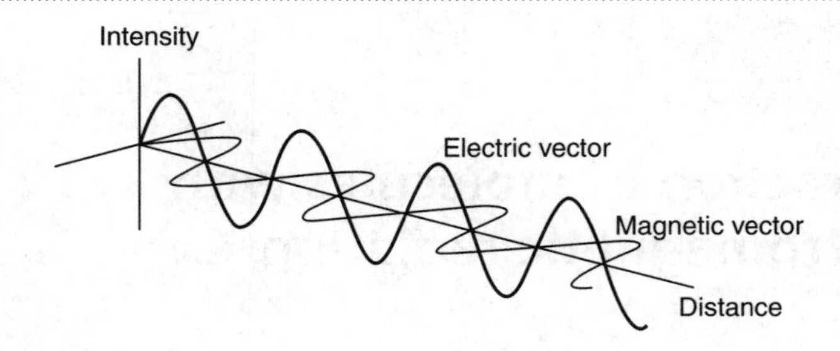

Fig. 17.1 Electromagnetic radiation consists of an oscillating electric field and an oscillating magnetic field at right angles. It can induce oscillations in any species that contains an electric or magnetic dipole.

an oscillating dipole creates an oscillating field; a wave is just a representation of this oscillating field. This may sound complex, but again it is an idea we are used to in another form. If you take a rope and move one end up and down in a regular way (i.e. cause one end to oscillate) then the whole rope starts to move. If you look at a single point in the middle of the rope, it also oscillates up and down. Most points on the rope oscillate out of phase with the end you are moving (they are not going up and down at the same time as the end but lag behind to different degrees). If you look along the length of the rope you get a wave. If you stop moving the end, other parts of the rope still oscillate and the wave appears to move along the rope.

Just like other waves, a continuous (infinitely long) electromagnetic wave has a perfectly defined frequency (ν) and wavelength (λ). These are related by the equation

$$\nu = c/\lambda$$

where c is the velocity of propagation of the wave (the speed at which the maxima appear to move). It also has an energy (E) related to the frequency by

$$E = h\nu$$

If the electromagnetic wave is not continuous but localised to a small region (such as a short pulse caused by a dipole that oscillates for a short period of time) then it is better represented by a wave-packet (in this case called a *photon*) whose frequency and energy can no longer be perfectly defined (see Heisenberg's uncertainty principle, Chapter 14).

An oscillating dipole generates a field that oscillates in the same direction (if you move the end of the rope up and down, the points in the middle can only move up and down, not side to side). In most cases, the dipoles that generate electromagnetic waves can oscillate in any direction. However, if we cause the dipoles to oscillate in a single direction then the resulting electromagnetic waves will only oscillate in that direction (figure 17.2). This is called *linearly polarised* or *plane polarised* radiation.

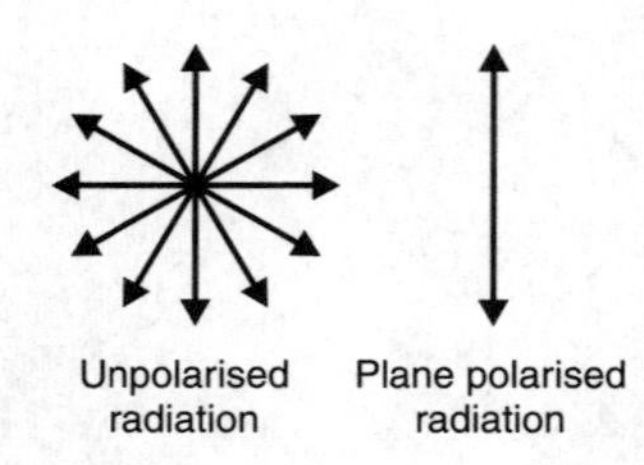

Fig. 17.2 Schematic representation of the direction of oscillations of the electric field component for unpolarised and plane polarised electromagnetic radiation.

For convenience, electromagnetic radiation is usually classified by its wavelength or frequency (see Table 17.1). Alternatively, to avoid the use of

TABLE 17.1

Type of radiation	Wavelength (m)	Energy (J)[a]	Properties
Gamma rays	$< 10^{-11}$	$> 2 \times 10^{-14}$	Very penetrating, can ionise molecules leading to chemical changes (e.g. DNA damage)
X-rays	10^{-11}–10^{-8}	2×10^{-14}–2×10^{-17}	
Far UV	10^{-8}–10^{-7}	2×10^{-17}–2×10^{-18}	
Near UV	10^{-7}–4×10^{-7}	2×10^{-18}–5×10^{-19}	Not very penetrating, interaction with molecules produces heat
Visible	4×10^{-7}–$6{\cdot}5 \times 10^{-7}$	5×10^{-19}–3×10^{-19}	
Infrared	$6{\cdot}5 \times 10^{-7}$–10^{-3}	3×10^{-19}–2×10^{-22}	
Microwaves	10^{-3}–1	2×10^{-22}–2×10^{-25}	Interact weakly with molecules, used for communications
Radio waves	> 1	$< 2 \times 10^{-25}$	

[a] The energy quoted is per photon. The energy in J mol^{-1} is given by multiplying this value by the Avogadro number.

very large or small numbers, *wavenumbers* can be used (defined as the inverse of the wavelength in centimetres (cm^{-1})). However, the fundamental nature of all forms of electromagnetic radiation (γ-rays through to radiowaves) is the same.

Absorption of electromagnetic radiation by matter

Absorption of electromagnetic radiation by matter occurs by a process called *resonance*. As we have seen, electromagnetic radiation at a given frequency is generated by an electric or magnetic dipole oscillating at that frequency. Conversely, electromagnetic radiation will cause electric and magnetic dipoles to try to oscillate, to *resonate* with the oscillating fields. If the electric or magnetic dipole does start to oscillate then it must have gained energy, i.e. absorbed energy from the oscillating field.

In atoms and molecules, the species that make up the electric or magnetic dipoles are the electrons and nuclei and so electromagnetic radiation will make them oscillate. The electrons or nuclei can only have specific energies because they have quantised energy levels. The induced oscillations can make the electrons or nuclei jump from one energy level to another, either up (absorption) or down (stimulated emission). There are two criteria that need to be met for this to occur. First, the two energy levels must be related by a dipolar movement of the electrons or nuclei, since a dipolar field can only induce dipolar changes. This means that when the atom or molecule moves from one energy level to the other its dipole moment must change. Second, the energy supplied by the electromagnetic radiation ($h\nu$) must match the difference in energy levels (ΔE).

$$\Delta E = h\nu$$

If the induced oscillations do not match the energy difference between two different energy levels then no absorption or emission of energy can take place.

Thus, for absorption of electromagnetic radiation to occur there have to be

- Two or more energy levels.
- A change of an electric or magnetic dipole associated with a change between these energy levels (this change is called the transition dipole).
- A match between the difference in energy of these levels and the energy of the radiation.

WORKED EXAMPLE

Which of the following molecules, O_2, CO and CO_2, can absorb electromagnetic radiation in the region $\lambda = 10^{-5}$ to 10^{-6} m?

SOLUTION: The energy of electromagnetic radiation is given by

$$E = \frac{hc}{\lambda}$$

where c is the speed of light and h is the Planck constant. Therefore, the wavelength range

$$\lambda = 10^{-5} \text{ to } 10^{-6}\ \text{m}$$

corresponds to the energy range

$$E = 2 \times 10^{-20} \text{ to } 2 \times 10^{-19}\ \text{J}$$

For small molecules, electronic energy levels usually have very large ΔEs. The energy level separations that match these wavelengths of radiation occur between vibrational energy levels (see Table 14.1). Thus, if this radiation is going to be absorbed it must be by the molecules undergoing vibrational energy transitions.

In order for electromagnetic radiation to cause a change in vibrational energy levels, the change in vibration must lead to a change in dipole moment. The only vibrations that can occur for O_2 and CO are bond stretching (figures 17.3a and 17.3b). O_2 does not have an electric dipole and stretching it does not cause one, i.e. the transition dipole is zero, and so O_2 cannot absorb electromagnetic radiation of this wavelength. CO does have an electric dipole. Stretching the bond changes the size of the dipole moment, i.e. the transition dipole is not zero, and thus CO can absorb this radiation.

CO_2 is rather more complex because more vibrations are possible. The molecule in the ground state does not have an electric dipole because of its symmetry. Vibrations that retain this symmetry, such as stretching both C–O bonds (figure 17.3c), do not give rise to an electric dipole and so cannot result in absorption of radiation. However, vibrations that break the symmetry, such as bending the molecule (figure 17.3d), will give rise to an electric dipole moment and so CO_2 can absorb radiation at these wavelengths.

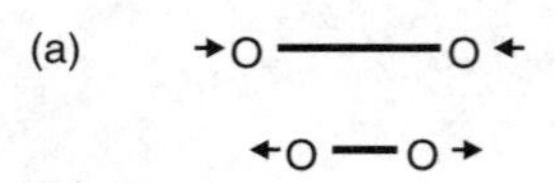

(b) δ+ C — δ− O

δ+ C — δ− O

(c) δ− O — δ+ C — δ− O

δ− O — δ+ C — δ− O

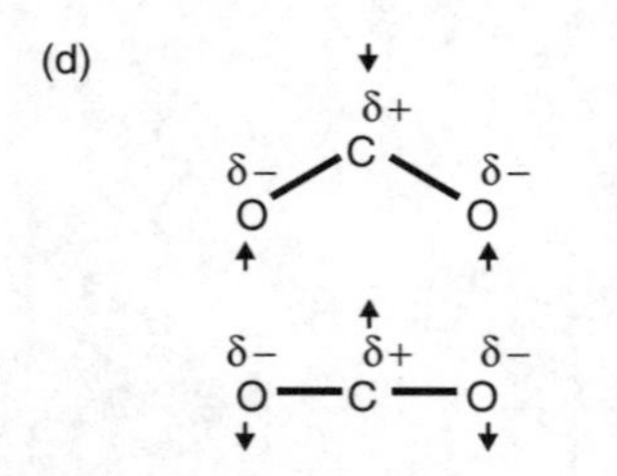

Fig. 17.3 Absorption of electromagnetic radiation requires a change in either the electric or the magnetic dipole moment of the chromophore. The vibrations shown in (a) and (c) do not lead to a change in dipole, the vibrations shown in (b) and (d) do lead to a change in dipole.

COMMENT: The normal components of the atmosphere, N_2 and O_2, do not absorb much IR radiation. However, CO_2 does, which is why it is classified as a 'greenhouse' gas; it lets through UV and visible radiation coming from the sun but absorbs IR radiation coming from the earth and thus reduces loss of heat from the earth.

This introduces the concept of *selection rules*. These are rules that tell us whether a transition between two energy levels can be caused by electromagnetic radiation.

Fate of excited atoms or molecules

Once an atom or molecule has absorbed energy from a photon, it has to lose this energy in order to return to its ground state, or for a collection of atoms or molecules to return to their Boltzmann equilibrium. The simplest way in which this can occur is by *spontaneous emission* of the extra energy as another photon, the reverse process to absorption. However, this does not happen instantaneously, so the atom or molecule spends a certain amount of time in the excited state and the average time spent in the excited state is called the excited state *lifetime*. During this time other things can happen to either some or all of the absorbed energy. These include

- Stimulated emission of a photon (just as absorption is caused by the molecule resonating with an external oscillating field, so emission can be caused by resonance with an external oscillating field).
- Conversion of the molecule from one excited state to another (such as conversion between an electronic excited state with all electron spins paired to one with two unpaired spins, called a singlet-to-triplet conversion).
- Non-radiative loss of energy (loss of energy to the environment as translational, rotational or vibrational energy without emitting any radiation).
- Resonance transfer of energy directly to another nearby molecule.
- Chemical reactions (using the extra energy to overcome an activation barrier).

WORKED EXAMPLE

Explain why

(a) molecules absorb photons at selected frequencies to give excited electronic states and then usually emit photons at lower frequencies, a phenomenon called fluorescence;

(b) atoms absorb photons at selected frequencies to give excited electronic states and then emit photons at the same frequencies.

SOLUTION:

(a) When a molecule absorbs a photon of the correct frequency, usually UV or visible radiation, an electron can be excited from an occupied orbital to a higher energy unoccupied orbital. As well as changing the electronic energy of the molecule, these transitions are usually accompanied by changes in vibrational energy and this vibrational energy can be lost by non-radiative means. On initial absorption of the photon, the atoms will remain the same distance apart as in the ground state since absorption happens too fast for the nuclei to move. This distance will be shorter than the equilibrium bond distance in the excited state since higher energy orbitals have less bonding character, and so the atoms will start to move apart. Having started to move they will not stop but will continue vibrating around the equilibrium distance for the excited state. Collisions with other molecules will dissipate this vibrational energy, returning the molecule to its vibrational ground state, before a photon is emitted and thus the emitted photon will have less energy and so be at a lower frequency. This process is shown schematically in figure 17.4a.

(b) Atoms do not have vibrational energy since there are no bonds, and so no energy can be lost in this way. Thus, all the absorbed energy has to be re-emitted (figure 17.4b).

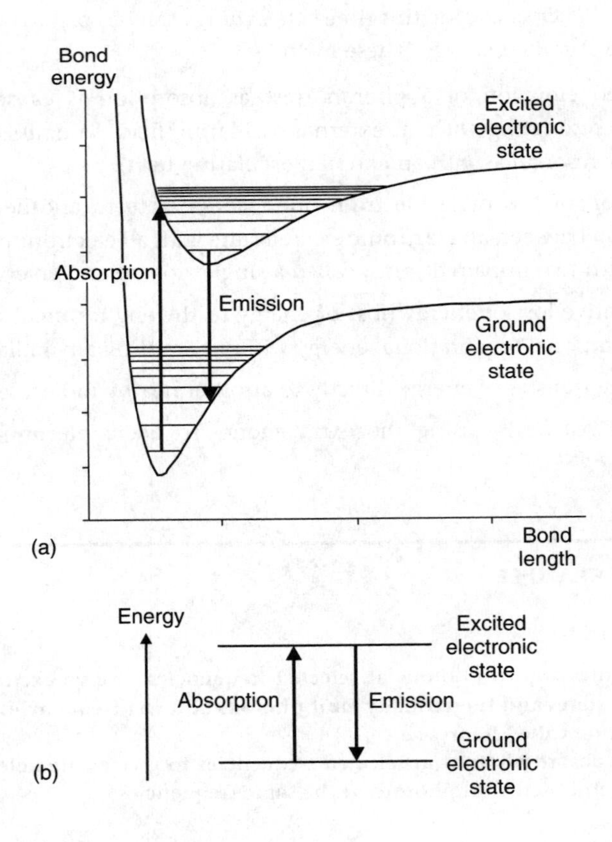

Fig. 17.4 Electronic transitions associated with absorption and emission by (a) a molecule and (b) an atom.

Basics of spectroscopy

Spectroscopy is an experimental technique for monitoring the interaction of electromagnetic radiation with matter by measuring changes in intensity of electromagnetic radiation. There are two basic protocols (figure 17.5).

- Absorption spectroscopy. Irradiate the sample at a specific frequency and measure the amount of radiation absorbed, i.e. the decrease in intensity of the transmitted radiation relative to the original source.
- Emission spectroscopy. Irradiate the sample and measure the intensity of the emitted radiation at a specific frequency.

A spectrum is obtained by scanning across a range of frequencies and plotting the intensity of the absorbed or emitted radiation versus frequency to give one or more peaks (figure 17.5). Spectroscopic techniques are classified according to the type of transition that is being monitored or the type of radiation used (see Table 17.2).

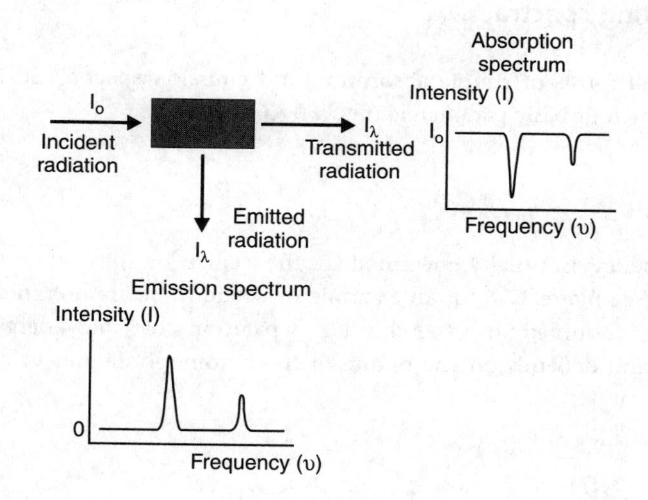

Fig. 17.5 Absorption spectroscopy involves measuring the reduction in intensity of transmitted radiation, whereas emission spectroscopy involves the direct measurement of emitted radiation following absorption. In both cases, intensity is plotted versus wavelength (λ) or frequency (ν).

TABLE 17.2

Type of spectroscopy	Nature of transition	Radiation used
Electronic	excitation of electrons from core orbitals	X-rays
Electronic	between valence electron orbitals	UV/Visible
Vibrational	between vibrational states of bonds	IR
ESR	between magnetic states of electrons	Microwaves
NMR	between magnetic states of nuclei	Radio waves

TABLE 17.3

Type of spectroscopy	Major absorbing/emitting components of proteins
Electronic	Trp and Tyr aromatic rings/Backbone peptide linkages/Metal ions
Vibrational	Carbonyl (>C=O) bonds/amide (N–H) bonds
NMR	Hydrogen atoms

In each case it is usually a specific part, or parts, of a molecule that absorbs or emits radiation at a specific wavelength and thus gives rise to the peaks in the spectrum. For electronic spectroscopy, a region of a molecule that absorbs or emits radiation is called a *chromophore*. Large complex molecules such as proteins often contain many different chromophores that absorb and emit at different frequencies (see Table 17.3).

Interpreting spectra

Peaks in all forms of simple absorption and emission spectra can be characterised by four basic parameters (figure 17.6).

Frequency

The frequency is usually measured at the centre or maximum point of the peak (see figure 17.6 for an example of where these are not coincident). This is determined by the precise separation of the energy levels involved and depends on the nature of the chromophore and its chemical environment.

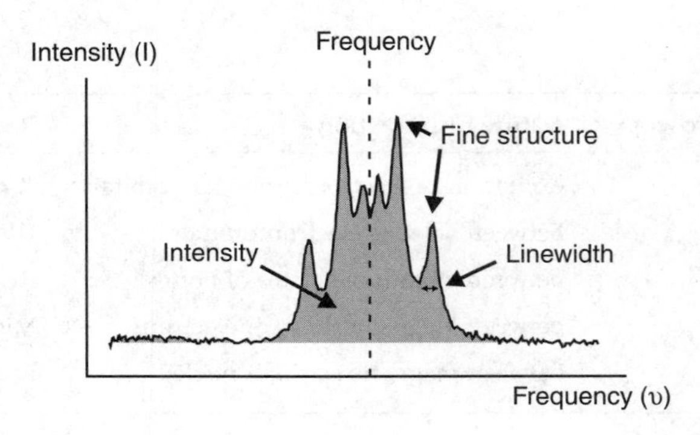

Fig. 17.6 A peak from a nuclear magnetic resonance (NMR) spectrum, illustrating the four parameters that can be measured.

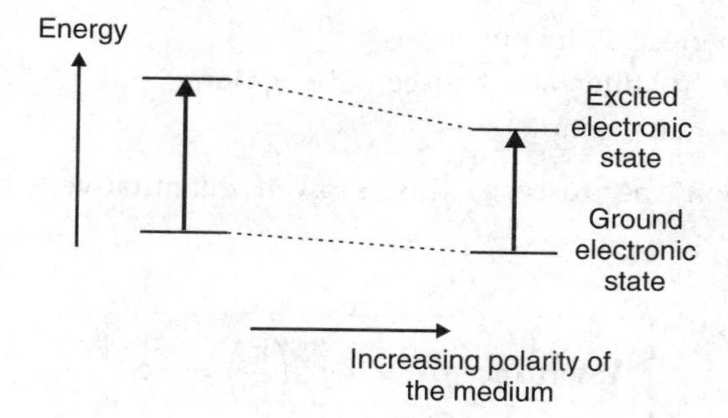

Fig. 17.7 Absorption and emission peak frequencies are often sensitive to the environment of the chromophore. For electronic transitions, changing the polarity of the medium has a different effect on the energies of the ground state and the excited state.

WORKED EXAMPLE

Some of the electronic absorption peaks of the amino acid tryptophan (Trp) in aqueous solution shift to higher frequency if a less polar solvent such as glycerol is added. This effect is often not seen for Trp residues in folded proteins. Explain.

SOLUTION: The frequency of the transition depends on the difference in energy between the ground and excited state of the molecule. As we have seen, absorption of radiation causes a change in electronic dipole moment of the molecule. In general, the electronic ground state of a molecule has the lowest possible dipole moment, corresponding to the least separation of charges, and the excited state has a higher dipole moment. The stability of a dipole depends on the polarity of the environment, the more polar the environment, the higher the dielectric constant, and the more stable the dipole is. Decreasing the polarity of the environment of the chromophore will have little effect on the ground state because of its small dipole moment but will destabilise the excited state (figure 17.7). Thus, the energy gap between the ground and the excited state will increase and the absorption is shifted to higher frequency.

In folded proteins, Trp residues are often buried in the core of the protein. Changing the polarity of the solvent does not affect the environment of the Trp and so no change in frequency is seen.

COMMENT: This is the basis of a technique called solvent perturbation spectroscopy, which can be used to establish whether a chromophore is exposed to the solvent.

Intensity

The intensity of a peak is the total peak area. This is a measure of the amount of radiation absorbed or emitted by the sample and it depends on

- The probability that radiation can cause the transition between the energy levels that gives rise to that peak, which is determined by how strongly the transition dipole interacts with the oscillating field of the radiation.

- How many molecules are present.
- The population difference between the ground and excited state (see Chapter 14 for an example).

For absorption spectroscopy, this is given quantitatively by the *Beer–Lambert law*

$$\text{Absorbance (A)} = \log_{10}\left(\frac{I_o}{I_\lambda}\right) = \varepsilon_\lambda cl$$

where I_o is the intensity of the incident radiation and I_λ is the intensity of the transmitted radiation (figure 17.5) at wavelength λ. c is the concentration of the chromophore in the sample and l is the path length of the radiation through the sample. cl is a measure of how many molecules are in the path of the radiation. ε_λ is a constant known as the *molar absorption coefficient* or *extinction coefficient* of the chromophore at wavelength λ. It is a measure of how strongly the chromophore interacts with radiation of that wavelength. The absorbance (A) is also referred to as the optical density (O.D.).

WORKED EXAMPLE

15·8% of the radiation at 340 nm incident on a sample of NADH was transmitted. The extinction coefficient of NADH at 340 nm is $6{\cdot}22 \times 10^6\ cm^2\ mol^{-1}$ (or $6{\cdot}22 \times 10^3\ M^{-1}\ cm^{-1}$). The path length was 1 cm. What is the concentration of NADH in this sample?

SOLUTION: As 15·8% of the incident radiation is transmitted

$$I_\lambda = 0{\cdot}158 \times I_o$$

The absorbance is given by

$$A = \log_{10}\left(\frac{I_o}{I_\lambda}\right) = 0.801$$

From the Beer–Lambert law, $A = \varepsilon_\lambda cl$ and $\varepsilon_\lambda = 6{\cdot}22 \times 10^6\ cm^2\ mol^{-1}$ and $l = 1$ cm. Thus

$$c = \frac{A}{\varepsilon_\lambda l} = \frac{0{\cdot}801}{(6{\cdot}22 \times 10^6) \times 1}\ mol\ cm^{-3}$$

$$c = 1{\cdot}29 \times 10^{-4}\ M$$

COMMENT: Monitoring NADH concentrations provides a powerful method for following enzyme kinetics as many enzyme reactions can be coupled to production or consumption of NADH.

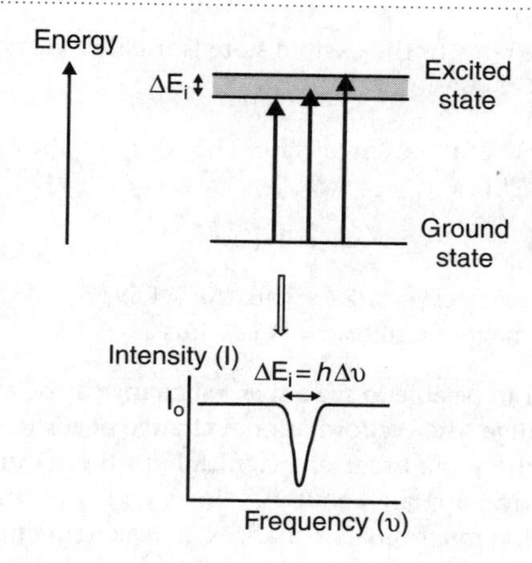

Fig. 17.8 The short lifetime of the excited state leads to an uncertainty in its energy and hence a broad absorption or emission peak.

Linewidth

Spectral peaks can appear broad for two reasons

- The observed peak may be a single peak that is inherently broad.
- The observed peak may be composed of a number of separate narrower peaks that overlap.

All peaks have a basic natural linewidth because the energy levels involved in the transition are not perfectly defined (figure 17.8). This can be explained by one version of the Heisenberg uncertainty principle

$$\Delta E_i \times \tau_i \geq \tfrac{1}{2}\hbar$$

where ΔE_i is the uncertainty in energy of a given energy level and τ_i is the time spent in that state. Molecules usually spend a very long time in the ground state, thus τ_i is large and so ΔE_i is small. The energy of the ground state is defined very well. If the molecule has a very short lifetime in the excited state then τ_i is small and so ΔE_i must be large. The energy of the excited state is poorly defined. This leads to a range of possible transitions, from E_1 in the ground state to $E_2 \pm x$ where x can have any value from 0 to $\tfrac{1}{2}\Delta E_2$.

WORKED EXAMPLE

For a Trp residue in a protein, the lifetime of its excited electronic states, as monitored by UV absorption spectroscopy, is about 10^{-7} s, and the lifetime of its excited nuclear magnetic states, as monitored by NMR spectroscopy, is about 0·1 s. How will this affect the widths of the absorption peaks in these two forms of spectroscopy?

SOLUTION: The width of an absorption peak depends on the excited state lifetime. A short lifetime gives a broad peak, a long lifetime gives a narrow peak.

The uncertainty in energy of the excited state is usually taken to be

$$\Delta E \sim \hbar/\tau$$

where τ is the lifetime of the excited state. The width of absorption peak ($\Delta\nu$) is given by $\Delta E = h\Delta\nu$. Thus

$$\Delta\nu \sim 1/(2\pi\tau)$$

For excited electronic states $\tau = 10^{-7}$ s, thus $\Delta\nu = 1\,591\,549{\cdot}4\ s^{-1}$ or 1·6 MHz.
For excited nuclear magnetic states $\tau = 0{\cdot}1$ s, thus $\Delta\nu = 1{\cdot}6\ s^{-1}$ or 1·6 Hz.

COMMENT: In order to be able to see a spectral change associated with a protein conformational change, the conformational change needs to cause a change in peak frequency of the same order of magnitude as the natural linewidth. The peaks in NMR spectroscopy are about 10^6 times narrower than in UV spectroscopy. This means that much smaller changes in peak frequency can be observed by NMR spectroscopy. Thus, whereas UV spectroscopy can only monitor relatively large protein conformational changes, NMR spectroscopy can be used to monitor very small, subtle changes.

The uncertainty principle imposes a minimum linewidth for a peak. Peaks can appear broader than this if they consist of a number of overlapping peaks at slightly different frequencies.

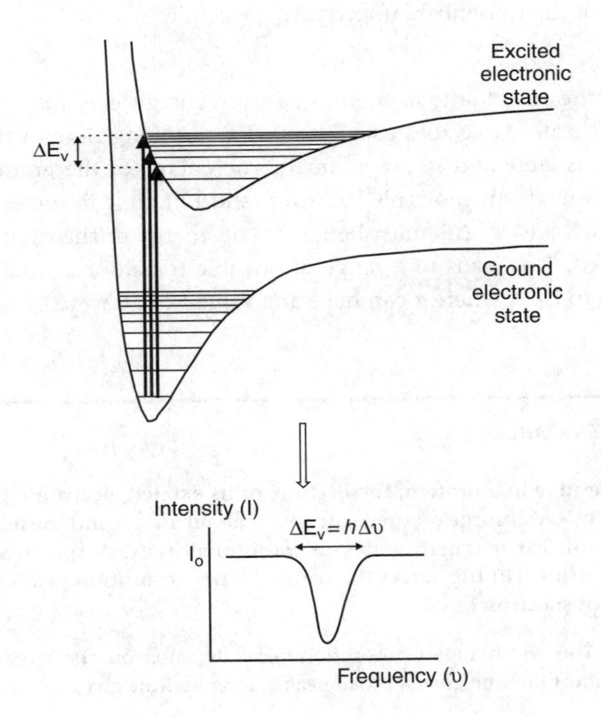

Fig. 17.9 Electronic transitions in molecules can be accompanied by different changes in vibrational energy, leading to additional peak broadening.

WORKED EXAMPLE

Electronic absorption spectra of molecules consist of much broader peaks than electronic absorption spectra of atoms. Explain the above observations.

SOLUTION: The lifetimes for the electronic excited states of molecules and atoms will be about the same and so the peaks will have the same minimum linewidth defined by the uncertainty principle. However, as we have already seen, electronic transitions in molecules are also accompanied by vibrational transitions. A single electronic transition can be accompanied by a range of vibrational transitions, each giving a slightly different change in total energy (figure 17.9). The resultant spectral peak then consists of a number of overlapping peaks. Atoms have no vibrational energy and so a single peak is observed with a linewidth close to that predicted by the uncertainty principle.

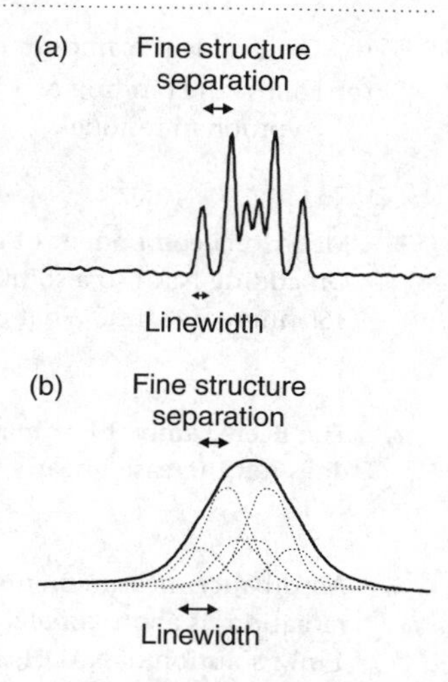

Fig. 17.10 Absorption or emission peak fine structure can only be observed if the linewidth is less than the fine structure separation (a). Otherwise a broad single peak is seen (b).

Fine structure

A single spectral peak may not always consist of one component but have several components, either fully resolved or partially overlapping. We have just seen in the previous worked example that an electronic transition in a molecule can be accompanied by several different vibrational transitions, leading to overlapping peaks and line broadening. However, if the separation between these peaks is larger than the inherent linewidth then the separate peaks will be resolved (figure 17.10). This is called the vibrational fine structure of an electronic transition. In practice, fine structure is usually only important in the types of spectroscopy that give very small inherent linewidths, such as ESR and NMR, where it can be very informative.

FURTHER READING

I.D. Campbell and R.A. Dwek, 1984, 'Biological spectroscopy', Benjamin Cummings—*an accessible account of various forms of spectroscopy with numerous biochemical examples.*

D. Sheehan, 2000, 'Physical biochemistry: principles and applications', Wiley—*an intermediate level treatment of biophysical techniques.*

K.E. van Holde, W.C. Johnson and P.S. Ho, 1998, 'Principles of physical biochemistry', Prentice Hall—*an advanced account of the full range of biochemically important physical methods.*

PROBLEMS

1. Explain why a sodium vapour lamp appears yellow whereas sodium vapour viewed by transmitted light appears blue.

2. Discuss, briefly, the principles behind using spectroscopy to
(a) Measure the concentration of a protein in solution.

(b) Determine the amount of copper bound to a membrane protein.

(c) Follow the binding of a ligand to the surface of a protein, the binding site being proposed to include a tryptophan residue.

3. Melittin, a component of bee venom, is a small amphipathic peptide that contains a single tryptophan residue. On adding NaCl to a solution of melittin in water, the frequency of the tryptophan emission peak shifts from 360 nm to 348 nm. What can you deduce from this?

4. The Beer–Lambert law may break down if the chromophore is present at very high concentrations, i.e. absorbance does not increase linearly with concentration. Why might this be?

5. Absorbance measurements can be made most accurately when I_λ, the intensity of the transmitted radiation, is approximately half of I_o, the incident radiation. By what factor would you dilute an approximately 1 mM solution of NADH, $\varepsilon = 6{\cdot}22 \times 10^3\ M^{-1}\ cm^{-1}$, to measure its concentration accurately in a cell with a 1 cm path length?

6. A solution of reduced cytochrome c, concentration 10 μM, in a 1 cm path length cell absorbs 44% of the incident radiation at 550 nm. On oxidation the solution absorbs 17% of the incident radiation. Calculate the molar extinction coefficients of the two forms of cytochrome c.

A mixture of oxidised and reduced cytochrome c in a 1 cm cell, total protein concentration of 10 μM, absorbed 38% of the incident radiation at 550 nm. Calculate the concentrations of the oxidised and reduced species present.

7. The absorbance, in a cell of 1 cm path length, of a solution containing NAD^+ and NADH is 0·21 at 340 nm and 0·85 at 260 nm. The extinction coefficient of both NAD^+ and NADH at 260 nm is $1{\cdot}8 \times 10^4\ M^{-1}\ cm^{-1}$, while the value at 340 nm for NADH is $6{\cdot}22 \times 10^3\ M^{-1}\ cm^{-1}$. NAD^+ does not absorb at 340 nm. Calculate the concentrations of NAD^+ and NADH in the solution.

8. The ionisation of *ortho*-nitrotyrosine produces a yellow colour

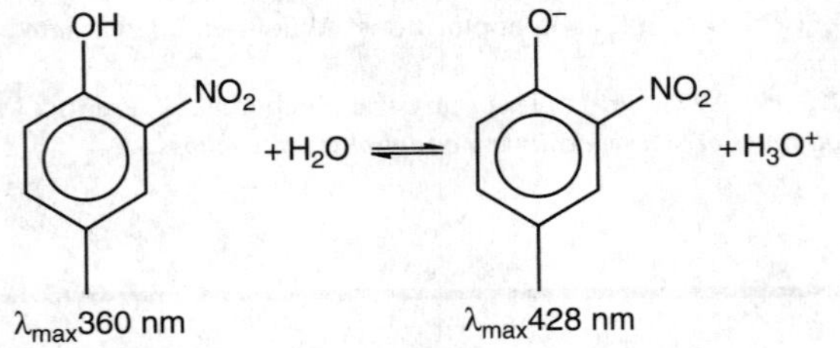

OH, NO_2 $+ H_2O \rightleftharpoons$ O^-, NO_2 $+ H_3O^+$

λ_{max} 360 nm $\qquad$ λ_{max} 428 nm

The following titration data were obtained for (a) nitrotyrosine and (b) the nitrotyrosine group in a sample of nitrated glutamate dehydrogenase.

(a) pH	$\varepsilon_{428} \times 10^{-5}$/cm^2 mol^{-1}	(b) pH	$\varepsilon_{428} \times 10^{-5}$/cm^2 mol^{-1}
4·6	1·6	5·6	2·0
5·4	3·6	6·4	3·8
6·0	7·0	7·2	7·6
6·6	16·0	7·8	16·0
7·0	27·6	8·2	28·0
7·5	37·0	8·8	34·6
8·6	41·0	9·6	37·6
10·0	42·0	10·8	39·0

Determine the pK_a values of the nitrotyrosine in these two cases and suggest why they might differ.

9. The haemoglobin (Hb) content of a solution can be determined by treating the solution with a reagent solution containing an excess of ferricyanide and cyanide. This converts both Hb and oxy-Hb to cyanomet-Hb, the concentration of which can be determined from its absorbance at 540 nm. The following results were obtained with a standard Hb solution of 0·6 g dm^{-3}.

Volume(cm^3) of standard Hb solution added to reagent (total volume = 5 cm^3)	Absorbance at 540 nm
0	0·025
1	0·090
2	0·160
3	0·230
4	0·290

When 0·01 cm^3 of a blood sample was added to 5 cm^3 of the reagent solution, the absorbance at 540 nm was 0·19. What is the concentration of Hb in the blood sample?

10. The graph below gives a plot of extinction coefficient versus wavelength of the incident radiation in the visible region for oxidised and reduced cytochrome c.

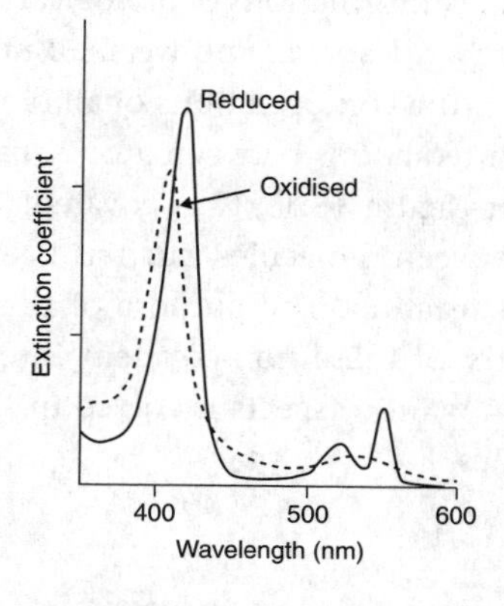

Why are there discrete absorption peaks? Give a brief qualitative explanation for (i) why the extinction coefficients are different at the different wavelengths and (ii) why the plots for oxidised and reduced cytochrome c are different.

18 Non-covalent interactions and macromolecular structure

KEY POINTS

- Non-covalent interactions include *dispersion forces, steric repulsion* and *hydrogen bonds.*
- A *molecular forcefield* is an equation that allows the energy of a molecule or a collection of molecules to be calculated and such calculations form the basis of molecular modelling.
- The *hydrophobic effect* describes the tendency of non-polar groups to aggregate in water and it arises from the disruption of the non-covalent interactions between water molecules.
- The energetics of the folding of macromolecules in water is dominated by the hydrophobic effect, but the specific conformation of the folded state is determined by the non-covalent interactions between different parts of the macromolecule.
- The folding and unfolding of macromolecules are cooperative processes.

There are many extensive textbooks that cover the three-dimensional structures and conformations of biological macromolecules, such as proteins, DNA, oligosaccharides, etc., and we shall not attempt to compete with these here. However, the conformations of all biological macromolecules are determined by the interactions between atoms, both covalent (see Chapter 16) and non-covalent (figure 18.1). Non-covalent interactions also determine all interactions between molecules, such as ligand binding to an enzyme and the subsequent reaction. The purpose of this chapter is to cover the basic physical chemistry of these non-covalent interactions and to deal with some of the thermodynamic aspects involved in the formation of well-structured macromolecules.

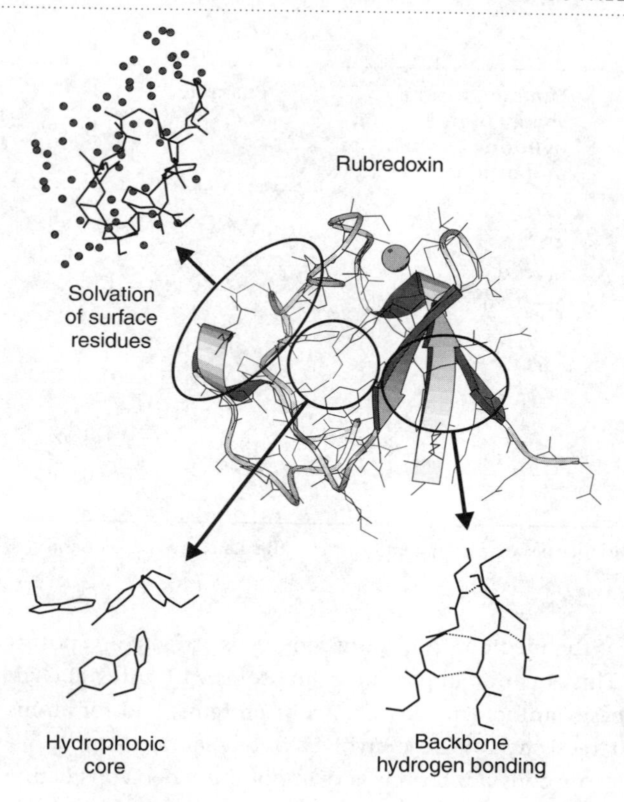

Fig. 18.1 Crystal structure of the electron transfer protein rubredoxin, highlighting some of the non-covalent interactions that determine the three-dimensional structure of the protein.

Non-covalent interactions between atoms

The basic types of simple non-covalent interactions that can occur between atoms are summarised in Table 18.1 and we will deal with each of these in turn.

Electrostatics

Two atoms with permanent electrical charges will interact electrostatically with each other. If they have the same charge they will repel each other, and if they have opposite charges they will attract each other. The energy of this interaction is given by

$$E = \frac{1}{4\pi\varepsilon_o\varepsilon}\frac{q_i q_j}{r_{ij}}$$

where q_i and q_j are the charges of the atoms, r_{ij} is the distance between them, ε_o is the vacuum permittivity and ε is the relative permittivity, or dielectric

TABLE 18.1

Type of non-covalent interaction		Equation for the energy of interaction (ignoring constants of proportionality)	Example	Typical strength (kJ mol^{-1})
Electrostatic	monopole : monopole	$\frac{q_i q_j}{\varepsilon r_{ij}}$	$-COO^- \cdots {}^+H_3N-$	−30
	dipole : dipole	$\sum_{\text{all pairs}} \frac{q_i q_j}{\varepsilon r_{ij}}$	$>C{=}O \cdots O{=}C<$	+1
Dispersion		$-\frac{B_{ij}}{r_{ij}^6}$	$C{-}H \cdots H{-}C$	−0·1
Steric repulsion		$+\frac{A_{ij}}{r_{ij}^{12}}$	$C{-}H \cdots H{-}C$	Very strong
Hydrogen bond		$\frac{q_i q_j}{\varepsilon r_{ij}} + \frac{C_{ij}}{r_{ij}^{12}} - \frac{D_{ij}}{r_{ij}^{10}}$	$>N{-}H \cdots O{=}C<$	−10

q_i is the whole or partial charge on atom i, r_{ij} the distance between atom i and atom j, ε the dielectric constant and A, B, C and D are constants

TABLE 18.2

Medium	ε
Vacuum	1
Centre of a protein	4
Water	80

constant, of the medium. A negative energy is attractive, a positive energy is repulsive. This equation applies both for atoms with integral charges, such as acidic or basic amino acid side chains in proteins, and for atoms with non-integral charges, such as those involved in a polar bond.

The dielectric constant term is very important when considering the effects of electrostatic interactions in macromolecules. Typical values for ε are given in Table 18.2. Electrostatic interactions on the surface of macromolecules are much less important than, and to a first approximation can be ignored when compared to, electrostatic interactions in the middle of macromolecules because of the high value of ε for the former and the low value of ε for the latter.

In the middle of macromolecules it is unusual to find charged atoms but it is very common to find atoms as part of dipoles, i.e. two bonded atoms where one has a slight positive charge and the other a slight negative charge (figure 18.2). The above equation can be used to calculate the energy of interaction of two dipoles by summing the electrostatic interactions over all the pairs of atoms that are not involved in covalent bonds.

WORKED EXAMPLE

Calculate the dipolar interaction energy of an −NH group and a >C=O group in an arrangement typical of a protein *β*-sheet (figure 18.3). The partial charges on the atoms in units of proton charge (*e*) are

$$q_N = -0{\cdot}46 \qquad q_H = +0{\cdot}25 \qquad q_O = -0{\cdot}50 \qquad q_C = +0{\cdot}62$$

and the distances between the atoms in nm are

$$r_{NO} = 0{\cdot}292 \quad r_{NC} = 0{\cdot}414 \quad r_{HO} = 0{\cdot}196 \quad r_{HC} = 0{\cdot}314$$

Assume the dielectric constant $\varepsilon = 1$ and $\varepsilon_o = 8{\cdot}854 \times 10^{-12}\,J^{-1}\,C^2\,m^{-1}$. Comment on this value compared to the electrostatic interaction between the H and O only.

SOLUTION: The electrostatic interaction between two charges is given by

$$E = \frac{1}{4\pi\varepsilon_o\varepsilon}\frac{q_i q_j}{r_{ij}}$$

For the interaction between two dipoles this becomes

$$E = \sum_{i<j}\frac{1}{4\pi\varepsilon_o\varepsilon}\frac{q_i q_j}{r_{ij}}$$

where the summation is taken over all the non-bonded pairs of atoms. In the example given,

$$E = E(N\cdots O) + E(N\cdots C) + E(H\cdots O) + E(H\cdots C)$$

Thus,

$$E = \frac{1}{4\pi\varepsilon_o\varepsilon}\left(\frac{q_N q_O}{r_{NO}} + \frac{q_N q_C}{r_{NC}} + \frac{q_H q_O}{r_{HO}} + \frac{q_H q_C}{r_{HC}}\right)$$

The absolute charges on the atoms are given by the partial charges times the charge on a proton ($1{\cdot}602 \times 10^{-19}$ C) and the distances need to be converted into metres. Putting in the numbers gives

$$E = -10{\cdot}4 \times 10^{-21}\,J$$

This value gives the energy of interaction for one pair of dipoles. To convert this into energy per mole of dipoles it is necessary to multiply by Avogadro's number.

$$E = -3{\cdot}2\,kJ\,mol^{-1}$$

COMMENT: If we only consider the interaction between the H and the O, $E(H\cdots O)$, and ignore all the other terms

$$E(H\cdots O) = -47{\cdot}8\,kJ\,mol^{-1}$$

The electrostatic influence of the N and C atoms reduces the overall energy by more than a factor of 10 compared to the monopolar $E(H\cdots O)$ value. Thus the dipolar interaction between a carbonyl group and an amide group cannot be treated as an electrostatic interaction between the oxygen and hydrogen alone.

It is also possible to derive equations for the energy of interaction of two dipoles. The equation for the energy of interaction of parallel dipoles is

$$E = \frac{\mu_i \mu_j}{4\pi\varepsilon_o\varepsilon}\frac{(3\cos^2\theta_{ij} - 1)}{r_{ij}^3}$$

where μ_i and μ_j are the two dipole moments, r_{ij} is the distance between the dipoles and θ_{ij} is the angle between the direction that the dipoles point and the line joining the two dipoles. If the dipoles are not parallel then the equation becomes more complex but still depends on r^{-3}. Thus dipolar interactions are much more sensitive to changes in distance than monopolar interactions, which depend on r^{-1}.

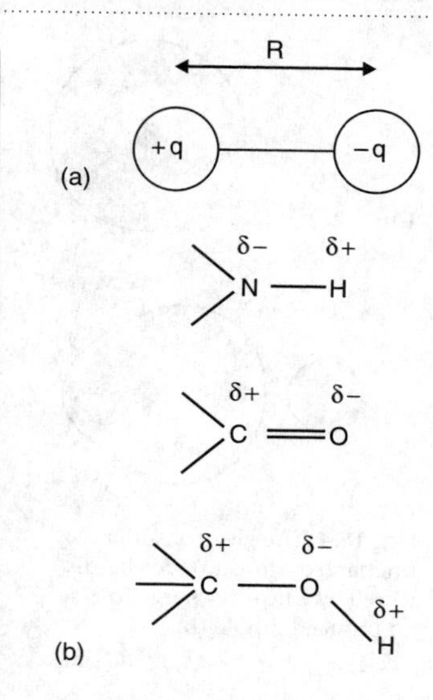

Fig. 18.2 (a) A dipole consists of a positive and a negative charge held at a fixed distance apart. The dipole moment is given by $q \times R$. (b) Typical dipoles found in biological macromolecules.

δ− δ+ δ− δ+
N—H······O═C

Fig. 18.3 A linear hydrogen bond. Typical hydrogen bonds in antiparallel β-sheets are not quite linear.

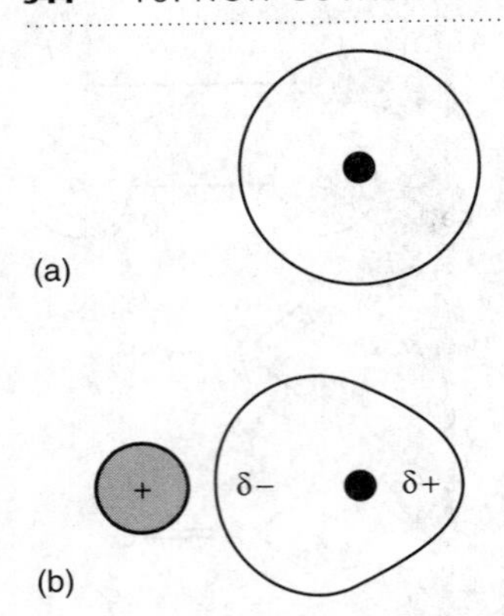

Fig. 18.4 The electrons in an uncharged atom (a) can be distorted by a nearby charge to give an induced dipole (b).

If an atom with a permanent charge or dipole is close to an atom with no charge then the former can induce a dipole in the latter by distorting its electron cloud (figure 18.4). This will give rise to an electrostatic interaction between the charge and the induced dipole. All interactions involving induced dipoles will be attractive because the permanent charge will always induce a region of opposite charge closest to itself. These effects will be much weaker than ordinary electrostatic interactions because the induced dipoles will be smaller.

Dispersion forces

For an atom, or group of atoms, with no permanent charge or dipole, random fluctuations in the electron cloud will produce temporary dipoles. Over time these dipoles will average to zero. However, temporary dipoles can induce complementary temporary dipoles in other non-polar atoms and molecules (figure 18.5). This leads to attractive induced-dipole:induced-dipole interactions called *dispersion forces*. The energy of this interaction is given by

$$E = -\frac{B_{ij}}{r_{ij}^{6}}$$

where B_{ij} is a constant for a given pair of atoms and depends on their polarisabilities. This effect is relatively weak and short range. It applies to all atoms and for non-polar atoms it is the only form of attractive interaction.

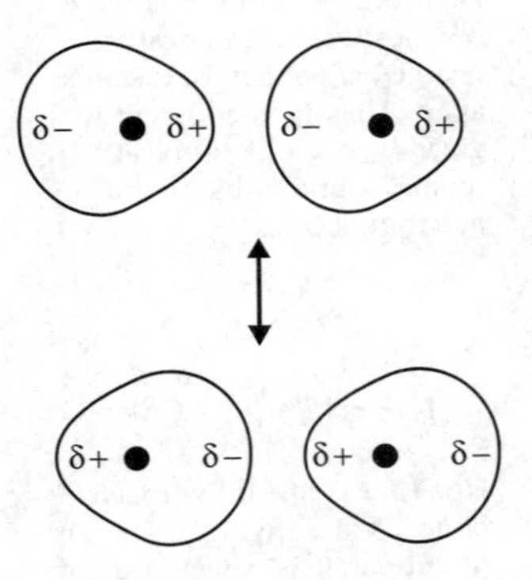

Fig. 18.5 Dispersion forces. Random fluctuations in the electrons in one atom can give a temporary dipole that induces a complementary dipole in another atom. The electrons in the two atoms fluctuate together.

Steric repulsion

When any two atoms, charged or uncharged, get very close together there is direct repulsion between their electron clouds. The energy of this interaction is given by

$$E = +\frac{A_{ij}}{r_{ij}^{12}}$$

where A_{ij} is a constant for a given pair of atoms and depends on the number of electrons on each atom. This effect is very short range and very strong.

Putting together the equations for steric repulsion and dispersion forces gives the following equation.

$$E = \frac{A_{ij}}{r_{ij}^{12}} - \frac{B_{ij}}{r_{ij}^{6}}$$

This equation is known as the Lennard-Jones potential and describes the interaction between two non-polar groups (figure 18.6). This can be extended

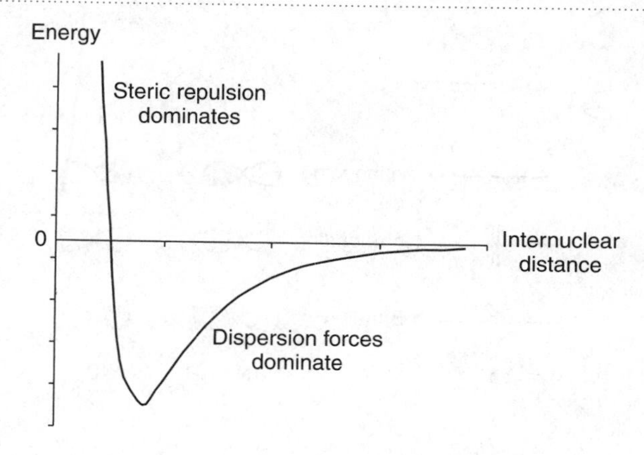

Fig. 18.6 Leonard-Jones potential curve for the interaction of two uncharged species.

to polar groups by including the electrostatic interactions as well.

$$E = \frac{1}{4\pi\varepsilon_o\varepsilon}\frac{q_i q_j}{r_{ij}} + \frac{A_{ij}}{r_{ij}^{12}} - \frac{B_{ij}}{r_{ij}^{6}}$$

This equation is known as the van der Waals potential. It applies to both polar and non-polar pairs of atoms and gives the total interaction energy between any pair of atoms that are not covalently bonded, with the exception of those involved in hydrogen bonds which are considered next.

Hydrogen bonds

A hydrogen bond occurs between a hydrogen atom that is bonded to an electronegative element called the donor, usually N or O, and a second electronegative element called the acceptor. This gives the hydrogen a high positive partial charge and the acceptor a high negative partial charge, so we would expect strong electrostatic interactions (figure 18.7a). However, hydrogen bonds are slightly stronger and slightly shorter than predicted by simple electrostatic interactions. The calculation of the dipolar energy in the worked example above gives the energy of an –N–H···O=C< hydrogen bond as $-3{\cdot}2\,\text{kJ mol}^{-1}$ whereas typical values are $-5\,\text{kJ mol}^{-1}$ or larger. This is because there is an additional small covalent contribution resulting from overlap of the hydrogen orbitals with those of the acceptor (figure 18.7b). Such a covalent interaction strengthens and shortens the bond but only contributes significantly at very short distances where orbital overlap is significant. It is thus only important in strong hydrogen bonds. A second consequence of this covalent interaction is that strong hydrogen bonds have a more marked preference for linearity (for example, a N–H···O angle of 180°) compared to other dipolar interactions.

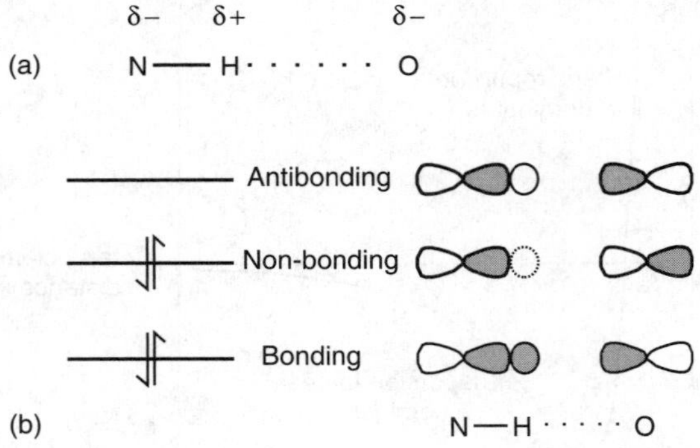

Fig. 18.7 A hydrogen bond consists of a strong electrostatic attraction (a) with an additional smaller covalent contribution (b) at short distances. [The dotted line indicates that the relative phase of the orbital is not important.]

WORKED EXAMPLE

Why are antiparallel β-sheets more common in folded proteins than parallel β-sheets?

SOLUTION: β-sheet structures in proteins are composed of extended peptide strands held together by interstrand hydrogen bonds between backbone –NH and backbone >C=O groups. In parallel β-sheets all the strands run in the same direction, in antiparallel β-sheets they run in alternate directions (figure 18.8). Both parallel and antiparallel β-sheets have the same number of interstrand backbone hydrogen bonds per residue; however, in antiparallel sheets these are linear whereas in parallel sheets they are bent. As linear hydrogen bonds are stronger than bent ones, the structure with linear bonds is more stable and thus more common.

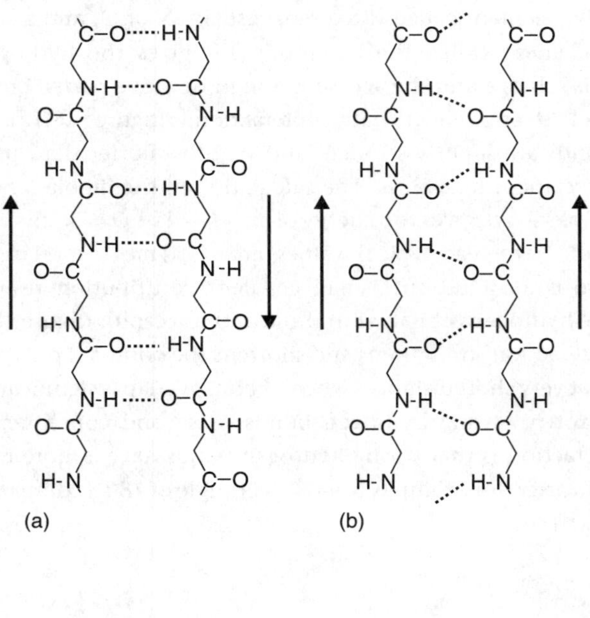

Fig. 18.8 Hydrogen bonding patterns in (a) antiparallel and (b) parallel β-sheets.

In practice, an empirical approach is used to calculate the strength of any given hydrogen bond. The electrostatic contribution can be calculated as above and additional terms are added to deal with the extra attraction and repulsion due to the covalent contribution at short distances. The energy of interaction between two atoms involved in a hydrogen bond is then

$$E = \frac{1}{4\pi\varepsilon_o\varepsilon}\frac{q_i q_j}{r_{ij}} + \frac{C_{ij}}{r_{ij}^{12}} - \frac{D_{ij}}{r_{ij}^{10}}$$

where C and D are constants for a given pair of atoms. These extra two terms typically make a contribution of 2 kJ mol^{-1} or less.

Calculating the energies of molecules

The total energy of a molecule can be calculated by summing all the interactions, covalent, non-covalent and hydrogen bonds, between all pairs of atoms in the molecule and between the atoms in the molecule and its environment, usually the solvent. An equation to calculate the total energy of a molecule is called a *molecular forcefield*. Just to see how one is constructed, a typical forcefield (the AMBER forcefield) is given in Table 18.3. It is suggested that students do not try and memorise this. The important point to stress is that each different type of energy can be calculated and these summed to give the total energy of the molecule.

An equation such as this involves a large number of parameters that have to be set, such as the force constants involved in the covalent interaction terms and the constants A, B, C and D involved in the non-covalent interaction terms. Sets of parameters have been produced from quantum mechanical calculations to optimise this type of calculation for specific classes of molecules, and the AMBER forcefield is generally used for calculations on proteins.

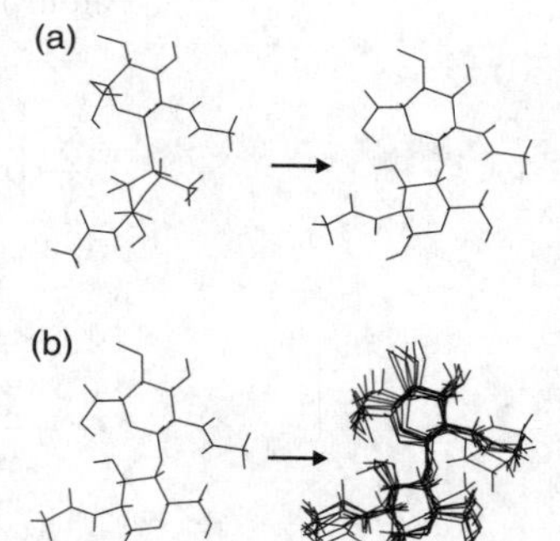

Fig. 18.9. (a) Energy minimisation. The left-hand structure is a distorted, high-energy conformation of chitobiose. By adjusting the structure and monitoring the total energy, the lowest energy and thus most stable conformation (right) can be found. (b) Molecular dynamics. Starting from a low-energy conformation of chitobiose (left) at a given temperature, the molecule is allowed to move. The picture on the right shows the overlay of several structures at different time points and gives a measure of the flexibility of different regions of the structure.

Applications of molecular forcefields

There are two major applications of energy calculations, energy minimisation and molecular dynamics.

- Energy minimisation. Most molecules can exist in many different conformations. The most stable of these is the conformation that gives the lowest overall energy. If we start from a given three-dimensional structure for a molecule, then we can use a molecular forcefield to calculate its total energy. We can then adjust the positions of the atoms slightly, recalculate the total energy and compare that to the starting structure. This procedure can be repeated, usually many thousands of times, until the lowest energy structure is found (figure 18.9a).

Table 18.3

Total energy	**Covalent energy terms**		
	Bond stretching	Bond bending	Bond rotation
$E_{tot} =$	$\sum_{bonds} K_b(b - b_{eq})$ +	$\sum_{angles} K_\theta(\theta - \theta_{eq})$ +	$\sum_{dihedrals} \frac{V_n}{2}[1 + \cos(n\phi - \gamma)]$
	b actual bond length b_{eq} equilibrium bond length K_b bond stretching force constant	θ actual bond angle θ_{eq} equilibrium bond angle K_θ bond bending force constant	ϕ actual dihedral angle γ reference dihedral angle V_n energy barrier to rotation n periodicity (no. of rotational minima, eg. 3 for rotation of a CH_3 group)

Non-covalent energy terms (includes interactions with solvent)	
Non-bonded interactions (summed over all pairs of atoms)	Additional hydrogen bond contribution (summed over all pairs involved in hydrogen bonds)
$+ \sum_{i<j} \left[\frac{A_{ij}}{r_{ij}^{12}} - \frac{B_{ij}}{r_{ij}^{6}} + \frac{1}{4\pi\varepsilon_o\varepsilon} \frac{q_i q_j}{r_{ij}} \right]$	$+ \sum_{i<j} \left[\frac{C_{ij}}{r_{ij}^{12}} - \frac{D_{ij}}{r_{ij}^{10}} \right]$
A, B constants for a given pair of atoms i and j q charge or partial charge on the atom r distance between the two atoms ε dielectric constant	C, D constants for a given pair of atoms i and j r distance between the two atoms

This procedure works well for organic molecules and small macromolecules, such as oligosaccharides, but only works for large macromolecules, such as proteins, if the starting structure is fairly close to the final structure. Energy minimisation techniques are frequently combined with experimental methods for determining the structures of macromolecules. Techniques such as X-ray crystallography and nuclear magnetic resonance spectroscopy are capable of determining the three-dimensional structures of large macromolecules but frequently give small errors in bond lengths, bond angles, etc. These can be eliminated by using energy minimisation as the last step in obtaining the structure.

- Molecular dynamics. At normal temperatures, molecules are not static but dynamic systems. All the atoms are moving to some degree. These movements may just be bond vibrations, but they can also include long-range cooperative motions, such as protein unfolding, or transitions between two or more equally stable conformations, as is observed for oligosaccharides. The flexibilities of molecules become particularly important when considering problems such as the conformational changes induced by ligand binding in ligands and proteins. There are few experimental techniques available for investigating these dynamic motions on a submicrosecond time scale but this can be done using theoretical calculations.

  We start with a given conformation of the molecule at a given total energy (determined by the temperature) and we can use a molecular forcefield to calculate the force on each individual atom in the molecule. We can then use Newton's laws of motion to calculate where each atom will be after a very small time increment (of the order of a picosecond). The whole process is then repeated using the new conformation. This gives us a picture of the molecule at each successive time point (just like the frames of a movie). We can run these together or overlay them to produce a picture of how the atoms in the molecule move (figure 18.9b).

  This technique is an example of where classical mechanics has to be used, because using quantum theory instead of Newton's laws makes the calculations far too complex to perform.

These two techniques can be applied to molecular complexes, as well as single molecules, and form the basis of rational drug design. A potential drug molecule can be designed on a computer and its conformation predicted. It can then be docked into the active site of a target enzyme and the strength of the interaction between the two calculated. Chemical modifications of the molecule can then be proposed and predictions made about whether these enhance binding or not. These predictions are never completely accurate but can speed up the process of drug development by suggesting the best compounds to test. This approach is now an integral part of most drug development programmes.

The role of solvent interactions: the hydrophobic effect

If a non-polar species, such as chloroform, is mixed with water then it tends to aggregate rather than dissolve. In the case of chloroform and water, the two solvents are completely immiscible. This tendency of non-polar molecules to aggregate in water is called the *hydrophobic effect*. Examples in biological systems include folded proteins, where the non-polar side chains cluster in the centre of the molecule, and formation of lipid vesicles and membranes, where the non-polar regions of lipids cluster away from water.

In general, non-polar molecules interact more strongly with water than with each other because dipole:induced-dipole interactions are stronger than dispersion forces. The non-polar molecules are also more ordered on aggregation. Thus, both the enthalpic and entropic terms for the non-polar molecules oppose aggregation. The origin of the hydrophobic effect is therefore not to do with interactions between the non-polar molecules, but in the molecular nature of water itself and the effect that non-polar species have on its structure.

Water molecules are polar, with the oxygen atom carrying a large partial negative charge and the hydrogen atoms partial positive charges. The molecules in pure water are strongly hydrogen bonded to each other. In order to solvate another species, some of these hydrogen bonds have to be broken. Polar solute molecules interact strongly with water molecules and so the loss of solvent hydrogen bonding can be offset. However, non-polar solute molecules interact more weakly with the water molecules, leading to an overall loss of bonding and an unfavourable change in enthalpy on solvation. In order to maximise the remaining hydrogen bonds, the water molecules surrounding the non-polar solute orient themselves to hydrogen bond with the rest of the solvent. This leads to an ordered cage-like structure around the solute molecule with an accompanying unfavourable change in entropy. Both the unfavourable enthalpic and entropic terms can be minimised by minimising the surface area of the non-polar solute in contact with the water. This leads to the observed aggregation.

The hydrophobic effect is not a force between atoms or molecules. It is the name given to a macroscopic phenomenon that occurs with some solutes in water as a result of the atomic level interactions dealt with above. It is nevertheless a useful concept, as long as it is only used as a shorthand description.

Thermodynamic stability of structured macromolecules

When considering the thermodynamic stability of a well-defined three-dimensional structure versus a random structure for a macromolecule, we need to consider the enthalpy and entropy changes for both the macromolecule and the solvent. For the following discussion we shall use the folding of proteins as an example. However, the same principles apply to the folding of all macromolecules.

A single polypeptide chain can often exist in many different conformations. However, we shall only consider three distinct states here (figure 18.10).

- Unfolded. The structure of the chain is completely random and highly flexible with all amino acid side chains fully exposed to the solvent.
- Molten globule. The peptide chain is not fully random but has all the hydrophobic amino acid side chains clustered in the centre away from the solvent. However, there is no well-defined structure.
- Folded. A single, well-defined conformation of the peptide that maximises non-bonded interactions.

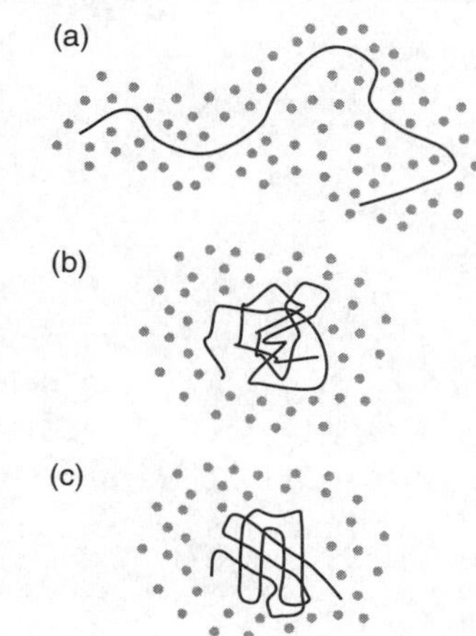

Fig. 18.10. Three possible folding states for a protein. (a) Fully unfolded. The whole chain is solvated. (b) Molten globule. The protein structure is not well defined but hydrophobic groups are on the inside and hydrophilic groups are on the outside. (c) Fully folded. The protein structure is now well ordered, maximising non-bonded interactions. The transition from (a) to (b) involves considerable reorganisation of both the solvent and the protein. The transition from (b) to (c) involves considerable reorganisation of the protein but relatively minor rearrangements of the solvent. Most proteins can exist in states (a) and (c) depending on the conditions. A few proteins (for example α-lactalbumin) show a well-defined molten globule state under specific conditions.

The molten globule is not an intermediate in the folding pathway of proteins. However, some proteins have stable molten globule states under certain conditions and comparisons with this state are useful for investigating the role that different types of interactions play in the stability of folded proteins. We shall take the unfolded molecule as our reference point and consider the changes in enthalpy and entropy that occur on folding to the molten globule or folded state.

To start with, we shall consider the protein in the absence of solvent (as we did in Chapter 2). The transition from the unfolded to the folded state has a relatively high negative ΔH, as the folded form maximises non-bonded interactions, and a high negative ΔS, as the folded form is very ordered. The transition to the molten globule form has a lower negative ΔH as there are few specific dipolar interactions or hydrogen bonds, just non-specific dispersion forces, and a lower negative ΔS as this form is not very well ordered. In the absence of solvent, ΔH favours the folded form and ΔS favours the unfolded form (see Chapter 2).

If we now consider the effects of hydration on the protein the situation becomes more complex. In the unfolded state the whole chain is solvated. All the hydrogen bond donors and acceptors of the protein interact with the solvent and so there are comparable numbers of hydrogen bonds formed by the protein residues in both the unfolded and folded states. The folded form also maximises dispersion forces and so ΔH for the transition from the unfolded to the folded state is negative but small. In the molten globule form, many of the hydrogen bonds with the solvent have been lost but very few new ones created and so ΔH for the transition from the unfolded to the molten globule state is positive. The ΔS terms for the different states do not change much relative to the unhydrated case above as these are dominated by the randomness of the peptide chain.

Lastly we need to consider the solvent itself and the hydrophobic effect. We shall take the solvent in the presence of unfolded protein as our reference point. In the unfolded form of the protein large numbers of water molecules are needed to solvate hydrophobic side chains of the protein, leading to weak hydrogen bonding and increased solvent organisation. In the molten globule and folded states, all the hydrophobic groups are buried and so there is an increase in the net hydrogen bonding of the solvent, giving a negative ΔH, and a decrease in solvent order, giving a positive ΔS. There is very little effect on the solvent going from the molten globule to the fully folded state.

These effects are summarised in Table 18.4. At room temperature in water, the unfolded state of the protein is unstable with respect to the molten globule state because of the changes in solvent enthalpy and particularly entropy. The hydrophobic effect drives the collapse to a molten globule state but not the adoption of a well-defined structure. The molten globule is unstable with respect to the folded state because of the favourable protein enthalpy change associated with maximising non-bonded interactions. These together overcome the unfavourable protein entropy term.

All these values are very strongly temperature dependent (see Chapter 3). In particular, the hydrogen bond network of water becomes weaker as the temperature increases and so the hydrophobic effect becomes less important. At high temperatures the dominant terms become the entropy associated with the protein chain and so the protein unfolds.

This discussion has only focused on the thermodynamics of protein folding by considering three possible folding states. Proteins usually fold through several intermediates whose structures are more complex than the molten globule and vary considerably from protein to protein. However, the above analysis comparing the unfolded and folded states still applies.

Similar effects are seen for other macromolecules and in general:

> Solvent entropy changes drive hydrophobic collapse to a partially disordered state. Macromolecular non-bonded interactions drive the formation of a single well-ordered structure.

TABLE 18.4

Change of state	In vacuo		In aqueous solution			
	Protein component		Protein component		Solvent component	
	ΔH	ΔS	ΔH	ΔS	ΔH	ΔS
Unfolded → Molten globule	– small	– medium	+ medium	– medium	– medium	**+ large**
Molten globule → Folded	– medium	– medium	**– medium**	– medium	zero	zero

[shaded] = Most significant terms for driving protein folding in solution at room temperature.

Cooperativity of macromolecular folding and unfolding

The above analysis of the thermodynamics of folding has compared the relative stabilities of three very different states for the protein. Many more partially or nearly fully folded states could also be proposed. Some of these might be expected to have free energies very similar to that of the folded state, for instance differing by one hydrogen bond. We might then expect to observe a Boltzmann distribution among these partially and fully folded states. However, most proteins fold to give a single well-defined structure. This is because protein folding and unfolding are cooperative processes.

- Protein folding. As a protein folds, more and more non-covalent interactions occur between different parts of the peptide chain. As the number of these interactions increase, each subsequent interaction becomes stronger. Thus, the folding process becomes more energetically favourable as folding progresses and drives the protein towards a fully folded state.
- Protein unfolding. When the first non-covalent interaction is broken, all the remaining interactions become weaker. It becomes easier to break the second non-covalent interaction and so on. Thus, once the process is started it will not stop until the chain has fully unfolded.

The result of this is that proteins are usually either fully folded or fully unfolded, depending on the conditions. Partially folded states mostly occur only as intermediates in the folding or unfolding process.

The origin of this behaviour is in the entropy changes of the protein chain during folding. An unfolded protein has a random chain conformation with very high entropy. On forming the first non-covalent interaction a large amount of this entropy is lost, the chain is no longer random but joined together at one point. On forming the second non-covalent interaction much less entropy is lost because the chain is much less random to start with (figure 18.11). Making the final few non-covalent interactions involves almost no loss of entropy because the chain is almost fully folded at this stage. Thus, each subsequent step is energetically easier than the last. Similarly, on unfolding, each subsequent step gives a greater gain in entropy.

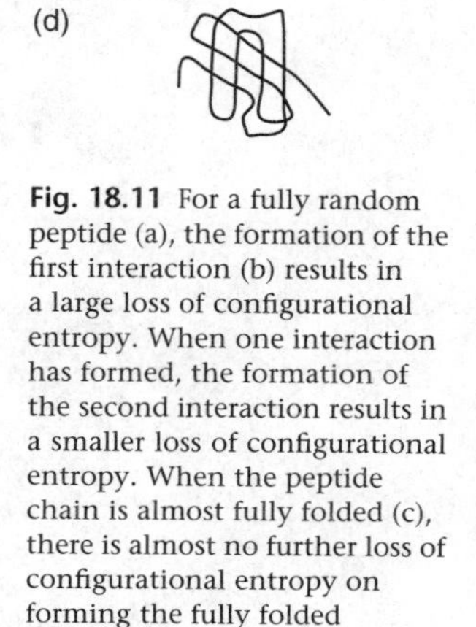

Fig. 18.11 For a fully random peptide (a), the formation of the first interaction (b) results in a large loss of configurational entropy. When one interaction has formed, the formation of the second interaction results in a smaller loss of configurational entropy. When the peptide chain is almost fully folded (c), there is almost no further loss of configurational entropy on forming the fully folded conformation (d).

An alternative approach, and one that can be used to generate a quantitative model, is to consider the intramolecular equilibrium constant for two groups involved in forming a non-covalent interaction (figure 18.12). The proximity of two groups present on the same chain will favour intramolecular interactions over intermolecular interactions because it results in an increase in the local concentration of one group with respect to the other. This can be described by the equation

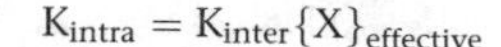

$$K_{intra} = K_{inter}\{X\}_{effective}$$

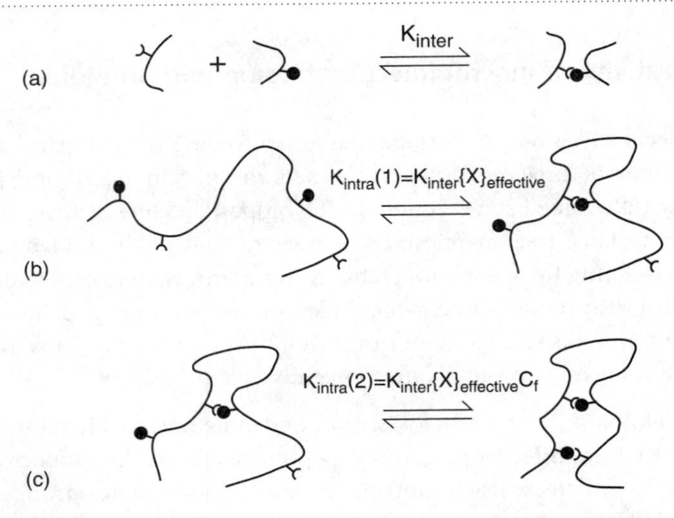

Fig. 18.12 (a) The interaction between two groups on different molecules is characterised by an equilibrium constant K_{inter}. (b) If these two groups are on the same chain, the equilibrium constant is changed because the groups are held closer together. This can be quantified as the effective local concentration of the groups relative to each other, $\{X\}_{effective}$. (c) As the chain folds, the average separations between the remaining groups decrease and so their effective concentrations increase. This can be modelled by using a cooperativity factor, C_f.

where K_{intra} is the intramolecular equilibrium constant for the interaction of two groups on the same molecule and K_{inter} is the intermolecular equilibrium constant for the interaction of the same two groups on different molecules. $\{X\}_{effective}$ is the factor describing the effect of the proximity of the two groups on their relative concentrations and is called the *effective concentration*. The effective concentration depends on the average distance between the groups, the further apart two groups are on average the lower their effective concentration. $\{X\}_{effective}$ usually has a value between 10^{-2} and 10^{-5} M depending on the size of the chain, the larger the chain the smaller the value of $\{X\}_{effective}$. In an extended peptide, the groups are at the maximum possible separation and so have the lowest possible effective concentration. As the protein folds the groups get closer together and so the effective concentration increases, increasing K_{intra} (figure 18.12). This can be modelled using the equation

$$K_{intra}(n) = K_{inter}\{X\}_{effective}C_f^{(n-1)}$$

where $K_{intra}(n)$ is the intramolecular equilibrium constant for the formation of the nth non-covalent interaction and C_f is the cooperativity factor. The overall equilibrium constant, $K_{intra}(\text{overall})$, for the formation of n interactions is given by the product of all the individual $K_{intra}(n)$ values.

$$K_{intra}(\text{overall}) = K_{intra}(1)K_{intra}(2)K_{intra}(3)\ldots K_{intra}(n)$$

WORKED EXAMPLE

A peptide chain can fold to form a maximum of 20 non-covalent interactions. Each interaction has an intermolecular association constant, K_{inter}, of 1×10^{-4} (M^{-1}) and the effective concentration of groups on the chain is 1×10^{-3} M. Determine whether the folded state is stable with a cooperativity factor of 1, 5 and 10. How would your conclusions change if the peptide chain contained more residues?

SOLUTION: $K_{intra}(n)$ can be calculated for values of n from 1 to 20 for the different values of C_f using the equation

$$K_{intra}(n) = K_{inter}\{X\}_{effective} C_f^{(n-1)}$$

and K_{intra}(overall) determined at each value of n by multiplying together the $K_{intra}(n)$ values. This gives

n	$C_f = 1$		$C_f = 5$		$C_f = 10$	
	$K_{intra}(n)$	K_{intra} (overall)	$K_{intra}(n)$	K_{intra} (overall)	$K_{intra}(n)$	K_{intra} (overall)
1	1×10^{-7}	1×10^{-7}	1×10^{-7}	1×10^{-7}	1×10^{-7}	1×10^{-7}
2	1×10^{-7}	1×10^{-14}	5×10^{-7}	5×10^{-14}	1×10^{-6}	1×10^{-13}
3	1×10^{-7}	1×10^{-21}	$2{\cdot}5 \times 10^{-6}$	$1{\cdot}25 \times 10^{-19}$	1×10^{-5}	1×10^{-18}
↓	↓	↓	↓	↓	↓	↓
19	1×10^{-7}	1×10^{-133}	$3{\cdot}81 \times 10^{5}$	$3{\cdot}34 \times 10^{-14}$	1×10^{11}	1×10^{38}
20	1×10^{-7}	1×10^{-140}	$1{\cdot}91 \times 10^{6}$	$6{\cdot}37 \times 10^{-8}$	1×10^{12}	1×10^{50}

$\Delta G°$ for folding is given by

$$\Delta G° = -RT \log_e K_{intra}(\text{overall})$$

For the folded form of the peptide to be stable, $\Delta G°$ must be negative and so K_{intra} (overall) after all 20 interactions have formed must be greater than one. Plots of $\Delta G°$ versus n are shown in figure 18.13. Thus, the folded form of the peptide would be unstable if $C_f = 1$ or 5 but would be stable if $C_f = 10$.

If we increase the size of the peptide chain, we increase the maximum number of non-covalent interactions that occur. For $C_f = 1$, where there is no cooperativity, $\Delta G°$ becomes increasingly positive as n increases and so increasing the maximum value of n still gives an unstable folded state. For $C_f = 5$ and 10, $\Delta G°$ starts by increasing as n increases, reaches a maximum and then decreases with increasing n. If we increase the maximum value of n, the folded state becomes increasingly more stable with respect to the unfolded state. For $C_f = 5$, $\Delta G°$ becomes negative after 23 interactions have formed and after this the folded state of the protein is more stable than the unfolded state.

COMMENT: This explains why proteins are large molecules. The active site of an enzyme may contain only 10 critical residues or less. It would take much less biochemical energy to build proteins that consisted of these 10 residues and a few others to provide structural support. However, such a small folded polypeptide would not be stable. Macromolecules need to be *macro-* to form stable folded structures.

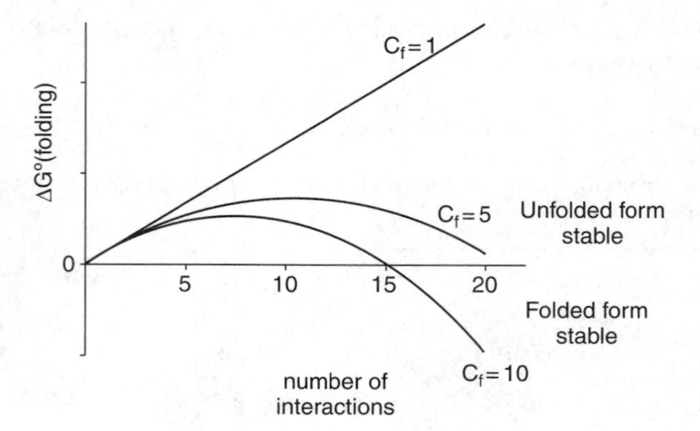

Fig. 18.13 Illustrative plots of $\Delta G°$(folding) versus number of interactions for the folding of a polypeptide with a cooperativity factor, C_f, of 1, 5 and 10. If $\Delta G° < 0$ then the folded form is more stable than the unfolded form.

A macromolecule in a folded conformation that is held together by weak forces, such as hydrogen bonds and van der Waals interactions, is only stable because there is a high degree of cooperativity between all the weak interactions. This cooperativity only occurs because the groups involved are on a single chain and so making one interaction brings other potentially interacting groups closer together. If the chain is broken, for instance by proteolysis, the degree of cooperativity decreases dramatically. In many proteins, it only needs the chain to be broken a few times for the whole protein to unfold.

FURTHER READING

P.W. Atkins, 1998, 'Physical chemistry', Oxford University Press—*a useful description of intermolecular forces.*

C. Branden and J. Tooze, 1999, 'Introduction to protein structure', Garland—*a well-illustrated descriptive account of protein structure.*

T.E. Creighton, 1993, 'Proteins: structure and molecular properties', Freeman—*a comprehensive account of the factors that determine protein structure, including an extensive treatment of the hydrophobic effect and cooperativity in folding.*

W. Saenger, 1984, 'Principles of nucleic acid structure', Springer-Verlag—*a good account of the thermodynamics of DNA and RNA structures.*

K.E. van Holde, W.C. Johnson and P.S. Ho, 1998, 'Principles of physical biochemistry', Prentice Hall—*includes a useful, more advanced, chapter on the thermodynamics of macromolecular structures and energy calculations.*

PROBLEMS

[Some of these questions require detailed information on protein secondary structures, which can be found in books such as *Introduction to protein structure* by Branden and Tooze.]

1. Which amino acids can use their side chains to

(a) form salt-bridges?
(b) form hydrogen bonds?
(c) give a large hydrophobic effect?

2. Why are α-helices more frequently found in proteins than π-helices or 3,10-helices?

3. An α-helix has a permanent electric dipole moment associated with its peptide backbone whereas a β-sheet does not. By considering the relative alignments of the backbone dipoles in these two types of secondary structure, show why this is so. Calculate the dipole moment for an α-helix 10 Å long and compare this value to that for an N–H bond. You can assume that the partial charges on the atoms of the backbone N–H and C=O groups are

$$q_N = -0{\cdot}5 \quad q_H = +0{\cdot}5 \quad q_O = -0{\cdot}5 \quad q_C = +0{\cdot}5$$

and that the N–H distance is 1 Å.

4. Calculate $\Delta G°$, $\Delta H°$ and $\Delta S°$ for the transfer of methane from a non-polar solvent to water at 298 K.

CH_4(non-polar solvent) $\rightarrow$ CH_4(g) $\quad \Delta G° = -14{\cdot}6\ \text{kJ mol}^{-1} \quad \Delta H° = +2{\cdot}1\ \text{kJ mol}^{-1}$
CH_4(aq) $\rightarrow$ CH_4(g) $\quad \Delta G° = -26{\cdot}3\ \text{kJ mol}^{-1} \quad \Delta H° = +13{\cdot}4\ \text{kJ mol}^{-1}$

Comment on the values that you obtain in relation to the proposed model of the hydrophobic effect.

5. $\Delta H°$ and $\Delta S°$ for the association of two enzyme subunits to form a dimer are $+16{\cdot}8\ \text{kJ mol}^{-1}$ and $+66{\cdot}9\ \text{kJ mol}^{-1}$ respectively at 293 K. Would association be favoured at this temperature? What would be the effect on the degree of association of lowering the temperature and do the data give any insight into the mechanism underlying the 'cold lability' of several multisubunit enzymes?

6. The following data give the free energies of transfer of various amino acids from water to a urea solution at 25°C.

Amino acid	$\Delta G°$transfer (kJ mol^{-1})
Gly	+0·42
Ala	+0·13
Leu	−0·13
Phe	−2·57

Do these data suggest that the action of urea as a denaturant is a consequence of destabilising the contribution of the hydrophobic effect to the folding of a protein? What additional data would you need to substantiate your conclusion?

7. A 20-residue peptide sequence forms a well-defined, solvent-exposed α-helix as part of a large, water-soluble protein. A 20-residue peptide with the same sequence has a random coil conformation in water. Explain this observation.

APPENDICES

APPENDIX 1

Note on units and constants

In this edition we have tried, wherever possible, to use SI units. These are based on the metre–kilogram–second system of measurement and are now almost universally accepted. SI units for the various physical quantities are listed below.

Quantity	SI unit	
Amount of substance	mole (mol)	this quantity contains 1 Avogadro number (L) of basic units (e.g. atoms, molecules, etc.)
Electric charge	coulomb (C)	
Length	metre (m)	1 Angstrom (Å) $= 1 \times 10^{-10}$ m or 0·1 nm
Mass	kilogram (kg)	multiple units are based on 1 gram (g) although the standard unit is kg
Molecular mass	dalton (Da)	1 Da = 1 atomic mass unit ≈ mass of 1 hydrogen atom.
Temperature	degree Kelvin (K)	0°C = 273·15 K
Time	second (s)	
Volume	cubic metre (m^3)	1 litre (l) = 1 dm^3 and 1 millilitre (ml) = 1 cm^3
Dipole moment	coulomb metre (C m)	1 debye (D) $= 3{\cdot}338 \times 10^{-30}$ C m
Electric potential	volt (V) (J C^{-1})	
Energy	joule (J) (1 m^2 kg s^{-2})	1 calorie = 4·18 J
Force	newton (N) (1 m kg s^{-2})	
Frequency	hertz (Hz) (1 s^{-1})	
Power	watt (W) (J s^{-1})	
Pressure	pascal (Pa) (1 N m^{-2})	1 atmosphere = 101·325 kPa
Radioactive decay	becquerel (Bq) (1 s^{-1})	one disintegration per second, 1 curie (Ci) $= 3{\cdot}7 \times 10^{10}$ s^{-1}
Concentration	mol kg^{-1}	moles per kg of solvent, called molality
		Concentration is also frequently measured as mol dm^{-3}, i.e. moles per litre of solution, called molarity (M). For water, 1 kg at room temperature has a volume of approximately 1 dm^3 and so 1 mol kg^{-1} ≈ 1 mol dm^{-3} = 1 M.

WARNING: equations and calculations only work if *all* the quantities involved (energy, temperature, etc.) are in the correct units. For instance, using temperature values in °C rather than K will always give the wrong answer.

For convenience, quantities are often given in multiples (in powers of 10) of the basic units. This avoids having lots of zeros before or after the decimal place and is indicated by putting prefixes before the standard SI unit. Compound prefixes are not allowed (e.g. mμm is not allowed).

Multiple	Prefix	Abbreviation	Multiple	Prefix	Abbreviation
10^1	deca	da	10^{-1}	deci	d
10^2	hecto	h	10^{-2}	centi	c
10^3	kilo	k	10^{-3}	milli	m
10^6	mega	M	10^{-6}	micro	μ
10^9	giga	G	10^{-9}	nano	n
10^{12}	tera	T	10^{-12}	pico	p

The implications of these units for the topics discussed in the text are

- Temperatures are quoted in degrees Kelvin. Thus, 25°C is 298·15 K.
- Enthalpy and free energy changes are given in $\mathrm{J\,mol^{-1}}$ or $\mathrm{kJ\,mol^{-1}}$.
- Entropies are given in $\mathrm{J\,K^{-1}\,mol^{-1}}$.
- Enzyme activities are given in katals (the amount of enzyme catalysing the transformation of 1 mol substrate per second). Specific activities are quoted in terms of katal $\mathrm{kg^{-1}}$.

Some non-SI units are retained in the text, partly for convenience but mostly to be consistent with other texts. These include

- Pressure atmosphere (1 atm = 760 mm Hg = 101·325 kPa)
- Concentration molar ($1\,M = 1\,\mathrm{mol\,dm^{-3}}$, for water $1\,M \approx 1\,\mathrm{mol\,kg^{-1}}$)

The following table gives the fundamental constants (together with their SI units) used in the text.

Constant	Symbol	Value	Units
Avogadro number	L	$6{\cdot}022 \times 10^{23}$	$\mathrm{mol^{-1}}$
Boltzmann constant	k	$1{\cdot}380 \times 10^{-23}$	$\mathrm{J\,K^{-1}}$
Charge on a proton (charge on an electron is $-e$)	e	$1{\cdot}602 \times 10^{-19}$	C
Faraday constant	F (= Le)	$9{\cdot}648 \times 10^{4}$	$\mathrm{C\,mol^{-1}}$
Gas constant	R (= Lk)	8·314	$\mathrm{J\,K^{-1}\,mol^{-1}}$
Planck constant	h	$6{\cdot}626 \times 10^{-34}$	J s
Speed of light in a vacuum	c	$2{\cdot}997 \times 10^{8}$	$\mathrm{m\,s^{-1}}$
Vacuum permittivity	ε_o	$8{\cdot}854 \times 10^{-12}$	$\mathrm{J^{-1}\,C^2\,m^{-1}}$

APPENDIX 2

Mathematical tools needed for this text

The purpose of this appendix is not to teach maths. It is merely a summary of the most important mathematical tools that are needed to understand the text and attempt the problems. It is probably most useful as a checklist to go through before you start the problems. Knowledge of basic algebra is assumed (rearranging and solving simple equations).

Notation

The following notation is used throughout the text.

Sums and products

Sum (addition) of terms $\sum_i a(i) = a(1) + a(2) + a(3) + \cdots$

Product (multiplication) of terms $\prod_i a(i) = a(1) \times a(2) \times a(3) \times \cdots$

Changes in quantities

Δ large change in a quantity (e.g. ΔX is a large change in X).

δ small change in a quantity (e.g. δX is a small change in X).

d continuous change in a variable (e.g. dX is a continuous change in the variable X).

Dealing with changes in quantities

A large change in a simple quantity can be calculated either by taking the difference between the final and initial states

$$\Delta X = X_{\text{final}} - X_{\text{initial}}$$

or by adding together lots of small changes

$$\Delta X = \sum \delta X$$

or by integrating a continuous change

$$\Delta X = \int_{\text{initial}}^{\text{final}} dX$$

A small change in a complex function can be obtained by considering the effect of a separate small change in each variable in that function. For example, if a complex function F is given by

$$F = a + b + cd$$

then δF is given by

$$\delta F = \delta a + \delta b + c\delta d + d\delta c$$

This does not work for large changes (Δ).

Solving quadratic equations

For a quadratic equation of the form

$$ax^2 + bx + c = 0$$

there are two solutions, given by

$$x = \frac{-b \pm \sqrt{b^2 - 4ac}}{2a}$$

Often the context of the problem will determine which of the two possible answers is correct (for instance you cannot have negative concentrations).

Logarithms

Logarithms mostly occur as natural logs, although base 10 logs occur in a few equations. The following standard notation is used throughout the text.

$\log_e(X)$ log to the base e (e = 2·71828) of x, also called 'natural logs' and can be written as ln(x)

$\log_{10}(x)$ log to the base 10 of x

These are defined by the following relationships.

$$\log_{10}(x) = y \quad \Rightarrow \quad x = 10^y$$
$$\log_e(x) = y \quad \Rightarrow \quad x = e^y \text{ which can also be written as } x = \exp(y)$$

The following properties of logs are illustrated for natural logs but work for base 10 logs as well.

$$\log_e(a) + \log_e(b) = \log_e(ab)$$
$$\log_e(a) - \log_e(b) = \log_e\left(\frac{a}{b}\right)$$
$$b\log_e(a) = \log_e(a^b)$$
$$\log_e(1 - x) \approx -x \quad \text{if x is very small}$$

Natural and base 10 logarithms can be interconverted by using the equation

$$\log_e(x) = 2{\cdot}303 \times \log_{10}(x)$$

Differentiation

Only very simple differentiation is needed, as follows.

General differentiation

$$\frac{d}{dx}(x^n) = nx^{(n-1)}$$

Differentiation of $\log_e(x)$

$$\frac{d}{dx}(\log_e(x)) = \frac{1}{x}$$

Integration

Only very simple integral calculus is needed, as summarised below. On a very few occasions, more complex integrals are required in the worked examples but in these cases the results are simply stated.

General definite (evaluated between limits) integration

$$\int_{\text{initial}}^{\text{final}} x^n dx = \left[\frac{1}{n+1}x^{(n+1)}\right]_{\text{initial}}^{\text{final}} = \frac{1}{n+1}\left((x_{\text{final}})^{(n+1)} - (x_{\text{initial}})^{(n+1)}\right)$$

Integration of 1

$$\int_{\text{initial}}^{\text{final}} dx = \int_{\text{initial}}^{\text{final}} x^0 dx = [x]_{\text{initial}}^{\text{final}} = x_{\text{final}} - x_{\text{initial}}$$

Integration of 1/x

$$\int_{\text{initial}}^{\text{final}} \frac{1}{x} dx = [\log_e(x)]_{\text{initial}}^{\text{final}} = \log_e\left(\frac{x_{\text{final}}}{x_{\text{initial}}}\right)$$

Probability

Only very simple concepts of probability are needed.

Dependent probabilities

Dependent probabilities are those where the events concerned are mutually exclusive. The sum of the probabilities of all possible outcomes must be one. Normalisation involves dividing each individual probability by the sum of all the probabilities to ensure that the overall probability is one. For example, if we toss a coin and it comes down N_h times heads and N_t time tails the normalised probabilities are

$$P(\text{heads}) = \frac{N_h}{N_h + N_t} \qquad P(\text{tails}) = \frac{N_t}{N_h + N_t}$$

and

$$P(\text{heads}) + P(\text{tails}) = 1$$

Independent probabilities

Independent probabilities are those where the events concerned do not influence each other. If there are two independent events with normalised probabilities P_1 and P_2, the probability of both of them happening is $P_1 \times P_2$. For example, if we toss a coin once the probability of it coming down heads is 0·5 (50%). If we toss two coins, the probability of both coming down heads is $0{\cdot}5 \times 0{\cdot}5 = 0{\cdot}25$ (25%).

Further reading

A. Cornish-Bowden, 1999, 'Basic mathematics for biochemists', Oxford University Press.

APPENDIX 3

Answers to problems

Chapter 2

1. The first law of thermodynamics, the law of conservation of energy, is a criterion that all chemical reactions must meet whether they are spontaneous or not. The second law provides a criterion for distinguishing between spontaneous and non-spontaneous reactions. Note that spontaneous implies that the reaction can proceed without an external input of energy, not that it will proceed at a noticeable rate.

2. $\Delta S^{o\prime} = 70{\cdot}2\,J\,K^{-1}\,mol^{-1}$. The association of the inhibitor with the enzyme is entropically driven. As we shall see in Chapter 18, this suggests that hydrophobic interactions between the two are dominant, the major contribution to $\Delta G^{o\prime}$ being the release of water from the binding faces.

3. Folding is entropically unfavourable, as the chain is more ordered, and is driven by the favourable enthalpy term, due to the dipole interactions between the different residues in the folded peptide (see Chapter 18). ΔG for folding is $-43{\cdot}9\,kJ\,mol^{-1}$ at 298 K, therefore folding will occur spontaneously. When the folded and unfolded forms are equally stable, ΔG for folding is zero and $\Delta H = T\Delta S$. This occurs at $T = 377\,K$. Above this temperature, ΔG for folding is positive and so folding is unfavourable. In practice, ribonuclease unfolds at a much lower temperature than this because the assumption that ΔH and ΔS do not change with temperature is invalid (see Chapter 3).

4. ΔH° for $CH_4(g) \rightleftharpoons C(g) + 4H(g)$ is $+1663.2\,kJ\,mol^{-1}$, thus the bond dissociation enthalpy is $415.8\,kJ\,mol^{-1}$. ΔH° for $C_2H_4(g) \rightleftharpoons 2C(g) + 4H(g)$ is $+2253{\cdot}1\,kJ\,mol^{-1}$. This involves breaking one C=C and four C–H bonds and so the dissociation enthalpy for C=C is $+589{\cdot}9\,kJ\,mol^{-1}$.

5. Benzene is $200{\cdot}7\,kJ\,mol^{-1}$ more stable than three isolated double bonds. This extra stabilisation is because the electrons are delocalised over all six atoms instead of just two in a double bond. The bonding in benzene is discussed in Chapter 16.

6. $\Delta H^{\circ}_{hydrolysis}(Gly{-}Gly) = -21{\cdot}5\,kJ\,mol^{-1}$. This involves breaking an O–H bond of water and the amide C–N bond and making a C–O and an N–H bond. Assuming that these are single bonds, the bond dissociation enthalpy terms contribute $+15\,kJ\,mol^{-1}$. This value is likely to be too small because the amide C–N bond has partial double bond character (see Chapter 16). The difference between this value

and $\Delta H^o_{hydrolysis}$ is due to the fact that the hydrolysis reaction in solution involves other contributions, notably interactions with the solvent. As two Glys have more charge than Gly–Gly, the former will interact more strongly with the solvent giving a large negative contribution to $\Delta H^o_{hydrolysis}$.

7. All three processes are entropically favourable, with $\Delta S > 0$. For the diffusion of Na^+ ions down a concentration gradient, ΔS is the only significant term. Dissolution of NaOH is both enthalpically and entropically favourable. Dissolution of $NaNO_3$ is enthalpically unfavourable, but in this case $T\Delta S > \Delta H$ and so ΔG is negative.

Chapter 3

1. $\Delta G°$ for the reaction is $+8{\cdot}4\,\text{kJ mol}^{-1}$. Using $\Delta G° = -RT\log_e K$ we obtain $K = 0{\cdot}034$ (M). We shall assume that water is in its standard state and so $[H_2O]_{eq}/[H_2O]° = 1$. The standard states for citrate and *cis*-aconitate are 1 M. Hence $K = [cis\text{-aconitate}^{3-}]/[\text{citrate}^{3-}]$. Substituting for K and [*cis*-aconitate] gives $[\text{citrate}^{3-}] = 11{\cdot}8\,\text{mM}$.

2. For this reaction, $K = 0{\cdot}0113$ (M). At equilibrium let $[\text{glycerol}] = 1 - x\,\text{M}$, $[P_i] = 0{\cdot}5 - x\,\text{M}$ and [glycerol 1-phosphate] $= x$ M. Thus

$$\frac{x}{(1-x)(0{\cdot}5-x)} = 0{\cdot}0113$$

and so $x = 0{\cdot}0056$. Therefore [glycerol 1-phosphate] $= 0{\cdot}0056$ M.

3. K' for this reaction is 0·31 (M). As glycogen is a very long homopolymer that is sequentially degraded by glycogen phosphorylase, for this reaction $[(\text{glycogen})_n] = [(\text{glycogen})_{n-1}] = [\text{glycogen}]_{total}$. Thus if $[P_i] = [\text{glucose 1-phosphate}]$ the mass action ratio, Γ, is 1·0. As $\Gamma > K'$, the reaction will proceed to the left, i.e. synthesis of glycogen is favoured. In muscle, $\Gamma = 0{\cdot}003$ and so as $\Gamma < K'$ the reaction will proceed to the right, i.e. degradation of glycogen is favoured.

This result shows that in the muscle cell, the levels of P_i and glucose 1-phosphate are such that phosphorylase acts to degrade glycogen. Confirmation of this role is provided by sufferers from McArdle's disease, where muscle phosphorylase is absent. These patients have high levels of muscle glycogen, confirming the link between muscle phosphorylase and glycogen degradation.

4. For phosphofructokinase, $K' = 1{\cdot}3 \times 10^3$ (M) and $\Gamma = 0{\cdot}031$ (M). As $\Gamma < K$ the system is not at equilibrium and will proceed to the right. $\Delta G' = -26{\cdot}6\,\text{kJ mol}^{-1}$. For adenylate kinase, $K' = 0{\cdot}44$ (M) and $\Gamma = 0{\cdot}42$ (M). As $\Gamma \approx K$ the system is at equilibrium, $\Delta G' = -0{\cdot}14\,\text{kJ mol}^{-1}$.

These types of analyses have been used to suggest enzymes that may act as regulation points in metabolic pathways. Phosphofructokinase is often considered to be the key regulatory enzyme in the glycolytic pathway because the reaction that it catalyses is so far from equilibrium. However, this type of argument is both simplistic and potentially misleading, as discussed in Chapter 13.

5. For the pyruvate kinase reaction in the direction given, $\Delta G^{\circ\prime} = -32\,kJ\,mol^{-1}$. Using the concentrations given, $\Delta G = -20\,kJ\,mol^{-1}$. For the alternative reaction, $\Delta G^{\circ\prime} = +2\,kJ\,mol^{-1}$ and $\Delta G = -38{\cdot}5\,kJ\,mol^{-1}$. Thus, the forward reaction converting phosphoenolpyruvate to pyruvate is spontaneous under these conditions and so the back reaction could not be used to synthesise phosphoenolpyruvate. The alternative reaction is spontaneous in the direction given and so it is thermodynamically feasible to synthesise phosphoenolpyruvate this way.

6. From a plot of $\log_e K$ versus 1/T, K at 303 K can be determined as $4{\cdot}59 \times 10^{-5}$ (M). ΔG° at 303 K can be calculated directly from K; ΔH° can be determined from the slope of the plot; and ΔS° can be calculated from ΔG° and ΔH°. These calculations give

$$\Delta G^\circ = 25{\cdot}2\,kJ\,mol^{-1} \qquad \Delta H^\circ = 20{\cdot}9\,kJ\,mol^{-1} \qquad \Delta S^\circ = -14{\cdot}2\,J\,K^{-1}\,mol^{-1}$$

The assumptions made are that ΔH° and ΔS° do not vary with temperature over the range used. As the plot of $\log_e K$ versus 1/T is a straight line, this assumption is reasonable.

7. The equilibrium constants for binding are

	293 K	313 K
Isoleucine tRNA	2.64×10^7	2.64×10^7
Valine tRNA	6.89×10^5	1.65×10^6

Thus binding of the correct tRNA is favoured by a factor of 38 at 293 K and a factor of 16 at 313 K.

Chapter 4

1. For a protein with n sites in total and i sites occupied

$$K_d^{L/PL_i} = \frac{\text{rate constant for dissociation of L from i sites on } PL_i}{\text{rate constant for binding of L to n} - (i-1) \text{ sites on } PL_{i-1}} = \frac{i \times k_d}{(n-i+1) \times k_a}$$

$$= \frac{i}{n-i+1} K_d$$

The product of all the K_d values is given by

$$K_d^{L/PL} \times K_d^{L/PL_2} \times \cdots \times K_d^{L/PL_{n-1}} \times K_d^{L/PL_n} = \frac{1}{n} K_d \times \frac{2}{n-1} K_d \times \cdots \times \frac{n-1}{2} K_d \times \frac{n}{1} K_d = K_d^n$$

2. (a) We can set up five equations involving [E], [S], [I], [ES] and [EI]

$$K_d^{S/ES} = 50 \times 10^{-6} = \frac{[E][S]}{[ES]} \qquad K_d^{I/EI} = 75 \times 10^{-6} = \frac{[E][I]}{[EI]}$$

$$[E]_{total} = 1 \times 10^{-6} = [E] + [ES] + [EI] \qquad [S]_{total} = 1 \times 10^{-6} = [S] + [ES]$$

$$[I]_{total} = 1{\cdot}5 \times 10^{-6} = [I] + [EI]$$

With five equations and five unknowns, the problem is soluble. However, it is very complex and usually requires a computer to calculate a numerical solution.

(b) In this case, both S and I are in large excess compared to the E and so we can assume that $[S] \approx [S]_{total}$ and $[I] \approx [I]_{total}$. This simplifies the equations above by removing two of the variables. The equations become

$$[ES] = \frac{[E][S]}{K_d^{S/ES}} \approx \frac{[E][S]_{total}}{K_d^{S/ES}} = \frac{[E] \times 150 \times 10^{-6}}{50 \times 10^{-6}} = 3 \times [E]$$

$$[EI] = \frac{[E][I]}{K_d^{I/EI}} \approx \frac{[E][I]_{total}}{K_d^{I/EI}} = \frac{[E] \times 150 \times 10^{-6}}{75 \times 10^{-6}} = 2 \times [E]$$

$$[E]_{total} = 1 \times 10^{-6} = [E] + [ES] + [EI] = 6 \times [E]$$

Solving these equations gives $[E] = 0{\cdot}17 \times [E]_{total}$, $[ES] = 0{\cdot}5 \times [E]_{total}$ and $[EI] = 0{\cdot}33 \times [E]_{total}$. (See Chapter 12 for further discussion of competitive inhibition.)

3. From a plot of fluorescence increase against NADPH added, we can deduce that the enzyme can bind 0·3 μM NADPH. As the enzyme concentration is 0·15 μM, there must be two sites on the enzyme.

4. A plot of $1/\bar{n}$ versus $1/[Mn^{2+}]_{free}$ gives a straight line showing independent binding to identical sites, with $n = 4$ and $K_d = 1{\cdot}48 \times 10^{-4}$ (M).

5. This can be solved in an identical fashion to question 4, to give $n = 4$ and $K_d = 6{\cdot}4 \times 10^{-4}$ (M). In this case, there is also a shortcut that can be taken. As ADP is present in large excess, $[ADP] \approx [ADP]_{total}$ and so a plot of $1/\bar{n}$ versus $1/[ADP]_{total}$ also gives a straight line from which n and K_d can be extracted. Pyruvate kinase contains four subunits, each of which binds one ADP molecule.

6. A plot of $1/\bar{n}$ versus $1/[NAD^+]_{free}$ gives a curve indicating cooperative binding. The shape of the curve is typical of positive cooperativity. Extrapolation to $1/[NAD^+]_{free} = 0$ gives the number of binding sites as $n = 4$. A plot of $\log_e(\bar{n}/n - \bar{n})$ versus $\log_e[NAD^+]_{free}$ gives a curve typical of positive cooperative binding. Measuring the gradients at 50% and 100% saturation gives $h = 2{\cdot}8$ and $h = 1{\cdot}0$ respectively.

7. $\bar{n}$ as a function of $[CTP]_{free}$ can be calculated for both sets of data and the results combined. A plot of $1/\bar{n}$ versus $1/[CTP]_{free}$ gives a curve indicating cooperative binding. The shape of the curve is typical of negative cooperativity. Extrapolation to $1/[CTP]_{free} = 0$ gives the number of binding sites as $n \approx 6$, thus each regulatory unit binds one CTP. A plot of $\log_e(\bar{n}/n - \bar{n})$ versus $\log_e[CTP]_{free}$ gives a curve typical of negative cooperative binding. Measuring the gradient at 50% gives $h = 0{\cdot}5$. The range of $\bar{n}$ for the data is from 0·6 to 5·2 and so extrapolating to 0% and 100% saturation is difficult. From the plot, it looks as though h is approaching 1·0 at 0% saturation but does not approach 1·0 at 100% saturation. This may be a problem with the second data set (experiment B) because very high protein and ligand concentrations were used, which can lead to aggregation or precipitation.

8. An antibody has two antigen-binding domains and is hence divalent, whereas the antigen is monovalent. If the antibody is linked to the solid support, the monovalent interaction between the antigen in solution and a single binding domain of the antibody is measured. If the antigen is linked to the solid support, the divalent interaction between the antibody in solution and the cross-linked, and hence multivalent, antigen is measured. The latter will be much higher affinity than the former.

This type of interaction, with one component immobilised on a solid support and the other in solution, is used in the technique of surface plasmon resonance (frequently referred to as BiaCore). This technique can be used to measure both the binding constant and the on and off rates for ligands binding to proteins but factors such as those above need to be considered when interpreting data.

9. As $\Delta G° = -RT\log_e K$, $\Delta\Delta G°$ is given by

$$\Delta\Delta G^o = \Delta G_2^o - \Delta G_1^o = -RT\log_e K_2 + RT\log_e K_1 = -RT\log_e \frac{K_2}{K_1}$$

where K_2 is the equilibrium constant for binding of the first ligand in the presence of the second ligand and K_1 is the equilibrium constant for binding of the first ligand in the absence of the second ligand. K must increase by a factor of 100 to go from <10% to >90% binding of the first ligand without changing the ligand concentration, i.e. $K_2/K_1 \approx 100$. Therefore, the maximum useful value of $\Delta\Delta G°$ is $\approx -11{\cdot}5$ kJ mol^{-1}. [For a more complete analysis, see S. Forsen and S. Linse (1997) *Trends in Biochemical Sciences*, **20**, 495.]

Chapter 5

1. By using the van't Hoff isochore, K_w at 310 K is $2{\cdot}57 \times 10^{-14}$. Therefore, neutral pH is 6.79.

2. (a) pH = 1·3; (b) pH = 2·88; (c) pH = 8·79; (d) pH = 2·72; (e) pH = 4.2. For (d), HCl will be fully dissociated and so if [acetate] = x M then [acetic acid] = (0·1 − x) M and $[H^+]$ = (0·001 + x) M. For (e) use the Henderson–Hasselbalch equation.

3. (a) At pH = 5·0, only Asp and Glu have pK_a values less than the pH and so these will be acidic, i.e. the protonated form will spontaneously dissociate at this pH. All the others will be basic.
(b) At pH = 10·0, only Arg has a pK_a value greater than the pH and so this side chain will be basic. For Lys and Tyr, the pH equals the pK_a and so these will be neither acidic nor basic, i.e. they will be 50% dissociated. All the others will be acidic.

4. A plot of peak frequency versus pH gives a standard titration curve, on which pK_a = pH at the mid-point of the titration. For His A the whole of the titration curve has been recorded, the mid-point giving pK_a(His A) = 7·2. For His B only half the titration curve can be observed, the peak becoming too broad to be measured below pH 5·75. If we assume that the change in peak frequency between the protonated and deprotonated states is the same for His B as for His A, i.e. change in δ is 1·0, then the mid-point of the curve is when the peak frequency has changed by 0·5, giving pK_a(His B) = 5·9.

Both His A and His B titrate with pH and so both must be exposed to the solvent, i.e. not buried in the hydrophobic core of the protein. Free histidine in solution has a pK_a of 6·9. The observation that the values for His A and His B differ from this indicates that they are both in structured regions of the protein, the local chemical environments changing the pK_a values.

5. The equilibria for glycine in solution are

$$NH_3^+{-}CHR{-}COOH \overset{K_a(1)}{\rightleftharpoons} NH_3^+{-}CHR{-}COO^- \overset{K_a(2)}{\rightleftharpoons} NH_2{-}CHR{-}COO^-$$

where

$$K_a(1) = \frac{[NH_3^+{-}CHR{-}COO^-][H^+]}{[NH_3^+{-}CHR{-}COOH]} \quad \text{and} \quad K_a(2) = \frac{[NH_2{-}CHR{-}COO^-][H^+]}{[NH_3^+{-}CHR{-}COO^-]}$$

When pH = pI, the average charge on the molecule is zero. $NH_3^+ - CHR - COO^-$ has a net charge of zero and so does not contribute to the average charge. For the average charge to be zero, $[NH_3^+ - CHR - COOH] = [NH_2 - CHR - COO^-]$. Thus

$$\frac{[NH_3^+{-}CHR{-}COO^-][H^+]}{K_a(1)} = \frac{K_a(2)[NH_3^+{-}CHR{-}COO^-]}{[H^+]}$$

which can be rearranged to give

$$[H^+]^2 = K_a(1) \times K_a(2)$$

Taking negative logs of both sides gives

$$-\log_{10}[H^+] \times 2 = -\log_{10} K_a(1) - \log_{10} K_a(2)$$

which can be rewritten as

$$pH = pI = \frac{pK_a(1) + pK_a(2)}{2}$$

Thus, the pI for glycine is the average of the pK_a values of the two titratable groups.

In general, the pI value is given by the average of the two pK_a values for protonation and deprotonation of the uncharged, zwitterionic, form. Glutamic acid has an amino group ($-NH_2$) with a pK_a of ≈9·5, a carboxyl group (–COOH) with a pK_a of ≈2·5 and a side chain carboxyl group with a pK_a of ≈4·4. The fully protonated form carries a+1 charge, deprotonation of the carboxyl group with a pK_a value of 2·5 gives the zwitterion and deprotonation of the carboxyl group with a pK_a value of 4·4 gives an overall charge of –1. Thus the pI value is given by

$$pI = \frac{2{\cdot}5 + 4{\cdot}4}{3} = 3{\cdot}45$$

6. Using the Henderson–Hasselbalch equation to calculate the ratio of [base] to [acid] to give a solution at pH = 7·0, 61·3 cm^3 of $H_2PO_4^-$ and 38·7 cm^3 of HPO_4^{2-} are required.

7. At pH = 8·3, the pH equals the pK_a of Tris and so the concentration of the base equals the concentration of the conjugate acid, thus [Tris] = [Tris–H^+] = 0.25 M. If we add 0·005 mol of H^+ ions to a litre of solution, 0·005 mol of Tris will be converted to Tris–H^+. The new concentrations in solution will then be $[\text{Tris–H}^-]=0{\cdot}255$ M and $[\text{Tris}]=0{\cdot}245$ M. From the Henderson–Hasselbalch equation, the new pH is given by

$$\text{pH} = \text{pK}_a + \log_{10}\left(\frac{[\text{Tris}]}{[\text{Tris–H}^+]}\right) = 8.3 + \log_{10}\left(\frac{0{\cdot}245}{0{\cdot}255}\right) = 8{\cdot}283$$

Thus, the pH change is 0·017 units. The buffering capacity is

$$\beta = \frac{\text{Amount of acid added}}{\text{Change in pH}} = \frac{0{\cdot}005}{0{\cdot}017} = 0{\cdot}294\,\text{mmol H}^+\ (\text{pH unit})^{-1}$$

From the worked example in the chapter, the buffering capacity of a 0·1 M solution of Tris at pH 8·3 is 0·057 mmol H^+ (pH unit)$^{-1}$, i.e. five times less than 0·294 to within the rounding errors of the calculation.

8. This question can be solved using the same method as in the previous question. At pH = 8·0, the ratio of Tris : Tris–H^+ is 1 : 2 and the reaction generates 1 mM H^+.
(a) In the presence of 0·1 M Tris, the new pH is 7·97.
(b) In the presence of 0·01 M Tris, the new pH is 7·78.
(c) If no buffer is present, then we might expect the final concentration of H^+ to be 1 mM, i.e. pH = 3·0. However, ADP and HPO_4^{2-} will both act as buffers. If the pK_a of both of these is assumed to be 7.2 and they are both present at 1 mM at the end of the reaction, the new pH is 6·96.

9. Using the van't Hoff isochore, for Tris $K_a(273)/K_a(298) = 0{\cdot}182$ and for HPO_4^{2-} $K_a(273)/K_a(298) = 0{\cdot}865$. The pK_a temperature dependence of the latter is much smaller and thus the pH of phosphate buffers are much less sensitive to changes in temperature.

10. The proton concentration, and hence the pH, can be calculated from the equation

$$[H^+] = \frac{-[\text{SID}] \pm \sqrt{[\text{SID}]^2 + 4K_w}}{2}$$

Addition of 2 mM HCl reduces the [SID] value by 0·002, since only the Cl^- ions contribute to [SID]. This gives the following results for the three solutions.

[SID]/M		$[H^+]$/M		pH	
Initial	Final	Initial	Final	Initial	Final
0·041	0·039	$1{\cdot}07 \times 10^{-12}$	$1{\cdot}13 \times 10^{-12}$	11·97	11·95
0·001	−0·001	$4{\cdot}4 \times 10^{-11}$	$1{\cdot}0 \times 10^{-3}$	10·36	3·00
−0·039	−0·041	$3{\cdot}9 \times 10^{-2}$	$4{\cdot}1 \times 10^{-2}$	1·41	1·39

Although equal amounts of HCl were added to each solution, the changes in $[H^+]$ are very different, and it is only in the case of the solution with an initial [SID] of −39 mM that the change in $[H^+]$ is directly related to the quantity of H^+ added to the solution.

Chapter 6

1. (a) $Zn^{2+}(aq) + 2e^- \rightarrow Zn(s)$
   (b) $H^+(aq) + e^- \rightarrow \frac{1}{2}H_2(g)$
   (c) $Co^{3+}(aq) + e^- \rightarrow Co^{2+}(aq)$
   (d) $AgBr(s) + e^- \rightarrow Ag(s) + Br^-(aq)$
   (e) $Hg_2Cl_2(s) + 2e^- \rightarrow 2Hg(l) + 2Cl^-(aq)$
   (f) $\text{fumarate}^{2-}(aq) + 2H^+(aq) + 2e^- \rightarrow \text{succinate}^{2-}(aq)$
   (g) $\text{Cyt-c}(Fe^{3+})(aq) + e^- \rightarrow \text{Cyt-c}(Fe^{2+})(aq)$
   (h) $CO_2(g) + H^+(aq) + 2e^- \rightarrow HCO_2^-(aq)$
   (i) $NAD^+(aq) + H^+(aq) + 2e^- \rightarrow NADH(aq)$

2. The electrochemical cell reaction is given by

$$\text{Left(reduced)} + \text{Right(oxidised)} \rightleftharpoons \text{Left(oxidised)} + \text{Right(reduced)}$$

and E° by E°(right) – E°(left).
   (a) $Zn^{2+}(aq) + Cu(s) \rightleftharpoons Zn(s) + Cu^{2+}(aq)$ $\quad E° = -1{\cdot}1\ V$
   (b) $H_2(g) + 2Ag^+(aq) \rightleftharpoons 2H^+(aq) + 2Ag(s)$ $\quad E° = 0{\cdot}80\ V$
   (c) $H_2(g) + 2AgCl(s) \rightleftharpoons 2H^+(aq) + 2Cl^-(aq) + 2Ag(s)$ $\quad E° = 0{\cdot}22\ V$
   (d) $H_2(g) + 2Fe^{3+}(aq) \rightleftharpoons 2H^+(aq) + 2Fe^{2+}(aq)$ $\quad E° = 0{\cdot}77\ V$
   (e) $NADH + \text{oxaloacetate}^{2-} + H^+ \rightleftharpoons NAD^+ + \text{malate}^{2-}$ $\quad E°' = 0{\cdot}15\ V$

3. (a) $Pb|Pb^{2+} \| Sn^{2+}|Sn$
   (b) $Pt|\text{pyruvate}^-, H^+, \text{lactate}^- \| NAD^+, H^+, NADH|Pt$

4. $\Delta G° = \Delta H° - T\Delta S°$ and $\Delta G° = -nFE°$ give $E° = (T\Delta S° - \Delta H°)/nF$. In this case, $n = 2$ and so

$$E° = \frac{298 \times 8{\cdot}8 + 148\,800}{2 \times 96\,500} = 0{\cdot}78\ V$$

5. The cell reaction is

$$Zn(s) + 2Fe^{3+}(aq) \rightleftharpoons Zn^{2+}(aq) + 2Fe^{2+}(aq)$$

Since $\Delta G° = -nFE°$ and $n = 2$

$$\Delta G°_{298} = -295{\cdot}3\ kJ\ mol^{-1}$$

From $\Delta G° = \Delta H° - T\Delta S°$

$$\Delta S° = -\frac{d(\Delta G°)}{dT} = nF\frac{d(E°)}{dT} = 2 \times 96\,500\frac{(1{\cdot}55 - 1{\cdot}53)}{(323 - 298)} = 154{\cdot}4\ J\ K^{-1}\ mol^{-1}$$

and

$$\Delta H° = \Delta G° + T\Delta S° = -249{\cdot}3\ kJ\ mol^{-1}$$

The assumptions made are that $\Delta H°$ and $\Delta S°$ are both independent of temperature over this range (see Chapter 3 for a discussion of these assumptions).

6. The two half-cell reactions are

$$CH_3CHO + 2H^+ + 2e^- \rightarrow CH_3CH_2OH$$
$$CH_3COCOO^- + 2H^+ + 2e^- \rightarrow CH_3CHOHCOO^-$$

(a) For the reaction

$$CH_3CHO + CH_3CHOHCOO^- \rightleftharpoons CH_3CH_2OH + CH_3COCOO^-$$

$E^{o\prime} = 0{\cdot}027\,V$. Thus, for products and reactants in their standard states this reaction will proceed to the right. For the concentrations given in the question

$$E = E^{o\prime} + \frac{RT}{nF}\log_e\left(\frac{[CH_3CHO][CH_3CHOHCOO^-]}{[CH_3CH_2OH][CH_3COCOO^-]}\right)$$

$$= 0{\cdot}027 + \frac{8{\cdot}314 \times 303}{2 \times 96\,500}\log_e\left(\frac{0{\cdot}001 \times 0{\cdot}01}{0{\cdot}1 \times 0{\cdot}1}\right) = -0{\cdot}06\,V$$

As E is negative, ΔG will be positive and so the reaction will proceed in the direction of acetaldehyde and lactate.

(b) Assume that at equilibrium, y M ethanol has been oxidised to acetaldehyde. The final concentrations are

ethanol	$0{\cdot}1 - y\,M$	pyruvate	$0{\cdot}1 - y\,M$
lactate	$0{\cdot}01 + y\,M$	acetaldehyde	$0{\cdot}001 + y\,M$

Substituting into the equation $E^{o\prime} = (RT/nF)\log_e K'$ gives

$$0{\cdot}027 = \frac{8{\cdot}314 \times 303}{2 \times 96\,500}\log_e\left(\frac{(0{\cdot}1 - y) \times (0{\cdot}1 - y)}{(0{\cdot}001 + y) \times (0{\cdot}01 + y)}\right)$$

which can be solved for y to give the final concentrations as

ethanol	0·0776 M	pyruvate	0·0776 M
lactate	0·0324 M	acetaldehyde	0·0234 M

7. From the Nernst equation for a half-reaction

$$E = E^o + \frac{RT}{nF}\log_e\frac{[\text{oxidised}]}{[\text{reduced}]} = 0{\cdot}21 + \frac{8{\cdot}314 \times 298}{1 \times 96\,500}\log_e\frac{[\text{Cyt-c}(Fe^{3+})]}{[\text{Cyt-c}(Fe^{2+})]}$$

and so

E/V	0·3	0·25	0·2	0·15	0·1
$[\text{Cyt-c}(Fe^{3+})]/[\text{Cyt-c}(Fe^{2+})]$	33·4	4·75	0·68	0·097	0·014

Because of the logarithmic nature of the Nernst equation, even quite large errors in the ratio [oxidised]/[reduced] lead to relatively small changes in E.

8. The half-reaction is

$$\text{oxaloacetate}^{2-}(aq) + 2H^+(aq) + 2e^- \rightarrow \text{malate}^{2-}(aq)$$

and from the Nernst equation

$$E = E^o + \frac{RT}{nF}\log_e\frac{[\text{oxaloacetate}^{2-}][H^+]^2}{[\text{malate}^{2-}]}$$

In the biochemical standard state $[H^+] = 10^{-7}$ M and $[malate^{2-}] = [oxaloacetate^{2-}] = 1$ M

$$E^{o\prime} = 0{\cdot}239 + \frac{8{\cdot}314 \times 300}{2 \times 96\,500} \log_e [10^{-7}]^2 = -0{\cdot}178\ \text{V}$$

9. The reaction of interest is

$$2NADH + O_2 + 2H^+ \rightarrow 2NAD^+ + 2H_2O$$

The potential (E) on a hypothetical inert electrode inserted into an equilibrium mixture of these reactants and products is given by

$$E = E^o_{O_2/H_2O} + \frac{RT}{4F} \log_e \frac{[O_2][H^+]^4}{[H_2O]^2}$$
$$= E^o_{NAD^+/NADH} + \frac{RT}{2F} \log_e \frac{[NAD^+][H^+]}{[NADH]}$$

Rearranging gives

$$E^o_{O_2/H_2O} - E^o_{NAD^+/NADH} = \frac{RT}{4F} \log_e \frac{[NAD^+]^2[H_2O]^2}{[NADH]^2[O_2][H^+]^2} = \frac{RT}{4F} \log_e K$$

$E^o_{O_2/H_2O}$ can be calculated from $E^o_{O_2/H_2O_2}$ and $E^o_{H_2O_2/H_2O}$ by using $\Delta G^\circ = -nFE^\circ$. Thus

$$E^o_{O_2/H_2O} = \frac{2FE^o_{O_2/H_2O_2} + 2FE^o_{H_2O_2/H_2O}}{4F} = 1{\cdot}229\ \text{V}$$

Hence

$$\log_e K = \frac{4 \times 96\,500}{8.314 \times 303}(1{\cdot}229 + 0{\cdot}11) = 51{\cdot}94$$

and so $K = 3{\cdot}6 \times 10^{22}$ (M).

10. (a) $\Delta G^{\circ\prime}$ for the coupled reaction

$$\text{Cyt-b}(Fe^{2+}) + \text{Cyt-f}(Fe^{3+}) \rightleftharpoons \text{Cyt-b}(Fe^{3+}) + \text{Cyt-f}(Fe^{2+})$$

can be calculated from the $E^{\circ\prime}$ values for the half-reactions using $\Delta G^{\circ\prime} = -nFE^{\circ\prime}$. Thus

$$\Delta G^{o\prime} = -F \times E^{o\prime}_{\text{Cyt-f } Fe^{3+}/Fe^{2+}} + F \times E^{o\prime}_{\text{Cyt-b } Fe^{3+}/Fe^{2+}} = -28{\cdot}95\ \text{kJ mol}^{-1}$$

(b) This reaction does not produce enough free energy to drive ATP synthesis under standard conditions.

(c) If two electrons were passed along the redox chain $\Delta G^{\circ\prime}$ would be -58 kJ mol^{-1}, which would be sufficient to drive ATP synthesis from ADP and P_i.

(d) This analysis is not meaningful in the context of a real cell for several reasons. First, we are not interested in $\Delta G^{\circ\prime}$ values but in ΔG values. The ΔG value for a chemical reaction within a cell is usually very different from the $\Delta G^{\circ\prime}$ value because the reactants and products are not in their standard states. ΔG for the hydrolysis of ATP within a cell is closer to $-50\,kJ\,mol^{-1}$ (as we have seen in Chapter 3) and so approximately $+50\,kJ\,mol^{-1}$ would be needed to drive ATP synthesis. Second, this analysis assumes a direct coupling of the free energy released by the redox reaction to ATP synthesis. In fact, as we shall see in Chapter 7, the mechanism for coupling electron transport to ATP synthesis involves setting up a proton gradient. Thus, the comparison between the free energy released by the redox reaction and the free energy required to drive ATP synthesis is invalid.

Chapter 7

1. The calculated freezing point depression for the serum of this Antarctic fish is 0·0011 K. The actual freezing point depression is some 550 times greater than this. The anti-freeze protein is thought to act by altering the hydrogen-bonded structure of water, and this lowers the freezing point by substantially more than would be predicted from a colligative effect alone. In contrast, the calculated freezing point depression for the lysozyme solution of 0·0013 K is in close agreement with the observed value.

2. The molecular weight of the protein is calculated to be 12 600 Da. In the presence of a smaller concentration of sodium chloride, the Donnan effect will lead to an unequal distribution of ions across the membrane unless the protein carries no net charge, i.e. unless it is at its isoelectric point. This would lead to an erroneous molecular weight for the protein. The Donnan effect is negligible if the concentration of sodium chloride is much larger than that of the protein.

3. The osmotic pressure difference is calculated to be 0·134 atm. This large osmotic pressure difference has pronounced effects in the diabetic patient. The high osmotic pressure of extracellular fluids necessitates a large urine volume, and would lead to loss of water from the tissues. A severely diabetic patient remains thirsty despite drinking large amounts of water and would suffer severe dehydration if untreated.

4. The equilibrium concentrations are

$$[Na^+]_{inside} = 79{\cdot}5\,mM \qquad [Na^+]_{outside} = 76{\cdot}5\,mM$$

$$[Cl^-]_{inside} = 73{\cdot}5\,mM \qquad [Cl^-]_{outside} = 76{\cdot}5\,mM$$

This result shows that the Donnan effect leads to a small difference between the ion concentrations on the two sides of the membrane. In the kidney, however, there is an accumulation of Na^+ and Cl^- on the plasma side of the membrane. This active transport requires the expenditure of energy, from ATP hydrolysis, and ensures the minimum loss of ions by excretion.

5. Initially $\Psi_o = P_o - \Pi_o = 0$, since $P_o = 0$ at atmospheric pressure and $\Pi_o = 0$ for pure water. Moreover since the cell is at equilibrium, $\Psi_o = \Psi_i$ and so $P_i = \Pi_i$ when $\Pi_o = 0$. As Π_o increases and Ψ_o decreases, P_i must decrease to maintain equilibrium. (Π_i is unable to respond at this point because the cell volume is unaffected.) Eventually $P_i = 0$, i.e. the hydrostatic pressure becomes zero, and the cell loses its turgor.

Increasing Π_o still further causes plasmolysis to occur and Π_i increases as water is lost from the cell. Graphically:

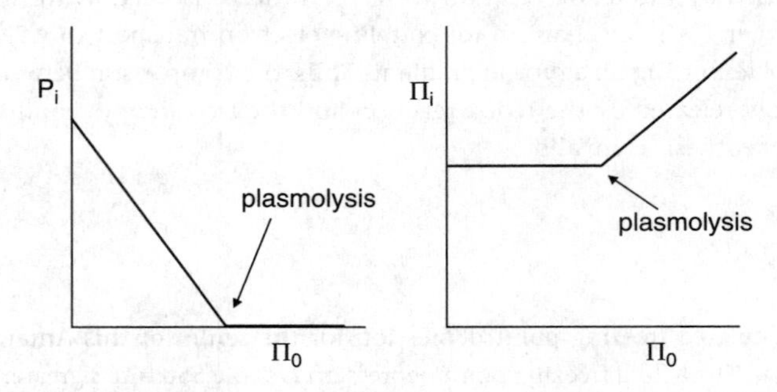

6. At equilibrium, $\mu_B(\text{in}) = \mu_B(\text{out})$, where μ_B is the chemical potential of the uncharged permeant weak base (B). Defining [Total B] = [BH⁺] + [B] and recalling that $K_a = [B][H^+]/[BH^+]$ for $BH^+ \rightleftharpoons B + H^+$ leads to

$$[B(\text{in})] = \frac{K_a + [\text{Total B(in)}]}{K_a + [H^+(\text{in})]}$$

Thus at equilibrium

$$[\text{Total B(in)}] = [\text{Total B(out)}]\frac{K_a + [H^+(\text{in})]}{K_a + [H^+(\text{out})]}$$

which leads to an intracellular concentration of 62·7 mM. Note that the weak base equilibrates to give a higher concentration in the more acidic (intracellular) compartment.

7. Applying the Nernst equation to the permeant ion leads to the following results

Permeant ion	$\Delta\Phi_{vc}$/mV	$[P_i]_{vac}$/mM
$H_2PO_4^-$	+20	2·5
	+30	3·7
	+40	5·5
HPO_4^{2-}	+20	722
	+30	1625
	+40	3250

Comparison with the measured vacuolar P_i concentration of 6·5 mM suggests that $H_2PO_4^-$ is the predominant form in which phosphate crosses the tonoplast.

8. According to the Nernst equation, the intracellular concentrations that would be in equilibrium with the given external concentrations are

$$[K^+] = 399\text{ mM} \qquad [Na^+] = 8{\cdot}77\text{ M} \qquad [Cl^-] = 27\text{ mM}$$

Comparison with the measured intracellular concentrations indicates that the thermodynamic driving force acts inwards on Na^+ and outwards on Cl^-. In contrast, K^+ is at equilibrium reflecting the presence of a K^+ leak channel in the plasma membrane. The inwardly directed driving force for Na^+ plays a critical role in the transmission of action potentials along the axon.

9. The energy required for ATP synthesis in the mitochondrial suspension is calculated using

$$\Delta G = \Delta G^{o\prime} + RT\log_e\left(\frac{[ATP]}{[ADP][P_i]}\right) = 57{\cdot}4\,kJ\,mol^{-1}$$

The energy stored in the proton gradient is calculated using $\Delta\bar{\mu}_{H^+} = F\Delta\Phi - 2{\cdot}303RT\Delta pH$. ΔpH can be calculated from the equation

$$[H^+(matrix)] = \frac{[\text{Total acetic acid(out)}][H^+(out)]}{[\text{Total acetic acid(matrix)}]}$$

on the assumption that K_a(acetic acid) $\gg [H^+(matrix)], [H^+(out)]$. This leads to

$$\Delta pH = pH(matrix) - pH(out) = \log_{10}\frac{[\text{Total acetic acid(matrix)}]}{[\text{Total acetic acid(out)}]} = 1{\cdot}70$$

$\Delta\Phi$ can be calculated by applying the Nernst equation to the equilibrium distribution of the Rb^+ ion. This gives

$$\Delta\Phi = \Phi(matrix) - \Phi(out) = -0{\cdot}119\,V$$

Substituting these values into the equation for the energy stored in the proton gradient gives $\Delta\bar{\mu}_{H^+} = -21{\cdot}25\,kJ\,mol^{-1}$. Hence 2·7 protons need to move down the proton electrochemical gradient to permit the synthesis of 1 molecule of ATP.

10. The movement of four protons into the matrix needs to generate 50 kJ mol^{-1} if it is to drive ATP synthesis. Thus, the minimum value needed for $\Delta\bar{\mu}_{H^+}$ is

$$\Delta\bar{\mu}_{H^+} = \frac{-50}{4} = -12{\cdot}5\,kJ\,mol^{-1}$$

12·5 kJ mol^{-1} is required to pump one proton out of the matrix against this electrochemical gradient. The transfer of one electron from Cyt-b to Cyt-f releases 29 kJ mol^{-1}, which would be sufficient to pump two protons out of the matrix.

It is worth making two further points. First, in practice the value of $\Delta\bar{\mu}_{H^+}$ may be higher than this; this is just a minimum value necessary to drive ATP synthesis. Second, the number of protons pumped out of the matrix by the transfer of one electron from Cyt-b to Cyt-f depends on the molecular machinery used to couple the redox process to proton pumping, and on whether sufficient free energy is available to drive the process.

Chapter 8

1. Non-ideality arises because the different species present in the solution interact with each other in different ways.
   (a) Ethanol and propanol are chemically very similar and are thus likely to form an ideal solution.
   (b) Ethanol and water are chemically different, for instance ethanol has only one hydrogen capable of forming hydrogen bonds whereas water has two, and so they will form a non-ideal solution. However, the departure from ideality is likely to be small because both species are uncharged and the primary intermolecular interactions for both are via hydrogen bonding.
   (c) NaCl will dissociate in water to give Na^+ and Cl^- ions. The deviation from ideality will be much larger than in (b) because of the additional electrostatic interactions involving these ions.
   (d) Lysozyme is a protein with both acidic and basic amino acid side chains that will be ionised in water. The concentration of charged species will be much greater than 1 mM. Thus the solution will be non-ideal and can be expected to deviate most from ideality.

2. Activity of water = 0·0434/0·0620 = 0·701, mole fraction of water = 0·773, activity coefficient of water = 0·701/0·773 = 0·906.

3. The plot of Π against protein concentration (x) suggested by the van't Hoff equation is non-linear, indicating that the solution is non-ideal. In this situation, the molecular weight of the protein can be estimated by measuring the limiting gradient of the curve at low concentration, i.e. as $x \rightarrow 0$ the solution becomes ideal. A more accurate approach is to note that, while for an ideal solution Π/x is a constant, i.e. Π/x = RT/M, for a non-ideal solution Π/x varies with x, i.e. Π/x = RT/M + f(x). Assuming the simplest possible function for f(x) then Π/x = RT/M + const. × x. A plot of Π/x versus x gives a straight line with an intercept on the Π/x axis of RT/M. This gives a molecular weight of 68 000 Da.

4. The ionic strength is given by $I = \frac{1}{2}\sum_j c_j z_j^2$, remembering that the summation is taken over all ions in solution.

   (a) $I = \frac{1}{2}\{(0{\cdot}1 \times 2^2) + (0{\cdot}2 \times (-1)^2)\} = 0{\cdot}3\,M$
   (b) $I = \frac{1}{2}\{(0{\cdot}01 \times 1^2) + (0{\cdot}005 \times (-2)^2)\} = 0{\cdot}015\,M$
   (c) $I = \frac{1}{2}\{(0{\cdot}001 \times 1^2) + (0{\cdot}0001 \times (-10)^2)\} = 0{\cdot}0055M$

   [Note: this is likely to be an underestimate of I. The net charge on the protein is − 10 but this is likely to be due to the presence of x positive charges and x + 10 negative charges and all charges contribute to I.]
   (d) [acetate] = $[H^+]$ = 1·31 mM and so $I = \frac{1}{2}\{(0{\cdot}00131 \times 1^2) + (0{\cdot}00131 \times (-1)^2)\} = 0{\cdot}00131\,M$

5. The activity coefficient of a specific ion in solution is given by $\log_{10}\gamma_x = -Az_x^2\sqrt{I}$. Assuming that $A = 0{\cdot}51\,M^{-1/2}$

   (a) $\gamma_{Mg^{2+}} = 10^{(-0{\cdot}51 \times 2^2 \times \sqrt{0.3})} = 0{\cdot}076$ and $\gamma_{Cl^-} = 10^{(-0{\cdot}51 \times (-1)^2 \times \sqrt{0{\cdot}3})} = 0{\cdot}526$

(b) $\gamma_{Na^+} = 10^{(-0{\cdot}51\times1^2\times\sqrt{0{\cdot}015})} = 0{\cdot}866$ and $\gamma_{SO_4^{2-}} = 10^{(-0{\cdot}51\times(-2)^2\times\sqrt{0{\cdot}015})} = 0{\cdot}563$
(c) $\gamma_{Na^+} = 10^{(-0{\cdot}51\times1^2\times\sqrt{0{\cdot}0055})} = 0{\cdot}917$ and $\gamma_{protein} = 10^{(-0{\cdot}51\times(-10)^2\times\sqrt{0{\cdot}0055})} = 1{\cdot}65\times10^{-4}$
(d) $\gamma_{H^+} = 10^{(-0{\cdot}51\times1^2\times\sqrt{0{\cdot}00131})} = 0{\cdot}958$ and $\gamma_{acetate} = 10^{(-0{\cdot}51\times(-1)^2\times\sqrt{0{\cdot}00131})} = 0{\cdot}958$

As can be seen, the value of γ at a given ionic strength is highly dependent on the charge of the ion.

We expect the simple Debye–Hückel equation to work up to I$\sim$0·04 M and so the values of γ for (a) will be unreliable. A value of γ of $1{\cdot}65\times10^{-4}$ for the protein in (c) would lead to very low effective concentration and so is also implausible, which suggests that the Debye–Hückel theory is not adequate to predict the behaviour of highly charged ions even at much lower ionic strengths.

6. (a) The only effect of adding $NaNO_3$ is to increase the ionic strength of the solution, and so change the activity coefficients. I = 0·01 M, thus for singly charged ions $\gamma = 0{\cdot}89$ and so $[Ag^+] = [Cl^-] = 1{\cdot}42\times10^{-5}$ M.
(b) The effect of adding NaCl is to increase the ionic strength of the solution, and so change the activity coefficients, and also to increase $[Cl^-]$, and so directly affect the equilibrium. Again, I = 0·01 M, thus for singly charged ions $\gamma = 0{\cdot}89$. $[Cl^-] = 0{\cdot}01$ M, due to the NaCl, and so $[Ag^+] = 2{\cdot}0\times10^{-8}$ M.
(c) Adding glucose at low concentrations has no effect on the ionic strength and so the solubility of AgCl is unaffected.

At much higher concentration, all three solutes result in an increase in γ because of effects on the structure of the solvent. This will decrease the solubility of AgCl, although in the case of NaCl this effect will be negligible compared to the common ion effect of adding Cl^-.

7. The equilibrium for the dissociation of acetic acid in water is

$$AcOH + H_2O \rightleftharpoons AcO^- + H_3O^+$$

Assuming that H_2O is in the standard state, the true equilibrium constant is given by

$$K_a = \frac{\gamma_{AcO^-}[AcO^-]\times\gamma_{H_3O^+}[H_3O^+]}{\gamma_{AcOH}[AcOH]} = K_a^{app}\frac{\gamma_{AcO^-}\times\gamma_{H_3O^+}}{\gamma_{AcOH}}$$

As AcOH is uncharged, $\gamma_{AcOH} = 1$. Using the simple Debye–Hückel equation to replace the other γ terms, substituting for A and rearranging gives

$$pK_a^{app} = pK_a - 1{\cdot}02\times\sqrt{I}$$

A plot of pK_a^{app} versus $\sqrt{I}$ should be a straight line with an intercept on the y-axis of pKa.

As NaCl is fully dissociated in water and is present at ten times the concentration of acetic acid or more, the ionic strength will be dominated by the NaCl. The plot of pK_a^{app} versus $\sqrt{I}$ is not a straight line over the whole range of $\sqrt{I}$ but is linear at small $\sqrt{I}$ and this region can be extrapolated back to give an intercept of 4·75.

8. The true equilibrium constant in the biochemical standard state is given by

$$K' = \frac{a_{ADP}a_{P_i}a_{H^+}}{a_{ATP}} = \frac{\gamma_{ADP}[ADP]\gamma_{P_i}[P_i]\gamma_{H^+}[H^+]}{\gamma_{ATP}[ATP]}$$

and the apparent equilibrium constant is given by

$$K'^{app} = \frac{[ADP][P_i][H^+]}{[ATP]}$$

Thus

$$K' = K'^{app} \times \frac{\gamma_{ADP}\gamma_{P_i}\gamma_{H^+}}{\gamma_{ATP}}$$

Taking $\log_{10}$ of both sides and substituting for $\log_{10}\gamma$ using the simple Debye–Hückel equation and the charges quoted in the question gives

$$\log_{10} K' = \log_{10} K'^{app} + 2A\sqrt{I}$$

Thus as I increases, K'^{app} decreases, i.e. [ATP] increases and [ADP] decreases.

Chapter 9

1. The reaction is first order, with $k = 0{\cdot}01205\ \text{min}^{-1}$.

2. For the data of the reaction between 1 mM *N*-acetylcysteine and 1 mM iodoacetamide, a plot of $\log_e$[*N*-acetylcysteine] versus t is a curve, so the reaction is not first order. A plot of 1/[*N*-acetylcysteine] versus t is a straight line, hence the reaction is second order overall with $k = 37\ M^{-1}\ s^{-1}$. Because the initial concentrations of *N*-acetylcysteine and iodoacetamide are equal, we cannot distinguish between the reaction being (i) second order with respect to one of the reactants and zero order with respect to the other, (ii) first order with respect to both reactants, or (iii) any combination of non-integral orders that add up to two.

For the data of the reaction between 1 mM *N*-acetylcysteine and 2 mM iodoacetamide, a plot of $\log_e$([*N*-acetylcysteine]/[iodoacetamide] versus t is a straight line, so the reaction is first order with respect to both reactants with $k = 30\ M^{-1} s^{-1}$.

3. The reaction is carried out with B in large excess and so [B] is approximately constant. The reaction is first order with respect to A and has a pseudo-first-order rate constant of $0{\cdot}018\ s^{-1}$. As the reaction is first order with respect to B, if the initial concentration of B is halved the initial rate will be halved. As long as B is still in large excess, the reaction remains pseudo-first order and so the half-time of A would double.

4. From the Arrhenius equation

$$\log_e k = \log_e A + \left(\frac{-\Delta H^{\ddagger}}{RT}\right)$$

Therefore

$$\frac{d(\log_e k)}{dT} = \frac{\Delta H^{\ddagger}}{RT^2}$$

Integration gives

$$\int_{k_{T_2}}^{k_{T_1}} d(\log_e k) = \int_{T_2}^{T_1} \frac{\Delta H^\ddagger}{RT^2} dT$$

$$\log_e k_{T_1} - \log_e K_{T_2} = -\frac{\Delta H^\ddagger}{R}\frac{1}{T_1} + \frac{\Delta H^\ddagger}{R}\frac{1}{T_2} \quad \text{or} \quad \log_e \frac{k_{T_1}}{k_{T_2}} = -\frac{\Delta H^\ddagger}{R}\left(\frac{1}{T_1} - \frac{1}{T_2}\right)$$

5. From the Arrhenius equation, $\Delta H^\ddagger = 69{\cdot}3\,\text{kJ mol}^{-1}$ and k at 283 K is $3{\cdot}43\,\text{M}^{-1}\,\text{min}^{-1}$.

6. From an Arrhenius plot, $\Delta H^\ddagger = 30{\cdot}9\,\text{kJ mol}^{-1}$.

7. $\Delta S^{o\ddagger} = 58{\cdot}6\,\text{J K}^{-1}\,\text{mol}^{-1}$. This indicates that the transition state is less ordered than the reactants, typical behaviour for the uncatalysed decomposition of one reactant to two products.

8. The rate of the reaction is given by

$$\text{Rate} = k_2[AB^\ddagger] = k_o \frac{\gamma_A \gamma_B}{\gamma_{AB^\ddagger}}[A][B] = k'[A][B]$$

Therefore

$$\log_{10}\frac{k'}{k_o} = \log_{10}\gamma_A + \log_{10}\gamma_B - \log_{10}\gamma_{AB^\ddagger}$$

The charge on the activated complex is $z_A + z_B$. Using the Debye–Hückel equation to replace the activity coefficients and assuming that $A = \frac{1}{2}$ gives

$$\log_{10}\frac{k'}{k_o} = -A(z_A^2 + z_B^2 - (z_A + z_B)^2)\sqrt{I} = -A(-2z_A z_B)\sqrt{I} = z_A z_B \sqrt{I}$$

For the reaction between $Cr(H_2O)_6^{3+}$ and SCN^-, a plot of $\log_{10} k'/k_o$ versus $\sqrt{I}$ is a straight line, with slope -3, as predicted by the equation. For the reverse reaction, one of the reactants is uncharged, i.e. $z_B = 0$, and so the rate constant is unaffected by the ionic strength.

9. $k_{^1H}/k_{^3H} = 12.5$. This is a maximum value if the C–H bond is fully broken in the rate-determining step. In practice, smaller values are observed corresponding to partial bond breakage.

Chapter 10

1. The kinetic equations for the [A] and [B] versus time are

$$-\frac{d[A]}{dt} = (k_1 + k_2)[A] \qquad [A]_t = [A]_o e^{-(k_1+k_2)t}$$

and

$$\frac{d[B]}{dt} = k_1[A] = k_1[A]_o e^{-(k_1+k_2)t} \qquad [B]_t = \frac{k_1[A]_o}{k_1 + k_2}(1 - e^{-(k_1+k_2)t})$$

Using these equations, the half-life of A is 3·3 min; [B] = 0·05 M at t = 5·73 min; and the maximum value of [B], at $t = \infty$, is 0·07 M.

2. These two decay processes are consecutive reactions.

$$^{209}\text{Tl} \xrightarrow{t_{\frac{1}{2}}=2\text{ minutes}} \beta + {}^{209}\text{Pb} \xrightarrow{t_{\frac{1}{2}}=3\text{ hours}} \beta + {}^{209}\text{Bi}$$

Radioactive decay is a first-order process and so the rate constant for each reaction is given by $k = \log_e 2/t_{\frac{1}{2}}$ giving values of $k_1 = 0{\cdot}006\ \text{s}^{-1}$ and $k_2 = 6{\cdot}4 \times 10^{-5}$ s^{-1} for the two processes respectively. The concentrations of the three species are given by

$$[^{209}\text{Tl}] = [^{209}\text{Tl}]_o \exp(-k_1 t)$$

$$[^{209}\text{Pb}] = \frac{[^{209}\text{Tl}]_o k_1}{k_2 - k_1}\{\exp(-k_1 t) - \exp(-k_2 t)\}$$

$$[^{209}\text{Bi}] = [^{209}\text{Tl}]_o - \frac{[^{209}\text{Tl}]_o k_2}{k_2 - k_1}\exp(-k_1 t) + \frac{[^{209}\text{Tl}]_o k_1}{k_2 - k_1}\exp(-k_2 t)$$

(a) At t = 300 seconds, $[^{209}\text{Tl}] : [^{209}\text{Pb}] : [^{209}\text{Bi}] = 0{\cdot}17 : 0{\cdot}82 : 0{\cdot}01$.
(b) At t = 7200 seconds, $[^{209}\text{Tl}] : [^{209}\text{Pb}] : [^{209}\text{Bi}] = 1{\cdot}7 \times 10^{-19} : 0{\cdot}64 : 0{\cdot}36$.

3. Applying the steady-state approximation to E′ gives

$$\frac{d[E']}{dt} = 0 = k_1[E][A] - k_{-1}[E'][P] - k_2[E'][B]$$

$$[E'] = \frac{k_1[E][A]}{k_{-1}[P] + k_2[B]}$$

The rate of formation of Q is given by

$$\frac{d[Q]}{dt} = k_2[E'][B] = \frac{k_1 k_2[E][A][B]}{k_{-1}[P] + k_2[B]}$$

The rate of formation of [P] is given by

$$\frac{d[P]}{dt} = k_1[E][A] - k_{-1}[E'][P] = k_1[E][A] - \frac{k_1 k_{-1}[E][A][P]}{k_{-1}[P] + k_2[B]}$$

which rearranges to the same equation as the rate of formation of Q, i.e. $d[P]/dt = d[Q]/dt$.

4. The basic rate equations, using the steady-state approximation for Br and H, are

$$\frac{d[HBr]}{dt} = k_2[H_2][Br] + k_3[Br_2][H] - k_{-2}[HBr][H]$$

$$\frac{d[Br]}{dt} = 0 = 2k_1[Br_2] + k_3[Br_2][H] + k_{-2}[HBr][H] - k_2[H_2][Br] - 2k_{-1}[Br]^2$$

$$\frac{d[H]}{dt} = 0 = k_2[H_2][Br] - k_3[Br_2][H] - k_{-2}[HBr][H]$$

Combining the equations for d[Br]/dt and d[H]/dt gives the steady-state value for [Br]

$$0 = 2k_1[Br_2] - 2k_{-1}[Br]^2 \qquad \text{or} \qquad [Br] = \sqrt{\frac{k_1[Br_2]}{k_{-1}}}$$

Rearranging the equation for d[H]/dt and substituting for [Br] gives the steady-state value for [H]

$$[H] = \frac{k_2[H_2][Br]}{k_3[Br_2] + k_{-2}[HBr]} = \frac{k_2[H_2]\sqrt{\frac{k_1[Br_2]}{k_{-1}}}}{k_3[Br_2] + k_{-2}[HBr]}$$

The steady-state values for [Br] and [H] can then be substituted into the equation for d[HBr]/dt giving the final rate equation, after some algebra

$$\frac{d[HBr]}{dt} = k_2[H_2]\sqrt{\frac{k_1[Br_2]}{k_{-1}}} + k_3[Br_2]\frac{k_2[H_2]\sqrt{\frac{k_1[Br_2]}{k_{-1}}}}{k_3[Br_2] + k_{-2}[HBr]} - k_{-2}[HBr]\frac{k_2[H_2]\sqrt{\frac{k_1[Br_2]}{k_{-1}}}}{k_3[Br_2] + k_{-2}[HBr]}$$

$$= k_2[H_2]\sqrt{\frac{k_1[Br_2]}{k_{-1}}}\left(\left[1 + \frac{k_3[Br_2]}{k_3[Br_2] + k_{-2}[HBr]} - \frac{k_{-2}[HBr]}{k_3[Br_2] + k_{-2}[HBr]}\right]\right)$$

$$= k_2[H_2]\sqrt{\frac{k_1[Br_2]}{k_{-1}}}\left(\left[\frac{k_3[Br_2] + k_{-2}[HBr] + k_3[Br_2] - k_{-2}[HBr]}{k_3[Br_2] + k_{-2}[HBr]}\right]\right)$$

$$= k_2[H_2]\sqrt{\frac{k_1[Br_2]}{k_{-1}}}\left(\left[\frac{2k_3[Br_2]}{k_3[Br_2] + k_{-2}[HBr]}\right]\right) = 2k_2[H_2]\sqrt{\frac{k_1[Br_2]}{k_{-1}}}\left(\frac{1}{1 + \frac{k_{-2}[HBr]}{k_3[Br_2]}}\right)$$

The rate equation for the reaction between H_2 and I_2 is much simpler. This is consistent with a single-step mechanism involving a bimolecular collision between one molecule of H_2 and one molecule of I_2 to give the products. However, the rate law does not prove this. For instance, the following scheme

$$I_2 \overset{K_1}{\rightleftharpoons} 2I$$

$$I + H_2 \overset{K_2}{\rightleftharpoons} H_2I$$

$$H_2I + I \overset{k_3}{\rightarrow} 2HI$$

gives

$$\frac{d[HI]}{dt} = 2K_1K_2k_3[H_2][I_2]$$

if the third step is slow and assumed not to perturb the first two equilibria. This reaction has been shown to proceed by a complex multistep mechanism. In general, a complex rate law indicates a complex mechanism but a simple rate law may be consistent with a simple or a complex mechanism.

5. A plot of $\log_{10} k_{relative}$ versus pH is linear with a slope of $+1$ up to pH $\approx 8{\cdot}5$ and then levels off above pH $= 10$. This indicates that a single group is titrating over this pH range and that the deprotonated form is reactive. From the plot, the pK_a of this group is 9·5. Looking at the reactants, the trinitrobenzene sulphonate does not titrate in this pH region. The carboxyl group will be deprotonated at pH $= 6{\cdot}5$ and the amino group will titrate in this pH range with an expected pK_a of 9·5 (see Chapter 5). This explains the pH dependence, the reactive form of the amino acid being $NH_2CHRCOO^-$.

6. We can calculate $[H^+]$ from the pH and [acetic acid] and [acetate] for each buffer using the Henderson–Hasselbalch equation (see Chapter 5). Direct inspection shows that the rate does not depend on $[H^+]$ or [acetic acid] and depends linearly on [acetate]. This reaction is an example of general base catalysis.

Chapter 11

1. The Michaelis–Menten equation breaks down when

   (i) The kinetic scheme used to derive the equation is not correct. This can occur when there is reactant or product inhibition or when the back reaction becomes significant.

   (ii) $K_d^{S/ES}$ is not constant. This occurs for a multisubunit enzyme when there is cooperative binding of S to the separate subunits.

   (iii) The steady-state assumption breaks down. This would occur if there is a build-up of the ES complex but this assumption is usually valid.

   There is also an experimental problem that many enzymes lose activity with time and so V_{max} decreases during the course of a reaction. Many of these problems can be eliminated by using initial rate measurements.

2. (i) $v = \frac{1}{2}V_{max}$ when $[S] = K_m$ and so $V_{max} = 90\ \mu M\ min^{-1}$. (ii) $K_m = 2{\cdot}4 \times 10^{-5}$ M.

3. From appropriate straight-line plots,

WT	$K_m = 54\ \mu M$	$k_{cat} = 0{\cdot}013\ s^{-1}$
Y63F	$K_m = 79\ \mu M$	$k_{cat} = 0{\cdot}010\ s^{-1}$
Y63L	$K_m = 16\ mM$	$k_{cat} = 0{\cdot}006\ s^{-1}$
D52A	$K_m = 56\ \mu M$	$k_{cat} = 0{\cdot}0005\ s^{-1}$

Mutation of Tyr 63 to Leu increases K_m but does not significantly affect k_{cat} and so Tyr 63 is involved in substrate binding but not transition state stabilisation. The mutation of Tyr 63 to Phe has little effect and so it must be the aromatic ring of the Tyr that is important for substrate binding and not the hydroxyl group. Crystallography shows that Tyr 63 is part of the B site of lysozyme and hence not close to the cleavage site, which is between sites D and E.

Mutation of Asp 52 does not affect substrate binding but reduces k_{cat} significantly, so this residue is involved in transition state stabilisation. Crystallography shows that Asp 52 is in site D and is in a position to stabilise the oxonium ion intermediate formed after proton donation by Glu 35 and bond cleavage.

Mutation of Asn 46 results in a similar decrease in k_{cat} to the D52A mutation. The double mutation results in very little further decrease in k_{cat} and so the effects of these two residues are not additive. Asn 46 and Asp 52 must be acting in a synergistic, or cooperative, manner with the loss of either one effectively eliminating the role of the other as well. Again this is confirmed by crystallography, which shows that these two residues and the substrate residue in site D form a hydrogen bond network. Removing either amino acid breaks this network.

A sequence comparison between human and hen egg white lysozymes shows that Asn 46 and Asp 52 are fully conserved whereas Tyr 63 is replaced by a Trp residue.

4. At a glucose concentration of 3 mM, the velocity of the reaction for hexokinase is 0·691 μmol min^{-1} g^{-1} and the velocity of the reaction for glucokinase is 0·992 μmol $min^{-1}g^{-1}$, i.e. glucokinase contributes 59% of the total activity. At a glucose concentration of 9·5 mM, the velocity of the reaction for hexokinase is 0·697 μmol min^{-1} g^{-1} and the velocity of the reaction for glucokinase is 2·095 μmol min^{-1} g^{-1}, i.e. glucokinase contributes 75% of the total activity.

   Thus hexokinase is effectively working at its maximum rate under these conditions, whereas glucokinase is not and so can respond to carbohydrate intake.

5. The plot of $\log_e$(activity) versus 1/T is not linear over the whole temperature range. If it is assumed to be linear up to 303 K, then $\Delta H^{\ddagger} = 68\,kJ\,mol^{-1}$.

6. From the Arrhenius plot it is seen that there are three distinct phases: (i) for $T < 293$ K, the activation energy is $220\,kJ\,mol^{-1}$; (ii) for $293\ K < T < 315$ K, the activation energy is $65\,kJ\,mol^{-1}$; (iii) for $T > 315$ K, the enzyme loses activity presumably because of thermal denaturation. As nitrogenase is a membrane-bound-enzyme, the transition at 293 K may be due to structural changes in the lipid bilayer.

7. A plot of $\log_{10}V_{max}$ versus pH gives the pK_a of the ionising group as 6·25. The temperature dependence of K_a gives $\Delta H^o_{ionization} = 29{\cdot}9\,kJ\,mol^{-1}$. From the data in the worked example, this residue is almost certainly a histidine, the deprotonated form being active.

Chapter 12

1. Assuming independent binding of A and B, the fractional saturation for sites A and B is given by

$$\theta_A = \frac{[A]}{K_d^{A/PA} + [A]} \quad \text{and} \quad \theta_B = \frac{[B]}{K_d^{B/PB} + [B]}$$

   The fraction of protein with both sites occupied is given by $\theta_A \times \theta_B$ and thus the rate is given by

$$v = V_{max} \times \theta_A \times \theta_B = V_{max} \times \left(\frac{[A]}{K_d^{A/PA} + [A]}\right) \times \left(\frac{[B]}{K_d^{B/PB} + [B]}\right)$$

   which gives the desired equation on expanding the brackets.

2. From plots of 1/v versus 1/[dGDP] at different values of [GTP], the reaction follows a ping-pong mechanism with $K_m^{A/EA}(GTP) = 107\,\mu M$, $K_m^{B/E'B}(dGDP) = 49\,\mu M$ and $V_{max} = 0{\cdot}79\,katal\,kg^{-1}$. The mechanism is likely to involve transfer of a phosphate group from GTP to the enzyme, release of the GDP, binding of dGDP and transfer of the phosphate group from the enzyme to give dGTP.

3. $K_m^{A/EA}(ATP) = 1{\cdot}7\,mM$, $K_m^{A/EAB}(ATP) = 0{\cdot}28\,mM$, $K_m^{B/EAB}(creatine) = 8{\cdot}5\,mM$ and $V_{max} = 3{\cdot}03\,katal\,kg^{-1}$. The binding of ATP to the free enzyme is less favourable

than the binding of ATP to the enzyme:creatine complex. Thus, the binding of the two substrates shows positive cooperativity (see Chapter 4).

4. Eadie–Hofstee plots: For a competitive inhibitor, the gradient becomes less negative and the intercept on the x-axis is unchanged. For a non-competitive inhibitor, the gradient is unchanged and the intercept on the x-axis is decreased.

   Hanes plots: For a competitive inhibitor, the gradient remains the same and the intercept on the x-axis becomes more negative. For a non-competitive inhibitor, the gradient becomes larger and the intercept on the x-axis is unchanged.

5. *N*-butyl deoxynojirimicin is a competitive inhibitor of ceramide and a non-competitive inhibitor of UDP-glucose, so is likely to bind to the ceramide binding site but not to the UDP-glucose binding site. Increasing the ceramide concentration will displace the inhibitor from the ceramide binding site, but increasing the UDP-glucose concentration will have no effect on the inhibitor binding as they bind at different sites.

6. If ES cannot be formed, $K_d^{I/EI}$ must be infinite. The general equation for enzyme inhibition then reduces to

$$v = \frac{V_{max}[S]}{K_d^{S/ES} + [S]\left(1 + \frac{[I]}{K_d^{I/ESI}}\right)}$$

which can be rewritten as

$$v = \frac{V_{max}^{app}[S]}{K_m^{app} + [S]}$$

where

$$K_m^{app} = \frac{K_m}{\left(1 + \frac{[I]}{K_d^{I/ESI}}\right)} \quad \text{and} \quad V_{max}^{app} = \frac{V_{max}}{\left(1 + \frac{[I]}{K_d^{I/ESI}}\right)}$$

Thus, both K_m and V_{max} are reduced by the same factor and so the ratio V_{max}/K_m remains the same. At low values of [S], where $K_m^{app} \gg [S]$, the rate is unaltered compared to the uninhibited reaction. At high values of [S], where $K_m^{app} \ll [S]$, the rate is reduced compared to the uninhibited reaction. As the inhibitor binds to the ES complex but not to the enzyme, it is likely that the binding site for the inhibitor includes part of the substrate.

7. For the uninhibited reaction, $K_m = 8\,\mu M$ and $V_{max} = 0{\cdot}185$ katal kg^{-1}. For the inhibited reaction, $K_m = 8\,\mu M$ and $V_{max} = 0{\cdot}102$ katal kg^{-1}. Thus, ATP is a non-competitive inhibitor. $K_d^{I/EI} = 9{\cdot}8\,\mu M$.

8. Pyruvate acts as a competitive inhibitor towards lactate and a non-competitive inhibitor towards NAD^+.

9. For the uninhibited reaction, $K_m = 0{\cdot}4$ mM and $V_{max} = 2{\cdot}95$ katal kg^{-1}. 2-Phosphoglycerate acts as a competitive inhibitor towards PEP, with $K_d^{I/EI} = 11{\cdot}5$ mM. Phenylalanine acts as a mixed inhibitor towards PEP with $K_d^{I/EI} = 9{\cdot}3$ mM and

$K_d^{I/ESI} = 19{\cdot}1$ mM. It is likely that 2-phosphoglycerate binds at the same site as PEP, as they are structurally similar. Phenylalanine probably binds at another site, altering the conformation of the catalytic site and so altering both K_m and V_{max}.

10. The equation for mixed enzyme inhibition can be rearranged to give

$$\frac{1}{v} = \frac{1}{V_{max}}\left(1 + \frac{[I]}{K_d^{I/ESI}}\right) + \frac{K_d^{S/ES}}{V_{max}}\left(1 + \frac{[I]}{K_d^{I/EI}}\right)\frac{1}{[S]}$$

When $-[I] = K_d^{I/EI}$, $(1 + [I]/K_d^{I/EI}) = 0$ and so the equation reduces to

$$\frac{1}{v} = \frac{1}{V_{max}}\left(1 + \frac{[I]}{K_d^{I/ESI}}\right)$$

1/v is independent of [S] and so plots of 1/v versus [I] at different values of [S] will intersect at this point.

When $-[I] = K_d^{I/ESI}$, $\left(1 + [I]/K_d^{I/ESI}\right) = 0$ and so the equation reduces to

$$\frac{[S]}{v} = \frac{K_d^{S/ES}}{V_{max}}\left(1 + \frac{[I]}{K_d^{I/EI}}\right)$$

[S]/v is independent of [S] and so plots of [S]/v versus [I] at different values of [S] will intersect at this point.

For a competitive inhibitor, the lines at different values of [S] intersect in plot (a) and are parallel in plot (b). For an uncompetitive inhibitor, the lines at different [S] are parallel in plot (a) and intersect in plot (b).

For the inhibition of xanthine oxidase, $K_d^{I/EI} = 1{\cdot}5 \times 10^{-9}$ M and $K_d^{I/ESI} = 8{\cdot}5 \times 10^{-9}$ M.

Chapter 13

1. For glucose 1-phosphate ⇌ glucose 6-phosphate, K = 18·8 (M), thus

$$\frac{[\text{glucose 6-phosphate}]}{[\text{glucose 1-phosphate}]} = 18.8 \quad \text{and} \quad \frac{[\text{fructose 6-phosphate}]}{[\text{glucose 6-phosphate}]} = 0.43$$

and

$$[\text{glucose 1-phosphate}] + [\text{glucose 6-phosphate}] + [\text{fructose 6-phosphate}] = 0.1\,\text{M}$$

At equilibrium, [glucose 1-phosphate] = 0·0036 M, [glucose 6-phosphate] = 0·0674 M and [fructose 6-phosphate] = 0·029 M.

2. A plot of $\log_e$(relative gluconeogenic flux) against $\log_e$(relative ICL activity) gives a straight line with a gradient, corresponding to the flux control coefficient, of 0·66. This result indicates that the ICL activity is quantitatively important in the control of gluconeogenic flux, and so developmental changes in the amount of enzyme could be an important factor in determining the conversion of lipid to sugar in young castor bean seedlings.

3. The relationship between f and r is

$$\frac{1}{f} = 1 - \frac{r-1}{r} C_E^J$$

For f = 2, the values of r are: (a) ∞; (b) 3; and (c) 2. Metabolic engineers wishing to increase the flux through a pathway by increasing the expression of an enzyme are constrained by this relationship. Setting $r = \infty$ gives the maximum value of f as $1/(1 - C_E^J)$ and this is only significant if C_E^J is close to 1.

Chapter 14

1. (a) The de Broglie relationship gives a mathematical relationship between the momentum of a classical particle and its associated wavelength when described as a wave-packet.
   (b) The uncertainty principle states that we can never know both the absolute position and the absolute energy of a particle. This is a direct consequence of the wave-packet nature of particles and how well we can define the properties of a wave-packet.
   (c) Because of the uncertainty principle molecules must always have some energy, they can never be completely stationary. This energy is inversely proportional to mass and so lighter atoms have a higher zero-point energy. This is exploited in studying the change in reaction rates on isotopic substitution (see Chapter 9).
2. Quantisation of energy occurs whenever waves are spatially localised. This only becomes important for very small masses localised to very small regions, otherwise the separation between the energy levels is so small as to be insignificant.
3. For a one-dimensional box, the energy level of the nth level is given by

$$E_n = \frac{n^2 h^2}{8ma^2}$$

Substituting into this equation gives the energies of the first three levels as $2{\cdot}41 \times 10^{-21}$ J, $9{\cdot}65 \times 10^{-21}$ J and $2{\cdot}17 \times 10^{-20}$ J respectively.

Normalised populations are given directly by the Boltzmann equation

$$P(E_n) = \frac{\exp\left(\frac{-E_n}{kT}\right)}{\sum_j \exp\left(\frac{-E_j}{kT}\right)}$$

This gives the relative populations of the first three levels as 0·845, 0·147 and 0·008 respectively. Note, all levels above n = 3 can be ignored for the denominator as being unpopulated.

4. The transition energy between two levels is given by

$$\Delta E = E_n - E_{n'} = \frac{(n^2 - n'^2)h^2}{8ma^2}$$

With four CH=CH units, there are eight π-electrons and so the first four energy levels are occupied. The first two transitions are from the fourth level to the fifth level and from the third level to the fifth level (this is at lower energy than fourth to sixth). The energies of these transitions are $1{\cdot}11 \times 10^{-18}$ J and $1{\cdot}97 \times 10^{-18}$ J respectively.

ΔE is inversely proportional to the length of the box squared, thus as the length increases ΔE decreases.

5. This is dealt with in Chapter 15, but try to do it as an exercise before you get there.

Chapter 15

1. Localisation of electrons around a nucleus leads to specific waves being allowed. These waves have very well-defined shapes, based on the allowed waves on a circle. Each wave also has a well-defined energy, based on the electrostatic attractions and repulsions with the nucleus and other electrons.

2. (a) Principal quantum number—n, radial quantum number, a large value of n indicates a large orbital with the electron having considerable density a long way from the nucleus.
   (b) Pauli exclusion principle—no two electrons can have the same four quantum numbers. Thus, only two electrons can occupy any one orbital and they must have opposite spins. A further consequence is that electrons with the same spin cannot occupy the same orbital and so have reduced electron–electron repulsion.
   (c) Aufbau principle—the ground-state electronic configuration of an atom or ion can be obtained by placing each electron sequentially into the lowest energy available orbital.

3.

H	1 electron	$1s^1$
C	6 electrons	$1s^2\ 2s^2\ 2p^2$
P	15 electrons	$1s^2\ 2s^2\ 2p^6\ 3s^2\ 3p^3$
Ni	28 electrons	$1s^2\ 2s^2\ 2p^6\ 3s^2\ 3p^6\ 4s^2\ 3d^8$
Mo	42 electrons	$1s^2\ 2s^2\ 2p^6\ 3s^2\ 3p^6\ 4s^2\ 3d^{10}\ 4p^6\ 5s^2\ 4d^4$

4. The general trend across both short periods is an increase in ionisation potential as the charge on the nucleus increases.

   The ionisation potentials of the second short period are generally lower than those of the first short period because, although the nuclear charges are higher, the outermost electrons are further from the nucleus.

   Instead of a steady, uniform increase in ionisation potential with atomic number there are two discontinuities.

   (i) Discontinuity between Be and B: the order of orbital energies for these atoms is $2s < 2p$ because of the differences in electron repulsion terms. For Li and Be, electrons are added to the 2s orbital. For B, the next electron goes into the 2p orbital and so is higher in energy, leading to a lower ionisation potential.

(ii) Discontinuity between N and O: from B to Ne, electrons are added to the 2p orbitals. These orbitals decrease steadily in energy as the nuclear charge increases. The first three of these electrons can go into the three 2p orbitals with their spins parallel. The fourth to sixth electrons have to pair up in the 2p orbitals and hence experience considerably more electron–electron repulsion leading to an increased energy and a decreased ionisation potential.

The same discontinuities are seen in the second short period, for the same reasons.

5. Increasing first ionisation potential: $K < Na < Ca < Cu < S < Br < Cl < H < Ne$.
Increasing first electron affinity: $Ca < Ne < K < Na < H < Cu < S < Br < Cl$.

In general, a high ionisation potential is associated with a high electron affinity because electrons are added to and removed from the same orbital. The exceptions for this are Ne, where ionisation involves removal of an electron from a 2p orbital but gain of an electron involves adding the electron to the 3s orbital, and Ca, where ionisation involves removal of an electron from a 3s orbital but gain of an electron involves adding the electron to the 3p orbital. H also has a lower than expected electron affinity because of the very high electron–electron repulsion in the small 1s orbital.

K, Na, Ca, Cu and H all commonly form positive ions in compounds. Even though the ionisation potential of H is very high, the H^+ ion is very small and so a large amount of electrostatic energy is released on forming an ionic compound. S, Br and Cl all commonly form negative ions in compounds. Ne has a high ionisation potential and a low electron affinity and so tends not to form compounds at all.

Chapter 16

1. $n = 4$ (t_{2g}^4 or $t_{2g}^3e_g^1$), $n = 5$(t_{2g}^5 or $t_{2g}^3e_g^2$), $n = 6$(t_{2g}^6 or $t_{2g}^4e_g^2$) and $n = 7$($t_{2g}^6e_g^1$ or $t_{2g}^5e_g^2$). Magnetic and spectroscopic measurements depend on the precise electron configuration and so can be used to determine spin state. Magnetic measurements give results that are much easier to interpret.

2. N is a small atom and is relatively high in the spectrochemical series, S is larger and relatively low in the spectrochemical series. Td coordination gives less steric repulsion between ligands and so is favoured by large ligands. Oh coordination gives a larger crystal field splitting and hence a larger crystal field stabilisation energy and so is favoured by strong field ligands.

3. In the $+2$ state, the ligands are closer to the metal than in the $+1$ state, because of the greater electrostatic attraction, and so cause greater repulsive effects for the 3d electrons. Thus, Δ is larger.

4. He_2^+, Li_2, Be_2^-, C_2^+ and O_2^{2-} exist. C_2^+ has the strongest bond with a bond order of 3/2.

5. $N_2 > N_2^+ \sim N_2^- > N_2^{2-}$ from bond order based on net number of bonding electrons.

6. Planarity is due to overlap of 2p orbitals to give delocalised π-systems. If these systems were not planar, π-orbital overlap could not occur. Although some properties of these systems can be explained in terms of alternating single and double bonds, others cannot, such as the thermodynamic stability of the delocalised π-system as seen in the problems for Chapter 2. Molecular orbital theory tells us that the delocalised model is correct.

7. The most favourable conformation of a peptide linkage is the one with the lowest total energy. The orbital energy of the electrons is the same for both *cis* and *trans* peptide bonds but it is also necessary to consider other interactions such as steric repulsion. For most amino acids, the *cis* form gives steric interactions between the preceding residue and the Cα carbon whereas the *trans* form gives steric interactions between the preceding residue and the amide proton. As the latter is much smaller, the *trans* form is more stable. For proline, the amide proton is replaced by the Cδ and so the difference in steric repulsion between the *cis* and *trans* forms is much smaller.

Chapter 17

1. Sodium vapour absorbs yellow light and so when viewed by transmitted radiation appears blue. A sodium vapour lamp emits yellow light. [cf. Difference between absorption and emission spectroscopy.]

2. (a) Protein concentrations in solution can be determined by UV absorption spectroscopy. The absorbance of the solution is measured and the Beer–Lambert law used to calculate the concentration. However, this requires a pure protein sample as most proteins absorb at similar frequencies and so the absorbance will depend on all proteins present. An alternative is to label an antibody with a specific chromophore, bind the antibody to the protein, separate the complex from the free antibody and then measure the concentration of the chromophore.
   (b) The copper content of a membrane protein can also be determined by UV absorption spectroscopy. In this case we want to see the spectra of the atoms present and so the sample is vaporised and the absorption spectrum of the vapour measured, each atom giving a unique spectrum (see Chapter 15). Alternatively, X-ray emission spectroscopy can be used on the intact metalloprotein (similar to the method used for elemental analysis of nettle hairs discussed in Chapter 15).
   (c) The Trp residue is likely to have a different UV absorption spectrum when exposed to solvent than when involved in interactions with the ligand and so monitoring changes in the UV absorption spectrum may allow the binding to be followed directly. If the absorption spectrum of the Trp does not alter on ligand binding, then solvent perturbation spectroscopy can be used, as the Trp will be exposed to solvent and the perturbant in the absence of ligand but not exposed in the presence of ligand.

3. Adding NaCl makes the solvent more polar. A shift to shorter wavelengths implies a less polar environment for the Trp. Thus, either a change in peptide conformation or aggregation (actual behaviour) is occurring.

4. If the absorbance is not linear with concentration, it implies that chemical changes are occurring involving the chromophore, altering its extinction coefficient. These can include (i) alteration of the chemical environment of the chromophore, for instance by oligomerisation/aggregation, and (ii) direct interactions between the transition dipoles of different chromophores, which lead to altered transition probabilities. The latter effect can give rise to an increase in absorption, hyperchroism, or a decrease in absorption, hypochroism (as is observed for the base pairs in double-helical DNA).

5. If $I_\lambda = \frac{1}{2}I_o$ the absorbance is 0·3. In this case, the concentration needs to be $4{\cdot}8 \times 10^{-5}$ M to give A = 0·3 and so a 20-fold dilution is necessary. Alternatively, rather than diluting the sample, a shorter path-length cell could be used or measurements taken at a different wavelength where the extinction coefficient is smaller.

6. For reduced cytochrome c,

$$\varepsilon = \frac{A}{cl} = \frac{\log_{10}\left(\frac{100}{56}\right)}{1 \times 10^{-5}} = 2{\cdot}5 \times 10^4\ \mathrm{M}^{-1}\ \mathrm{cm}^{-1}$$

Similarly for oxidised cytochrome c, $\varepsilon = 0{\cdot}8 \times 10^4\ \mathrm{M}^{-1}\ \mathrm{cm}^{-1}$. For the mixture, $A = \varepsilon_1 c_1 l + \varepsilon_2 c_2 l$ and so

$$\log_{10}\left(\frac{100}{62}\right) = (2{\cdot}5 \times 10^4 \times c \times 10^{-6}) + (0{\cdot}8 \times 10^4 \times (10 - c) \times 10^{-6})$$

where c is the concentration of the reduced cytochrome c in μM. Thus, [Cyt-c(Fe^{2+})] = 7·5 μM and [Cyt-c(Fe^{3+})] = 2·5 μM.

7. [NADH] can be calculated from the absorbance at 340 nm to give [NADH] = 33·8 μM. [NADH] + [NAD^+] can be calculated from the absorbance at 260 nm to give [NAD^+] = 13·5 μM.

8. A plot of extinction coefficient versus pH gives a titration curve (see Chapter 5) from which the pK_a can be determined. This gives pK_a(nitrotyrosine) = 6·8 and pK_a(nitrated enzyme) = 7·9. The two values differ because the chemical environment of the nitrotyrosine in the enzyme is different from that in free solution (see Chapter 5).

9. By constructing a standard curve of absorbance versus amount of Hb, it can be seen that the 0·01 cm^3 of blood contain 1·4 mg Hb, i.e. the Hb content is 140 g dm^{-3}. Normal ranges for males are 130 – 180 g dm^{-3} and for females 110 – 160 g dm^{-3}.

10. Absorption peaks in the visible region correspond to the excitation of electrons from one energy level to another. Each absorption peak corresponds to a transition between two specific energy levels where the energy difference between the levels matches the energy of the radiation. As several different transitions with appropriate energy separations are possible, several peaks are observed in this range of wavelengths.

    (i) Some wavelengths do not match the energy of any possible transitions and so at these wavelengths no absorption is seen and the extinction coefficients are close to zero. The separate peaks have different extinction coefficients because the electronic transitions involved have different transition dipoles and so have different probabilities.

(ii) Oxidised and reduced cytochrome c give different plots because the electron energy levels are different in the two species. The electronic transitions for cytochrome c in the visible region involve electrons in the Fe d orbitals. The energies of these d orbitals depend on the surrounding ligands. In oxidised cytochrome c the Fe is in the +3 oxidation state, compared to +2 in reduced cytochrome c. In the +3 state, the ligands will be closer to the Fe and so alter the d-orbital energies more than in the +2 state. (See Chapter 16 for a discussion of crystal field theory.)

Chapter 18

1. (a) Asp, Glu, Lys, Arg, His. (b) Asn, Gln, Ser, Thr, Tyr. (c) Ile, Leu, Phe, Trp, Val.

2. α-Helices have linear backbone hydrogen bonds whereas the hydrogen bonds in all other helices are non-linear and so slightly less stable.

3. Each H–N and C=O group constitutes a dipole, with the H and the C being the positive ends. These two dipoles have approximately the same dipole moment. In a β-strand, the H–N and C=O dipoles of a single residue point in opposite directions, cancelling each other out. As there is no dipole associated with each residue in a β-strand, a β-sheet cannot have an overall backbone dipole. In an α-helix, the H–N and C=O dipoles of each residue point in the same direction, along the axis of the helix towards the C-terminal end. Thus, an α-helix has an overall backbone dipole with the N-terminal end positive and the C-terminal end negative. Because of this, antiparallel α-helices are slightly more stable than parallel α-helices.

For a series of parallel dipoles along a helix, the negative end of one dipole cancels with the positive end of the next dipole, except at the ends of the helices. Thus, the dipole moment of the helix can be calculated by assuming that there is a $+0{\cdot}5$ charge at the N-terminus and a $-0{\cdot}5$ charge at the C-terminus. The magnitude of the dipole moment is given by

$$\text{dipole moment} = q \times e \times R$$

where e is the charge on a proton, q is the charge at the end of the dipole in units of e and R is the distance between the two charges. This gives a dipole moment of

$$0{\cdot}5 \times 1{\cdot}602 \times 10^{-19} \times 10 \times 10^{-10} = 8 \times 10^{-29}\,\text{C m} = 24\,\text{D}$$

for a 10 Å helix. The magnitude of the dipole moment for an N–H bond is 2.4 D, i.e. 10 times smaller.

4. From the data given, for the process

$$CH_4(\text{non-polar solvent}) \rightarrow CH_4(\text{aq})$$

$\Delta G° = 11{\cdot}7\,\text{kJ mol}^{-1}$ and $\Delta H° = -11{\cdot}3\,\text{kJ mol}^{-1}$ and so $\Delta S° = -77{\cdot}2\,\text{J mol}^{-1}\,\text{K}^{-1}$. Thus, in this case transfer of a non-polar molecule from a non-polar solvent to a polar solvent is enthalpically favourable and entropically unfavourable. This fits with the dominant component of the hydrophobic effect being the unfavourable ordering of water around non-polar molecules.

5. At 293 K, $\Delta G^\circ = -2{\cdot}8\ \text{kJ mol}^{-1}$ and so association is favourable. The association is enthalpically unfavourable and entropically favourable, suggesting that the hydrophobic effect is the major factor in dimer formation. Assuming that ΔH° and ΔS° are independent of temperature over the temperature range of interest, association would become less favourable as the temperature is lowered. This behaviour is observed in some multisubunit enzymes, which become less active at lower temperatures due to subunit dissociation.

6. All these amino acids have charged backbone groups and non-polar side chains. As the non-polar side chain becomes larger, the transfer from water to urea becomes more favourable. This suggests that the hydrophobic effect is a significant factor in the interactions between urea and amino acids. To confirm this, ΔH° and ΔS° for the transfer need to be determined. A significant contribution from the hydrophobic effect would be associated with a positive value for ΔS°. If the hydrophobic effect is the major factor in determining ΔG° transfer, then you would expect ΔH° to be positive as well. If competition between urea and water for hydrogen bonding with the backbone groups is also significant, then you would expect ΔH° to be either zero or negative.

7. Folded proteins are stable because of the high degree of cooperativity on forming weak, non-covalent interactions between different parts of the peptide chain. As we have seen in figure 18.13, a significant number of interactions have to be formed before the folded form becomes more stable than the unfolded form. A 20-residue peptide forming a helix does form enough non-covalent interactions for this to occur. A 20-residue helix within a larger protein can form additional non-covalent interactions with the rest of the protein, increasing the stability of the helix.

Index